T0327510

FLUID–STRUCTURE INTERACTION

FLUID–STRUCTURE INTERACTION

AN INTRODUCTION TO FINITE ELEMENT COUPLING

Jean-François Sigrist

DCNS Research, France

Registered office
John Wiley & Sons Ltd, The Atrium, Southern Gate, Chichester, West Sussex, PO19 8SQ, United Kingdom

For details of our global editorial offices, for customer services and for information about how to apply for permission to reuse the copyright material in this book please see our website at www.wiley.com.

Library of Congress Cataloging-in-Publication Data

SIGRIST, Jean-François.
  Fluid-structure interaction : an introduction to finite element coupling / Jean-François SIGRIST, DCNS Research, France.
      pages cm
  Includes bibliographical references and index.
  ISBN 978-1-119-95227-5 (hardback)
  1. Fluid-structure interaction. 2. Finite element method. I. Title.
  TA357.5.F58.S54 2015
  624.1′71 – dc23
                      2015022631

A catalogue record for this book is available from the British Library.

Cover Image: Courtesy of the Author

ISBN: 9781119952275

Typeset in 10/12pt, TimesLTStd by SPi Global, Chennai, India.

1   2015

Contents

Foreword

The dynamical behaviour of flexible structures in the presence of fluids stands as a challenging and rapidly evolving part of the mechanical engineering sciences. Its mastering is of major importance for designing a large variety of artefacts produced in several industrial domains, including in particular the energy and transport sectors, where the financial and safety stakes may be especially high.

The relatively abundant literature already available in the field remains largely focused on the physics of the basic mechanisms involved in the fluid–structure coupled problems, which indeed reveal often as very appealing from an academic standpoint. However, the analytical and numerical models used in such research-oriented works often turn out to be of limited efficiency to deal with the complexity usually met in the industrial context.

In most cases, numerical models require the development of numerical tools able to deal with large sized discrete models, typically more than hundred thousand degrees-of-freedom, together with a high level of computational efficiency making it possible to perform versatile parametric and even probabilistic studies. Furthermore, it is usually desired to treat the fluid–structure coupled problem of concern paying equal attention to the structure part of the mechanical system as to the fluid part of it, both in the linear and non-linear domains of response.

In this respect, the designer is faced, already at the starting point, with the alternative of choosing between a so-called fluid–structure computer code, in which the coupled problem is formulated as a whole, and the coupling by an appropriate interface of two distinct codes, one dealing with the solid part of the problem and the other with the fluid part of it. Accordingly, from the engineering standpoint, the topical subject of fluid–structure interaction clearly deserves to be complemented by a book aimed at presenting a didactical introduction to the numerical methods made available nowadays to carry out design-oriented numerical simulations of a large variety of fluid–structure coupled systems.

It must be emphasised that one essential task of the engineer when performing design analyses based on computational simulations is to get a clear judgement about the tools suited to build a numerical model of his problem and the practical value of the numerical data so produced. This necessarily requires a sufficient acquaintance of the analyst with the numerical techniques, including standard and more advanced tools used in the finite element and code coupling methods.

This book is aimed at meeting such a need, partially at least, by providing the reader with a didactical introduction to a fairly large number of methods, some basic in nature and other far more advanced, which are described in the context of the finite element discretisation of fluid–structure coupled systems as restricted to the case of motions arising about a static and stable state of equilibrium, that is, no steady motion of the solid and no steady flow of the fluid. The book is intended for beginners as well as for practitioners in the computational fluid–structure dynamics. The reader will find a presentation of the formalism of the finite element discretisation methods which make accessible a global vision together with a logical progression in the degree of sophistication of the problems treated, starting from the case of incompressible fluids in confined configuration to the case of infinite extents of fluid.

This gives the author the opportunity to outline several advanced methods concerning, on one hand, the coupling between finite element models and boundary element models and the delicate problem of modelling anechoic conditions to mock-up a physically unbounded extent of fluid.

To uninitiated beginners, the book is intended to make the travel to destination easier by complementing at each important step the presentation of the mathematical tools of general value, by a thorough discussion of illustrative examples, judiciously selected for their physical and didactical interest. The reader is so rewarded for the effort paid to master a bunch of some abstract and rather arid concepts by the access it gains to the actual solution of a large variety of engineering problems of practical importance, governed by an underlying physics which would be difficult to apprehend in the absence of well-suited numerical tools.

To experienced practitioners, the book may also be quite valuable, because besides the basic concepts they are already well acquainted with, they will be provided with several enlightening outlines concerning various topical methods which still pertain more to the research than to the industrial community, nowadays at least. Though not exhaustive, the references included are sufficient to direct the interested reader towards the more advanced literature concerning the specific subjects introduced in this book.

I feel sure that this work by Jean-François SIGRIST will reward those students and practitioners who are concerned with learning and applying numerical techniques to the solution of fluid–structure coupled problems in a large variety of situations.

François AXISA – Paris, January 2015.

Images Credits

- Cover image – Propeller and vibrating shell – Source: Jean-François SIGRIST, DCNS Research, Nantes, France, 2013–2014. Reproduced with permission of DCNS.
- Figure 1.1 – *Time and Tide*, Quiberon, France, 2011 – Source: © Jean-François SIGRIST.
- Figure 1.3 – *L'Aviso Patrouilleur de Haute Mer Commandant Blaison sort de l'abri de la rade vers la mer d'Iroise le 25 février 2014* – Source: © A.Monot/Marine Nationale.
- Figure 1.14 (b) – Sloshing in a reservoir – Source: Jean-Sébastien SCHOTTÉ, ONREA, Chatillon, France, 2009. Reproduced with permission of Jean-Sébastien SCHOTTÉ.
- Figure 1.14 (c) – Swimming eel – Source: Alban LEROYER, Ecole Centrale, Nantes, France, 2005. Reproduced with permission of Alban LEROYER.
- Figure 1.14 (d) – Yacht sail (the computation is performed with the code ARAVANTI, property of K-EPSILON) – Source: Patrick BOT, Ecole Navale, Brest, France, 2014. Reproduced with permission of Patrick BOT.
- Figure 1.21 – Hydrodynamics and Fluid–Structure Interaction – Source: André ASTOLFI, Ecole Navale, Brest, France and Fabien GAUGAIN, DCNS Research, Nantes, France, 2013. Reproduced with permission of André ASTOLFI.
- Figure 2.1 – *Bel Ami*, Nantes, France, 2009 – Source: © Jean-François SIGRIST.
- Figure 2.9 – Finite element model of a surface ship – Source: DCNS Lorient, France, 2008. Reproduced with permission of DCNS.
- Figure 2.16 – Modal analysis of a propeller – Source: DCNS Nantes, France, 2005. Reproduced with permission of DCNS.
- Figure 3.1 – *Viscosity*, Nantes, France, 2010 – Source: © Jean-François SIGRIST.
- Figure 3.4 – Finite volume mesh (the mesh is generated with the code STAR-CCM+) – Source: Julien MANERA, CD-adapco, Bobigny, France, 2014. Reproduced with permission of CD-adapco.
- Figure 3.5 – Computational fluid dynamics and hydrodynamics – Source: Jean-François SIGRIST and Jean-Jacques MAISONNEUVE, DCNS research, Nantes, France, 2012 and 2014. Reproduced with permission of DCNS.
- Figure 3.8 – Simulation with the smoothed hydrodynamics particles technique (the computation is performed with the code SPH-flow) – Source: Erwan JACQUIN, HydrOcéan, Nantes, France, 2012. Reproduced with permission of HydrOcéan.
- Figure 3.11 – Finite element model of a pulsating sphere (the numerical model is generated within the framework of the project SiNuS from IRT Jules VERNE) – Source: Cédric LEBLOND, DCNS Research/IRT Jules VERNE, Nantes, France, 2014. Reproduced with permission of IRT Jules VERNE.

- Figure 4.2 – *Y+*, Saint-Nazaire, France, 2013 – Source: © Jean-François SIGRIST.
- Figure 5.1 – *While my Guitar...*, Nantes, France, 2014 – Source: © Jean-François SIGRIST.
- Figure 6.1 – *October Mood*, Pornic, France, 2012 – Source: © Jean-François SIGRIST.
- Figure 6.12 – Comparison of computational techniques – Source: Lucie ROULEAU, CNAM, Paris, France, 2013. Reproduced with permission of Lucie ROULEAU.
- Figure 6.13 – Vibro-acoustics of a cruise ship – Source: Sylvain BRANCHEREAU, STX, Saint-Nazaire, France, 2014. Reproduced with permission of STX France.

Preface

"Ce que nous n'avons pas eu à déchiffrer, à éclaircir par notre effort personnel, ce qui était clair avant nous, n'est pas à nous."

Marcel PROUST – *Le Temps Retrouvé* (1927)

Fluid–structure interaction has received a growing interest over the past decades, as it comes into play in numerous industrial applications, also being of paramount importance as far as the safety and reliability issues of many mechanical systems are concerned. Many reference textbooks have already been published on this topic as a result of intense researches carried out in order to understand the physics involved in the interaction, to propose mathematical models which account for fluid and structure coupling and to develop appropriate numerical methods which solve the equations of the physical models.

As a research and development engineer, I have been involved in the development of a few specific methods aimed at solving problems of industrial concern arising in naval shipbuilding and nuclear engineering; as a consultant teacher at École Centrale de Nantes, I have also had the opportunity to share views with students on such engineering applications of numerical techniques for coupled systems.

This book originates from this 15 years experience, both as a practitioner and as an observer of numerical techniques of engineering relevance. It aims at providing an introduction to the general concepts of finite and boundary element methods for dealing with fluid–structure interaction, with illustrations of a few fundamental coupling effects, highlighted by examples simple enough to be computed with general-purpose programming languages, such as MAT-LAB. It may be viewed as an invitation to the practice of such numerical methods and may also serve as a tool for self-learning and for code validation and verification.

Taking time to write a book is a unique opportunity which would not have been possible to undertake, carry out and complete without the aid of devoted friends and colleagues. May I take here the opportunity to acknowledge my thankfulness to those who helped me through the way in achieving the project, with kind advice, wise remark, warm encouragement – as well as by providing some illustrative material. Special thanks go to Daniel BROC, Aziz HAMDOUNI, Cédric LEBLOND, Frédérique LE LAY and Eric VÉRON for their precious contribution to the early versions of the manuscript through fruitful comments and ideas. I am grateful to François AXISA for his enlightened comments and his constant assistance during the final stage of the writing process; I am also deeply honoured by his kind introduction to the present edition.

I do wish that, despite the inevitable simplifications and reductions arising from its pedagogical character, the material proposed in this introduction of finite element coupling for fluid–structure interaction will prove useful to the reader – as far as its author is concerned, conceiving and writing this book has simply and truly been a rewarding experience.

Jean-François SIGRIST – Nantes.

1

Fluid–Structure Interaction

A short overview of some analytical models and numerical methods is proposed in this chapter as an introduction to fluid–structure interaction (FSI). The analytical models derived in what follows focus on the description of the so-called inertial effect on the one hand and of vibro-acoustic coupling on the other hand, both being investigated from various points of view in the subsequent chapters. The principles of a coupled simulation are also concisely exposed and illustrated with a few specific examples borrowed from academic researches and industrial applications, thereby highlighting the variety of possible approaches on FSI. The interactions between a vibrating structure and a stagnant fluid are described by a set of partial differential equations whose numerical solution can be obtained from finite element or boundary element discretisation, both numerical techniques being presented and studied in a detailed manner in this book.

Figure 1.1   Fluids and Solids. The interactions between mechanical systems are of various nature and intensity. Structural and fluid dynamics have long been considered in a separate manner, which holds only when the fluid flow and the structure motion are weakly coupled, that is, when their evolutions occur within different characteristic times. When it is more pronounced, FSI has to be accounted for in numerical simulations: for instance, vibrations of an elastic structure in contact with a quiescent fluid are commonly described with finite element-based techniques. *Source*: © Jean-François Sigrist

Fluid–Structure Interaction: An Introduction to Finite Element Coupling, First Edition. Jean-François Sigrist.
© 2015 John Wiley & Sons, Ltd. Published 2015 by John Wiley & Sons, Ltd.
Companion Website: www.wiley.com/go/sigrist

1.1 A Wide Variety of Problems

FSI is concerned with the coupled dynamics of structures in contact with a fluid. As sketched in Figure 1.2, the basic mechanism of fluid and structure coupling may be described as follows: the motion of the structure modifies the flow conditions at the interface with the fluid, which in turn induces a fluctuation in the pressure and/or viscous forces; the loading applied to the fluid–structure interface subsequently changes the structure motion.

A wide variety of industrial problems, ranging from civil to naval and offshore engineering, from transportation to medical applications, from power nuclear to aerospace industries, are concerned by FSI. From an engineering perspective, taking FSI into account is often of major importance since the structural response to fluid loading enters into consideration for safety, reliability or durability issues of mechanical systems – for instance, in shipbuilding, the so-called hydrodynamic impact is a rather spectacular illustration of FSI; see Figure 1.3.

FSI modelling usually assumes that both the structure and the fluid parts of a coupled system are studied within the theoretical framework of the continuum mechanics. As a consequence, their motion is governed by a set of partial differential equations associated with some appropriate boundary conditions. However, in certain circumstances, specified later when needed, it may be found advantageous to formulate the differential equations of the fluid motion into an integral form.

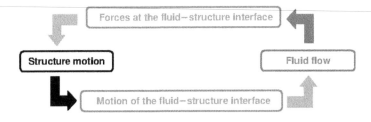

Figure 1.2 Fluid–structure interaction: coupling mechanism

Figure 1.3 Hydrodynamic impact. Among many phenomena involving fluid and structure motions in extreme conditions, *slamming* is one of the most spectacular examples of fluid–structure interaction. When sailing at high speed and/or in rough sea condition, the impact of a ship on water induces high loads on the structure: slamming is, among other cases, of primary concern when designing ships. *Source*: © A.Monot/Marine Nationale

Analytical solutions to the equations of motion can be made available for simple geometries of the solid and fluid domains, scarcely met in practice. Such particular cases are of academic interest, for instance, in order to highlight some essential features of FSI, as first considered in Section 1.2.

In most applications of engineering relevance, numerical methods are resorted to in order to produce approximate solutions. A short overview of the large variety of FSI problems in relation to the numerical methods made available to the designer to solve them is proposed in Section 1.3.

1.2   Analytical Modelling of Fluid–Structure Interactions

Beyond the definition of energy transfer from one medium to another lies a large number of situations: the *dimensional analysis*, as developed for instance by De Langre (2001), offers a general framework for the classification of various coupling phenomena. For a given problem, a set of non-dimensional numbers characterises the intensity of the coupling (whether 'strong' or 'weak'): it is usually derived from the fluid–structure mathematical model.

An illustration of some manifestations of FSI is proposed hereafter, starting from the vibrations of a spring-mass system, as represented in Figure 1.4.

Without fluid coupling, vibrations of the cylinder are accounted for by the equation of motion $m\ddot{u}(t) + ku(t) = 0$. It is a second-order ordinary differential equation, endowed with initial conditions $u(t = 0) = u_o$ and $\dot{u}(t = 0) = \dot{u}_o$. Using non-dimensional variables $u^* = u/R$, with R scaling displacement, and $t^* = \omega_o t$, with $\omega_o = \sqrt{\dfrac{k}{m}}$ scaling time, the non-dimensional equation of motion may be recast as follows:

$$\ddot{u}^*(t^*) + u^*(t^*) = 0 \qquad u^*(t^* = 0) = u_o^* \qquad \dot{u}^*(t = 0) = \dot{u}_o^*$$

As made conspicuous in the following subsections, analytical solutions to the fluid flow equations are conveniently derived using the Laplace transform.[1] In order to enable a

---

[1] It is recalled that when it exists, the Laplace transform of function $\psi(t)$ (denoted here $\hat{\psi}(s) = \mathcal{L}(\psi(t))$ with s the Laplace variable), and its inverse, are defined by the integrals:

$$\hat{\psi}(s) = \int_0^\infty \psi(t)\exp{(-st)}dt$$

$$\psi(t) = \mathcal{L}^{-1}(\hat{\psi}(s)) = \frac{1}{2i\pi}\int_{\sigma-i\infty}^{\sigma+i\infty} \hat{\psi}(s)\exp{(+st)}ds$$

σ is such that the integral converges (which is ensured for any σ greater than the real value of any singularity of $\hat{\psi}(s)$) and if $|\hat{\psi}(s)| = o(|s|^2)$ for $|s| \to \infty$. As illustrated in what follows, the Laplace transform is well suited for the description of time-dependent problems with initial conditions. The elementary properties of the Laplace transform, such as recalled below, make it useful for solving partial differential equations.

- The Laplace transform of a function derivative is $\mathcal{L}(\psi'(t)) = s\mathcal{L}(\psi(t)) - \psi(0)$;
- The Laplace transform of a Dirac function is $\mathcal{L}(\delta(t)) = 1$, $\forall s$;
- The Laplace transform of a function product is $\mathcal{L}(\psi_a * \psi_b(t)) = \mathcal{L}(\psi_a(t))\mathcal{L}(\psi_b(t))$, where the convolution product of ψ_a and ψ_b is defined by:

$$\psi_a * \psi_b(t) = \int_{-\infty}^{+\infty} \psi_a(t-\tau)\psi_b(\tau)d\tau$$

when the integral exists. It is recalled that a function remains unchanged when convoluted with the Dirac function: $\psi * \delta(t) = \psi(t)$.

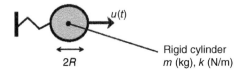

Figure 1.4 Spring-mass system. Computer simulation of mechanical systems can handle numerical models of increasing complexity. Analytical models still prove useful to understand some 'elementary' physics and to validate numerical methods

comparison with the non-coupled case, the Laplace transform of the former equation is written as follows:

$$\hat{\ddot{u}}^*(s^*) + \hat{u}^*(s^*) = 0$$

with the non-dimensional Laplace variable $s^* = \omega_o s$.

When the cylinder is coupled to a fluid initially at rest, various flow models may be considered, under the assumption that the motions of the solid are small regarded to the size of the fluid domain. Small perturbations of the pressure and/or velocity fields are considered – such assumptions will be clarified in Chapter 3. For the sake of simplicity, the equations of the fluid flow are taken for granted in what follows.

1.2.1 Potential Flow. Inertial Coupling

As depicted in Figure 1.5, the vibrations of the cylinder coupled to an inviscid and incompressible fluid are considered, assuming small amplitude perturbation of the flow. The equation of motion is $m\ddot{u} + ku = \phi$, where ϕ stands for the force exerted by the fluid. The latter is calculated as follows:

$$\phi = -\int_0^{2\pi} p(R, \theta, t)\cos\theta R\, d\theta \tag{1.1}$$

where p stands for the pressure fluctuation in the fluid.

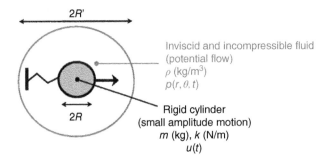

Figure 1.5   Cylinder coupled to an incompressible fluid in annular confinement (potential flow model)

Under potential flow assumptions, the fluctuation of the pressure field in the fluid is described by the Laplace equation, stated in the cylindrical coordinate system as follows:

$$\frac{\partial^2 p}{\partial r^2} + \frac{1}{r}\frac{\partial p}{\partial r} + \frac{1}{r^2}\frac{\partial^2 p}{\partial \theta^2} = 0$$

The coupling of the fluid with the moving inner wall and the fixed outer wall are expressed by the following boundary conditions:

$$\left.\frac{\partial p}{\partial r}\right|_{r=R} = -\rho \ddot{u}(t)\cos\theta \qquad \left.\frac{\partial p}{\partial r}\right|_{r=R'} = 0$$

According to Fritz (1972), the pressure field is found to be

$$p(r,\theta,t) = -\frac{\rho}{1-\alpha^2}\left(r + \frac{\alpha^2 R^2}{r}\right)\ddot{u}(t)\cos\theta$$

with $\alpha = \frac{R'}{R}$.

The pressure is proportional to the acceleration of the structure, indicating that the perturbations of the fluid flow resulting from the vibration of the structure are instantaneously propagated throughout the fluid domain. Accordingly, the pressure force on the cylinder is also found to be proportional to its acceleration:

$$\phi(t) = -\rho\pi R^2 \frac{\alpha^2+1}{\alpha^2-1}\ddot{u}(t) = -m_a\ddot{u}(t)$$

where m_a is the so-called fluid *added mass*.

The vibrations of the cylinder are therefore governed by

$$(m + m_a)\ddot{u} + ku = 0$$

which is also stated in a non-dimensional form in the Laplace domain:

$$(1 + \mathcal{M}_a)\hat{\ddot{u}}^* + \hat{u}^* = 0 \tag{1.2}$$

$\mathcal{M}_a$ is the *mass number*, defined as the ratio of the fluid added mass and the structure mass:

$$\mathcal{M}_a = \frac{\rho\pi R^2}{m}\frac{\alpha^2+1}{\alpha^2-1}$$

In the Laplace domain, the coupling is represented by the FSI function $\hat{\mu}(s^*) = \mathcal{M}_a\delta(s^*)$, so that in the time-domain, the pressure force is proportional to the structure acceleration. FSI is of *inertial nature*, and it is quantified by $\mathcal{M}_a$, as illustrated in Figure 1.6.

The fluid added mass is a *virtual* mass which accounts for the inertial effect, and, as depicted in Figure 1.7, it is usually different from the physical mass of the fluid – in the present case, the latter is $\rho\pi R^2(\alpha^2 - 1)$ while the former is $m_a = \rho\pi R^2\frac{\alpha^2+1}{\alpha^2-1}$.

For a strong confinement ($\alpha \ll 1$), the added mass largely exceeds the actual mass of the fluid and tends to be 'infinite' in extreme confinements ($\alpha \to 1$). For an infinite extent of

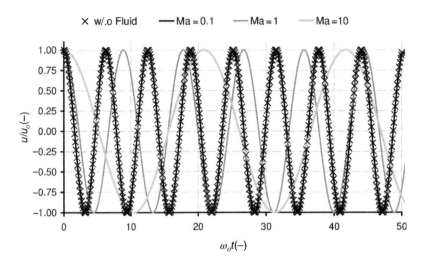

Figure 1.6 FSI effect for potential flow. The free vibrations of the cylinder without and with fluid are represented using non-dimensional values, for initial conditions $u_o^* = 1$ and $\dot{u}_o^* = 0$ and for different mass numbers. In the present case, FSI is of inertial nature: according to Equation (1.2), the cylinder vibrates with an additional inertia, so that the period of its natural vibrations increases with $\mathcal{M}_a$

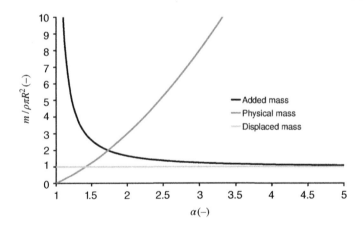

Figure 1.7 Added mass and displaced mass for a cylinder in cylindrical confinement and physical mass of fluid within the annular space

the fluid domain ($\alpha \gg 1$), it reaches a finite value $\rho \pi R^2$, which is the mass of the fluid displaced by the solid.

As the fluid is set into motion by the vibration of the structure, it gains kinetic energy. The latter is evaluated as follows:

$$\mathcal{E} = \int_R^{\alpha R} \int_0^{2\pi} \frac{1}{2}\rho(\dot{\xi}_r^2 + \dot{\xi}_\theta^2) r \, dr d\theta$$

where $\xi = (\xi_r, \xi_\theta)$ is the fluid displacement field. For potential flows, ξ is shown to be derived from the displacement potential φ, that is, $\xi = \nabla\varphi$. φ may be calculated from the fluid pressure according to $p = -\rho\frac{\partial^2\varphi}{\partial t^2}$, so that in the present case, the following relation is arrived at:

$$\varphi(r, \theta, t) = \frac{1}{1-\alpha^2}\left(r + \frac{\alpha^2 R^2}{r}\right) u(t)\cos\theta$$

Hence:

$$\xi_r = \frac{\partial\varphi}{\partial r} = \frac{1}{1-\alpha^2}\left(1 - \frac{\alpha^2 R^2}{r^2}\right) u(t)\cos\theta$$

$$\xi_\theta = \frac{1}{r}\frac{\partial\varphi}{\partial\theta} = -\frac{1}{1-\alpha^2}\left(1 + \frac{\alpha^2 R^2}{r^2}\right) u(t)\sin\theta$$

and:

$$\mathcal{E} = \frac{1}{2}\rho\dot{u}^2\frac{\pi}{(1-\alpha^2)}\int_R^{\alpha R}\left(\left(1 + \frac{\alpha^2 R^2}{r}\right)^2 + \left(1 - \frac{\alpha^2 R^2}{r}\right)^2\right) r\, dr$$

The kinetic energy in the fluid is therefore found to be:

$$\mathcal{E} = \frac{1}{2}\rho\pi R^2\frac{1+\alpha^2}{1-\alpha^2}\dot{u}^2 = \frac{1}{2}m_a\dot{u}^2$$

which gives a straightforward physical signification of the added mass.

1.2.2 Viscous Flow. Viscous Damping

The vibrations of a cylinder coupled to a fluid is investigated for a viscous flow, as in the configuration depicted in Figure 1.8.

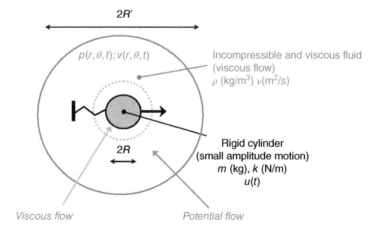

Figure 1.8 Cylinder coupled with an incompressible fluid in annular confinement (viscous flow model)

The force exerted by the fluid on the cylinder is expressed as follows:

$$\phi = \int_0^{2\pi} (\cos\theta e_r + \sin\theta e_\theta)\sigma(R,\theta)e_r R \, d\theta = \int_0^{2\pi} \sigma_{rr} R\cos\theta \, d\theta + \int_0^{2\pi} \sigma_{r\theta} R\sin\theta \, d\theta$$

where σ_{rr} and $\sigma_{r\theta}$ are the fluid stress tensor components. The latter are calculated as follows:

$$\sigma_{rr} = -p + \frac{\partial v_r}{\partial r} \qquad \sigma_{r\theta} = \frac{\rho v}{2}\left(\frac{1}{r}\frac{\partial v_r}{\partial\theta} + \frac{\partial v_\theta}{\partial r} - \frac{v_\theta}{r}\right)$$

where $p(r,\theta)$ and $\mathbf{v}(r,\theta) = (v_r(r,\theta), v_\theta(r,\theta))$ stand for the fluid pressure and velocity fields, and v denotes the fluid kinematic viscosity.

As exposed in a detailed manner in Chapter 3, a fluid flow may be described by the mass and momentum equations. In the present case, these equations are expressed as follows[2], using a cylindrical coordinate system:

$$\frac{1}{r}\frac{\partial(\rho r v_r)}{\partial r} + \frac{\partial(\rho r v_\theta)}{\partial\theta} = 0$$

and:

$$\frac{\partial v_r}{\partial t} = -\frac{1}{\rho}\frac{\partial p}{\partial r} + v\left(\frac{\partial^2 v_r}{\partial r^2} + \frac{1}{r}\frac{\partial v_r}{\partial r} - \frac{v_r}{r^2} + \frac{1}{r^2}\frac{\partial^2 v_r}{\partial\theta^2} - \frac{2}{r^2}\frac{\partial v_\theta}{\partial\theta}\right)$$

$$\frac{\partial v_\theta}{\partial t} = -\frac{1}{\rho r}\frac{\partial p}{\partial\theta} + v\left(\frac{\partial^2 v_\theta}{\partial r^2} + \frac{1}{r}\frac{\partial v_\theta}{\partial r} - \frac{v_\theta}{r^2} + \frac{1}{r^2}\frac{\partial^2 v_\theta}{\partial\theta^2} + \frac{2}{r^2}\frac{\partial v_\theta}{\partial\theta}\right)$$

The initial conditions of the fluid flow are represented by the following relations:

$$p(t=0) = p_o \qquad v_r(t=0) = v_r^o \qquad v_\theta(t=0) = v_\theta^o$$

while the boundary conditions, namely the coupling conditions with the moving cylinder at $r = R$ and with the fixed wall at $r = R'$, are expressed as:

$$v_r(R,\theta) = \dot{u}\cos\theta \qquad v_r(R',\theta) = 0 \qquad v_\theta(R,\theta) = \dot{u}\sin\theta \qquad v_\theta(R',\theta) = 0$$

An analytical solution to the fluid flow equations may be derived using a matched asymptotic development technique: in the vicinity of the cylinder, a pure viscous flow solution is derived, while in the remainder of the fluid domain, a potential flow solution is obtained. Matching these solutions describes the fluid flow throughout the fluid domain, as sketched in Figure 1.8. An application of this technique is proposed by Leblond et al. (2009) for the problem under concern here; using the Laplace transform, it can be shown that the fluid force on the cylinder is expressed as follows:

$$\hat{\phi}(s) = -\rho\pi R^2\left(\frac{\alpha^2+1}{\alpha^2-1} + \frac{4\alpha^2}{\alpha^2-1}\frac{1}{\sqrt{R^2 s/v}}\right)\hat{\ddot{u}}(s)$$

In this equation, the first term is related to the potential flow, while the second term is a viscous correction. Substituting this expression of the fluid force into the equation of motion for the

---

[2] As small amplitude motions of both the cylinder and the fluid are assumed, the convection terms are discarded in the momentum equation, see Chapter 3.

cylinder yields the following equation:

$$(1 + \mathcal{M}_a \hat{\mu}(s^*/S_t))\hat{\ddot{u}}^* + \hat{u}^* = 0 \tag{1.3}$$

The above expression is formulated in the Laplace domain using non-dimensional variables. The influence of the fluid on the cylinder motion is accounted for by the *FSI function* $\hat{\mu}(s^*/S_t)$, which is expressed here as follows:

$$\hat{\mu}(s^*/S_t) = \frac{\alpha^2 + 1}{\alpha^2 - 1} + \frac{4\alpha^2}{\alpha^2 - 1}\frac{1}{\sqrt{s^*/S_t}}$$

with:

$$S_t = \frac{v}{\omega_o R^2}$$

S_t is the so-called *viscosity number* or Stokes number: in terms of order of magnitude, it may be interpreted as the ratio between the propagation velocity of viscous shear waves in the fluid, which is v/R, and the vibration velocity of the cylinder, which is $\omega_o R$.

In the Laplace domain, the influence of the fluid on the solid motion is expressed by Equation (1.3) and consequently, in the time-domain, FSI is represented by the convolution product of μ^* and $\ddot{u}^*$, so that the fluid force at time t depends on the acceleration of the cylinder at time t, and also at any time $t' < t$. There is a *history effect* which is associated with the propagation of viscous shear waves in the fluid and its relative importance in comparison to the vibration velocity of the cylinder is quantified by the viscosity number, as further illustrated in Figure 1.9.

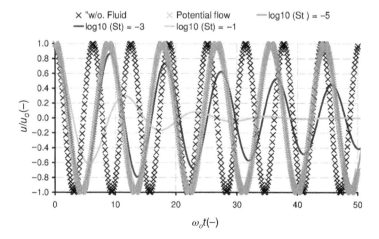

Figure 1.9   FSI effects for viscous flow. The free vibrations of the cylinder without and with fluid are represented using non-dimensional values, for initial conditions $u_o^* = 1$ and $\dot{u}_o^* = 0$ and for various viscosity numbers. For small values of S_t, FSI is mainly of inertial nature and the potential flow model suffices to account for the interactions. For larger values of S_t however, viscous effects are marked by a raise of the kinetic energy in the fluid shear layer (inertial effect is stronger than in the potential flow case), and by a dissipation of the kinetic energy which results from the friction between fluid layers

1.2.3 Compressible Flow. Radiation Damping

As represented in Figure 1.10, the vibrations of a cylinder coupled to a fluid are investigated here under the assumption of an acoustic flow in an infinite domain.

According to Equation (1.1), the fluid force ϕ is obtained by integrating the pressure around the cylinder circumference; for a compressible flow, p satisfies the acoustic wave equation, which is expressed in the present example as follows:

$$\frac{\partial^2 p}{\partial r^2} + \frac{1}{r}\frac{\partial p}{\partial r} + \frac{1}{r^2}\frac{\partial^2 p}{\partial \theta^2} - \frac{1}{c^2}\frac{\partial^2 p}{\partial t^2} = 0$$

using a cylindrical coordinate system. In the above expression, c stands for the speed of sound in the fluid.

The coupling condition with the cylinder motion at $r = R$ is stated as follows:

$$\left.\frac{\partial p}{\partial r}\right|_{r=R} = -\rho \ddot{u}(t)\cos\theta$$

A condition at infinity is also required in order for the problem to be well posed; as made conspicuous in Chapter 3, this condition states that waves travelling away from the structure are not reflected at infinity. In the Laplace domain, the former equation and boundary conditions are expressed as follows:

$$\frac{s^2}{c^2}\hat{p} - \Delta\hat{p} = 0 \qquad \left.\frac{\partial\hat{p}}{\partial r}\right|_{r=R} = -\rho\hat{\ddot{u}}\cos\theta \qquad \hat{p}|_{r\to\infty} < +\infty$$

The pressure field is shown to be

$$\hat{p}(r,\theta,s) = -\rho\frac{1}{s/c}\frac{K(rs/c)}{K'(Rs/c)}\hat{\ddot{u}}(s)\cos\theta$$

where K is the modified Bessel function of the second kind and the first order (Abramowitz and Stegun, 1970).

The fluid force may thereby be expressed as follows:

$$\hat{\phi}(s) = -\rho\pi R^2 \,\hat{\mu}(s)\hat{\ddot{u}}(s)$$

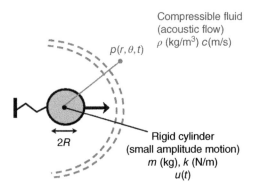

Compressible fluid
(acoustic flow)
ρ (kg/m³) c(m/s)

$p(r,\theta,t)$

$2R$

Rigid cylinder
(small amplitude motion)
m (kg), k (N/m)
$u(t)$

Figure 1.10 Cylinder coupled to a compressible fluid in an unbounded domain

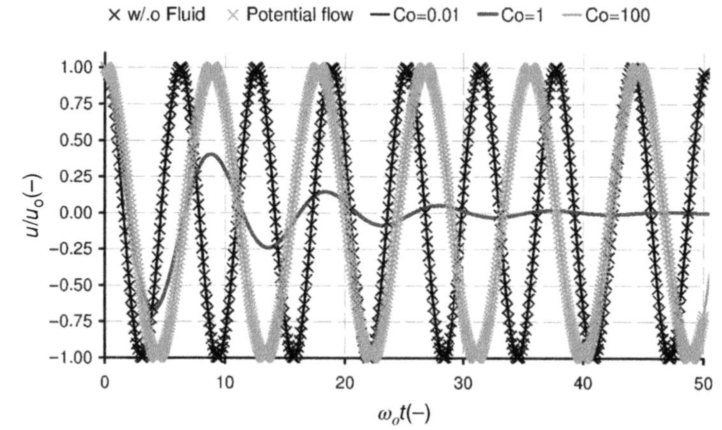

Figure 1.11 FSI effects for acoustic flow. The free vibrations of the cylinder without and with fluid are represented using non-dimensional values, for initial conditions $u_o^* = 1$ and $\dot{u}_o^* = 0$ and for various compressibility numbers. For large values of C_o on the one hand, FSI is mainly of inertial nature: the system behaves as if coupled to an incompressible fluid. For small values of C_o on the other hand, FSI may safely be neglected and the system behaves as if no fluid were present. For intermediate values of C_o, FSI combines inertial and radiative effects: some kinetic energy is conveyed by the vibrating cylinder to the fluid, which in turn propagates this energy throughout the external domain. This corresponds to an energy loss for the structure, hence its vibrations are damped

where the FSI function reads as

$$\hat{\mu}(s) = -\frac{1}{Rs/c}\frac{K(Rs/c)}{K'(Rs/c)}$$

The equation of motion of the cylinder in the Laplace domain using non-dimensional variables is then expressed as follows:

$$(1 + \mathcal{M}_a\hat{\mu}(s^*/C_o))\hat{\ddot{u}}^* + \hat{u}^* = 0 \tag{1.4}$$

with:

$$C_o = \frac{c}{\omega_o R}$$

C_o is the so-called *compressibility number*: it may be interpreted as the ratio between the propagation velocity of acoustic waves in the fluid and the vibration velocity of the cylinder.

In the same manner as for the viscous flow, the influence of the fluid on the solid motion is expressed by the FSI function $\hat{\mu}(s^*/C_o)$. The history effect is associated here with the propagation of acoustic waves in the fluid, and it is quantified by the compressibility number[3], as evidenced in Figure 1.11.

Remark 1.1 Radiative damping without inertial effect *In the example depicted in Figure 1.10 and discussed above, FSI is, as a general rule, a combination of inertial effect*

[3] It is stressed here that viscous and radiative damping are different in nature, since the latter corresponds to energy dissipation away from the structure by travelling waves, whereas the former corresponds to energy dissipation in the shear layer of the fluid, that is, in the vicinity of the structure.

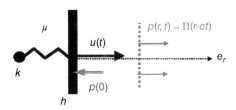

Figure 1.12   *Plate coupled to a compressible fluid in an unbounded domain*

and acoustic damping. In some particular cases however, only a single effect is observed, as evidenced by the following example represented in Figure 1.12.

The motion of the plate is governed by a second-order differential equation:

$$\mu h \ddot{u} + k u = -p|_{r=0}$$

where the pressure field complies with the wave equation, stated here as follows:

$$\frac{1}{c^2} \frac{\partial^2 p}{\partial t^2} - \frac{\partial^2 p}{\partial r^2} = 0$$

and endowed with a non-reflection condition at infinity and a coupling condition with the moving wall at r = 0:

$$\left. \frac{\partial p}{\partial r} \right|_{r=0} = -\rho \ddot{u}$$

As detailed in Chapter 3, the pressure field is found to be $p(r,t) = \Pi(r - ct)$: it describes a plane wave propagating away from the moving plate. The coupling condition is $\Pi'(-ct) = -\rho \ddot{u}$ so that the wall pressure is found to be $p|_{r=0} = \Pi(-ct) = \rho c \dot{u}$. The plate motion is thereby

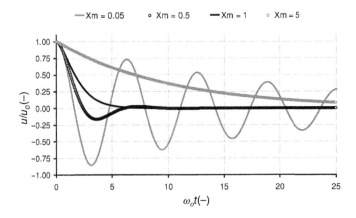

Figure 1.13   Radiative damping. *For a plate/plane wave coupled system, FSI is found to be of dissipative nature as a result of radiation. This is visible in the graph which plots the displacement of the system in the time domain with the initial conditions $u(0)/u_o = 1$, $\dot{u}(0)/\dot{u}_o = 0$, and which evidences the typical free oscillation regimes: (i) the pseudo-periodic regime ($\mathcal{X}_m < 1$); (ii) the aperiodic regime ($\mathcal{X}_m > 1$); (iii) the critical aperiodic regime ($\mathcal{X}_m = 1$)*

governed by the differential equation $\mu h \ddot{u} + \rho c \dot{u} + k u = 0$. In a non-dimensional form, the latter equation may be recast as follows:

$$\ddot{u}^* + 2\mathcal{X}_m \dot{u}^* + u^* = 0$$

with $\mathcal{X}_m = \frac{1}{2}\frac{\rho}{\mu}\frac{c}{\omega_o h}$.

As a remarkable feature of FSI in this case, radiative damping solely governs the interaction, which could be counter-intuitive at first glance.[4] $\mathcal{X}_m = 1/2\mathcal{M}_a\mathcal{C}_o$ combines mass number and compressibility number and quantifies radiative damping, as evidenced in Figure 1.13. ∎

Further discussions on FSI effects for diversified fluid flow configurations may be found in many textbooks, which provide a deeper insight into the topic: for instance, Axisa (2007) investigates various physical aspects of fluid–structure coupling for stagnant fluids, Païdoussis (2004) and Païdoussis *et al.* (2011) propose an extensive overview of fluid–structure interactions, particularly on stability issues for structures subjected to axial flows or cross-flows.

1.3   Numerical Simulation of Fluid–Structure Interactions

The continuous development of numerical techniques and the constant growth of computational capacities make it possible to perform complex simulations, which account for various multi-physic coupling, among which fluid and structure interactions.

Simulation of FSI tends to become a specific topic in Computational Mechanics, as exposed, for instance, in Bungartz and Schäffer (2006), Benson and Souli (2010), Bazilevs *et al.* (2013) and Bodnar *et al.* (2014), and opens new paths in structural analysis for exploring and evaluating new concepts or designs, especially in applications where the empirical approach is dominant, for instance, in the design of musical instruments (Derveau *et al.*, 2003).

As an illustration of some of the numerical techniques available to the researcher and the practitioner, a few specific examples of FSI simulations are reported in Figure 1.14.

The nature of the physics involved in fluid and structure interactions is so diverse, and the scope of the numerical methods which can be used to represent them is so large that it is difficult to propose a general classification of FSI computational techniques. However, the vast majority of existing methods may be found to belong to one of the approaches depicted in Figure 1.15, the choice of a particular numerical strategy being always a compromise between computational cost, accuracy, robustness and stability.

Using a fluid–structure code is possible when a mathematical model of the coupled problem is established in a form which is suited for numerical discretisation with a unique technique (in terms of space and time discretisations); see for instance Le Tallec and Mouro (2001). This approach is therefore adapted to strong coupling mechanisms and it achieves high accuracy and stability; in terms of modelling and computing, this methodology generally requires a dedicated code to be developed and it proves very costly, especially for non-linear

---

[4] This example is not only worthy of interest from the pedagogic standpoint, but it has also some practical relevance, in particular, for the pre-design of naval or offshore structures (Shin, 2004). The response of a submerged structure to a pressure impulse triggered by a distant underwater explosion may indeed be accounted for with the plate/plane wave interaction model as further discussed by Taylor (1941).

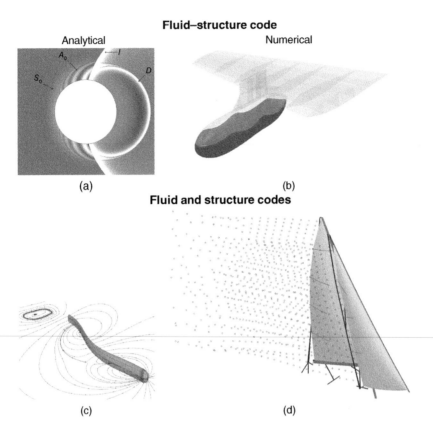

Figure 1.14 Examples of FSI simulations. (a) Leblond *et al.* (2009) develop a coupled model to tackle the vibro-acoustics of submerged shells. The partial differential equations involved in the model are solved in a single code, which allows the description of specific aspects of the interaction. The incoming acoustic wave (I) is diffracted by the shell (D) whose vibrations generate waves in the fluid (A_0 and S_0). The simulation is in good agreement with experimental observations discussed in Ahyi *et al.* (1998). (b) Schotte and Ohayon (2009) formulate a mathematical model accounting for free surface effects in deformable reservoirs of complex shape. The coupled problem is solved using the Finite Element Method (FEM), according to the general principles which will be detailed in Chapter 5. *Source*: Jean-Sébastien SCHOTTÉ, ONREA, Chatillon, France, 2009. Reproduced with permission of Jean-Sébastien SCHOTTÉ. (c) Leroyer and Visonneau (2005) investigate the self-propulsion of a fish-like body, using a numerical procedure which couples the resolution of the Navier–Stokes equation (describing the fluid flow) with a finite volume method (FVM) in Eulerian formulation and the resolution of the Newton law equations (accounting for the structure motion) with a finite element technique in Lagrangian formulation. *Source*: Alban LEROYER, Ecole Centrale, Nantes, France, 2005. Reproduced with permission of Alban LEROYER. (d) Augier *et al.* (2012) propose a numerical model for yacht sail design, by coupling a finite element model composed of beams, cables and membranes, which are standing for the constitutive parts of the boat (*e.g.* spars, rigging and sails), with a vortex lattice method (VLM), which is suited for external fluid flows where vorticity develops in the vicinity of lifting surfaces. *Source*: Patrick BOT, Ecole Navale, Brest, France, 2014. Reproduced with permission of Patrick BOT

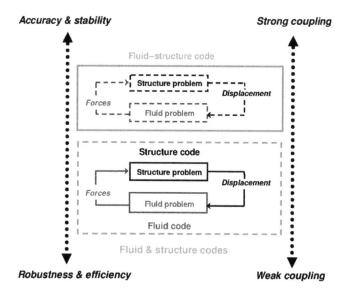

Figure 1.15 Numerical simulation of FSI: coupled fluid–structure code or fluid and structure codes coupling strategy

problems. As exposed in Chapter 5, this strategy is however suited for linear problems formulated on a fixed domain, among which coupled problems involved in vibro-acoustics.

Coupling fluid and structure codes takes advantage of the robustness of each numerical tool and provides efficient solutions for engineering purposes (Felippa *et al.*, 2001; Degroote and Vierendeels, 2011). This *partitioned approach* has widely been investigated since the pioneer researches of Farhat and Piperno (1997) for aero-elastic simulations. Such a methodology requires the development of specific formulations, such as Arbitrary Lagrangian–Eulerian (ALE), and a coupling procedure which allows for space and time coupling (Casadeï *et al.*, 2001).

- ALE (Donea *et al.*, 1982) is particularly suited to simulate FSI, since it combines the Lagrangian formulation, which is the standard framework for structural dynamics, and the Eulerian formulation, which is adapted to fluid dynamics; see Figure 1.16.

    With the Eulerian formulation, the equation of motion is written in a spatial domain so that the system moves through a fixed grid: this approach is therefore particularly suited to the description of fluid flows. In the Lagrangian formulation, the equation of motion is written in a material domain: the motion is tracked by a grid which deforms while the system moves. This approach is adapted for structures undergoing large displacements, as long as the grid deformation remains limited.

    Arbitrary Lagrangian–Eulerian formulation combines both descriptions by stating the equation of motion in a moving frame: it allows a control of the mesh geometry independently from the material geometry. ALE is of broad use and interest in the context of FSI simulation, since large deformations of the fluid–structure interface can be conveniently

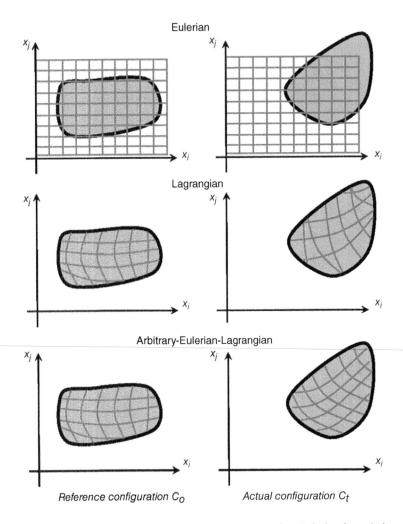

Figure 1.16   Lagrangian, Eulerian and Arbitrary Lagrangian–Eulerian formulations

handled – with a Lagrangian-dominated approach to adjust the structure motion, while the physics of the fluid flow is accounted for with an Eulerian-dominated approach (Souli *et al.*, 2000).

For small transformations about an equilibrium state, which is the framework adopted in this book to describe the vibrations of a structure or a fluid (as discussed further in Chapters 2 and 3), the Lagrangian and Eulerian formulations are equivalent.

- Coupling in space handles the information exchanges between the two codes involved in the simulation (Farhat *et al.*, 1998). The first one, referred to as the *CSD code*[5] solves the equation of motion of the structure subsystem, while the second one, referred to as the

[5] As discussed in Chapter 6, CSD stands for *Computational Structural Dynamics* and designates the set of numerical techniques available to solve the second-order ordinary differential equations which account for the motion of mechanical systems.

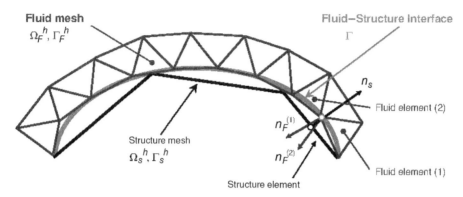

Figure 1.17 **Space coupling**. The finite volume method (FVM) and the finite element method (FEM) are typically used as discretisation techniques, respectively, for Computational Fluid Dynamics (CFD) and Computational Structural Dynamics (CSD). As different physics are represented by each code, different meshes may be required to solve the equations of each sub-problems: at the fluid–structure interface, where force and motion are transferred from one code to another, the compatibility between the fluid and structure meshes is not always possible. Consequently, the definition of the normal at the interface may be ambiguous, as depicted above. Various techniques may be used to convert the pressure computed on one node in the fluid mesh to the corresponding finite element in the structure mesh, for instance with interpolation finite elements (Guruswamy, 1989). As an approximation of the exchanges is inevitably introduced by the discretisation of the interface, the projection technique should be designed in order to limit energy losses (Piperno and Farhat, 2001)

CFD code,[6] each of the two codes may make use of different discretisation and approximation techniques.

The space coupling is therefore designed to be as accurate as possible in order to limit the loss of information, hence of energy, due to discretisation and approximation (Maman and Farhat, 1995; Piperno and Farhat, 2001), see for instance Figure 1.17.

- Coupling in time organises the information exchanges between two time iterations of the codes and achieves a strong or weak coupling: as a delay is introduced between the structure and fluid codes, some energy is lost between iterations. Inaccuracies and instabilities in the simulation may result from the use of algorithms which would fail to correctly account for the mechanical coupling (Piperno *et al.*, 1995), see for instance Figure 1.18.

From a practical point of view, CFD–CSD numerical strategies are relatively mature for many industrial applications; they are commonly based on general procedures, such as the one described in Figure 1.19: coupling techniques are accessible to the engineer with a combination of various CFD and CSD codes; see for instance Gaugain (2013) and Yvin (2014).

The architecture of code coupling depends on the problem under concern, whether FSI is driven by the fluid flow or by the structural response. Robust simulations are often based on staggered or iterated procedures: within a time loop, a coupling loop handles the exchanges between the two codes. The fluid force is passed from the CFD code to the CSD code and, conversely, the fluid–structure interface motion from the CSD code to the CFD code. When large

---

[6] CFD stands for *Computational Fluid Dynamics* and, as illustrated in Remark 3.1, it refers to the numerical techniques aimed at solving the equations of fluid flows.

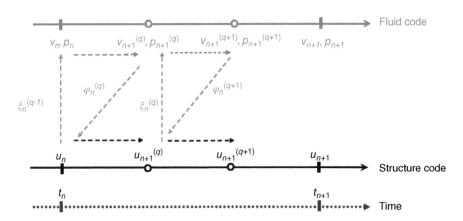

Figure 1.18   Time coupling. From time t_n to time t_{n+1}, CFD and CSD codes are coupled in a staggered manner: the structure displacement $\mathbf{u}_n$ and the fluid–structure interface motion $\boldsymbol{\xi}_n$ are first computed and serve as boundary conditions for the CFD code. The latter computes the fluid pressure and velocity fields p_n and $\mathbf{v}_n$, yielding the fluid force on the structure $\boldsymbol{\varphi}_n$. The CSD code computes the displacement which results from the loading on the structure. Iterations of this procedure within a time step $[t_n, t_{n+1}]$ produce a sequence of structure and pressure fields $\mathbf{u}_n^q, \boldsymbol{\xi}_n^q, p_n^q, \mathbf{v}_n^q$ and $\boldsymbol{\varphi}_n^q$. It gives the updated fields at time t_{n+1} when a convergence criterion is satisfied (Schäffer and Teschauer, 2001)

displacements of the structure are accounted for, the mesh of the fluid problem is modified: this is usually achieved with moving mesh and/or re-meshing techniques; see for instance Figure 1.20.

Iteration of the process is required when the fluid and the structure are 'strongly' coupled. Achieving accuracy and stability demands an important computational effort: coupled strategies based on ALE formulation are in general not affordable for 'highly non-linear' problems.

Although coupled simulations achieve satisfying levels of accuracy and reliability for industrial and academic purposes, it should be mentioned that for the sake of robustness, multi-physic algorithms offered by general-purpose codes are often adapted from coupling strategies presented in the scientific literature: a theoretical and practical involvement is therefore required from any code user to validate the coupling procedures.

Experimental tests on academic cases provide fruitful data for code verification, together with physical insights which guide numerical simulations, as illustrated for instance in Figures 1.21 and 1.22. Experiments and numerics often proceed in a joined manner: on the one hand, a mathematical model is an ideal representation of the physical world, so that a numerical method always exhibits finite accuracy; on the other hand, the reproducibility of experiments is often questionable and data acquisition is always limited. A meaningful numerical/experimental validation might stem from crossing error estimates in simulations with uncertainty quantification in experiments.

When experimental data are not available, it may be worthy of interest to turn to analytical solutions. As outlined in the previous section, analytical models generally deal with simple geometry and physics, but provide some deeper insights into a particular aspect and may serve as a reference for validation purposes; see for instance (Placzek *et al.*, 2009).

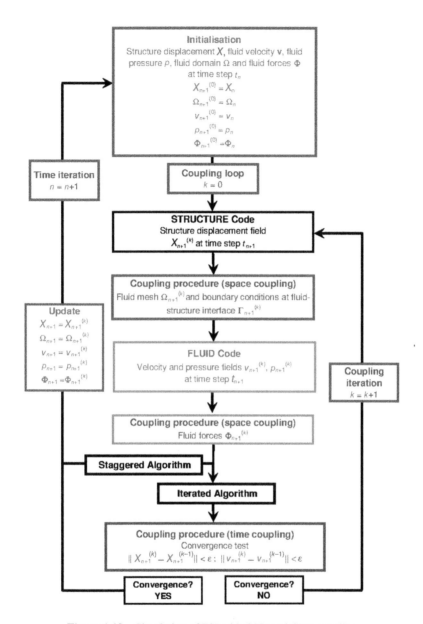

Figure 1.19 Simulation of FSI with CSD and CFD coupling

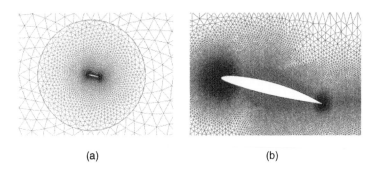

(a)   (b)

Figure 1.20   Moving mesh technique. The moving mesh techniques – for instance with sliding mesh (a) or deforming mesh (b) – are well suited to FSI simulation since they allow for the mesh to adjust to the motion of boundaries, as is the case in the present example, which is concerned with the pitching motion of a hydrofoil in a fluid flow (Ducoin *et al.*, 2009)

1.4   Finite Element and Boundary Element Methods

As suggested by the earlier examples, FSI may be addressed from various points of view, whether the fluid is confined in the structure or contains the structure, whether the fluid is stagnant or flowing. In this case, an additional complexity arises from the flow conditions, which may be compressible or incompressible, separated or attached, laminar or turbulent.

Among the many techniques which are available to the practitioner, Figure 1.23 gives a simplified overview of coupling methods which are of common use for engineering purposes.

Flow–Structure Interaction is generally described with *time-domain equations* in the *actual configuration* of the system: coupling CSD and CFD codes, as presented in the previous section, is, among various options, one of the most convenient strategies to tackle FSI.

Fluid–Structure Interaction is usually modelled with equations of motion written in the *frequency domain* (for linear problems, it is equivalent to a time-domain formulation, through the Laplace transform). Since small displacements of the fluid and structure are considered, these equations are stated in the *reference configuration* of the system. Coupling fluid and structure finite elements is adapted to problems involving a fluid contained in a structure. *Infinite elements* or *boundary elements* allows for FSI modelling when considering structures submerged in an unbounded fluid domain. In the former case, a unique finite element code solves the coupled problem, whereas in the latter two codes may need to be coupled.

This book is centred for the most part on the mathematical modelling and on the numerical simulation of FSI, for elastic *structures* coupled to a *quiescent* fluid, whether bounded or not, using *finite element* or *boundary element* methods. According to the FEM, both the continuous domain and its boundaries are discretised, while according to the Boundary Element Method (BEM), the discretisation is restricted to the boundaries, which may be advantageous in terms of computational effort when one has to deal with a large, or even infinite, extent of fluid – in the latter case, the unbounded character of the fluid domain may also be tackled with the infinite element method (IEM).

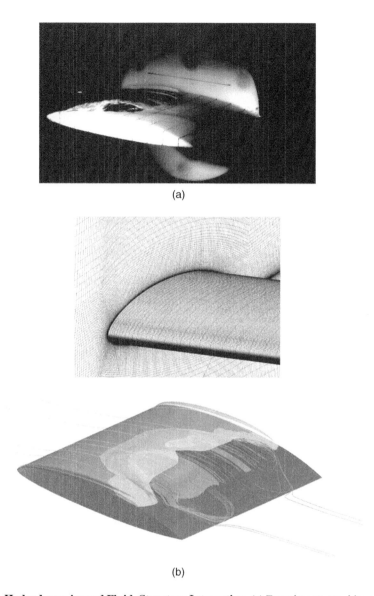

(a)

(b)

Figure 1.21 Hydrodynamics and Fluid–Structure Interaction. (a) Experiments provide meaningful data to understand the physics of complex interactions, such as the one evidenced in Ducoin *et al.* (2012) as far as the dynamic behaviour of an elastic hydrofoil in cavitating flow is concerned. (b) Coupling between the formation of sheet cavitation and the vibrations of the hydrofoil is accurately reproduced with CFD–CSD coupling procedures built with general-purpose numerical codes (Gaugain *et al.*, 2012). *Source*: André ASTOLFI, Ecole Navale, Brest, France and Fabien GAUGAIN, DCNS Research, Nantes, France, 2013. Reproduced with permission of André ASTOLFI

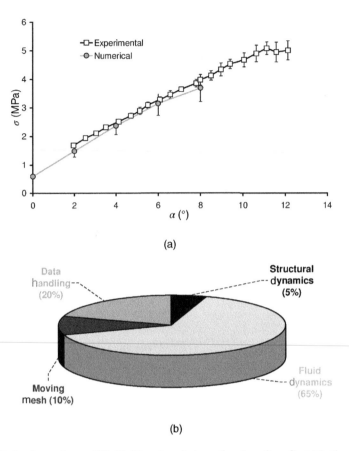

(a)

(b)

Figure 1.22   Hydrodynamics and Fluid–Structure Interaction (continued). (a) In the example proposed in (Gaugain, 2013), the agreement between experiments and simulations is quite remarkable and builds confidence in the future engineering applications of the numerical tool. (b) For such simulations, the computational cost is mostly driven by CFD, but moving mesh and data handling are also demanding. However, coupled simulations which capture some fine physics remain today out of reach for large industrial applications

The vibrations of structure and fluid systems, as well as the finite element method and the boundary element method, are introduced in Chapters 2 and 3. Finite element coupling is then described in Chapter 4 for the modelling of inertial effects. Fluid–structure coupling is addressed in Chapter 5, mainly focussing on vibro-acoustics; various mathematical formulations of the coupled problem are detailed and discussed. Structural dynamics with FSI, either for time-domain or for frequency-domain analyses, is finally presented in Chapter 6. Numerous application examples are also proposed to deepen the analysis; each of them has been programmed using MATLAB, which offers a convenient approach of numerical methods for engineers (Kiusalaas, 2005).

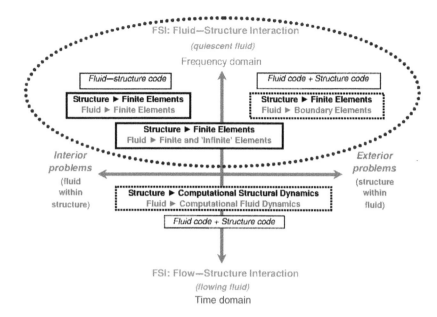

Figure 1.23   Engineering methods for FSI simulations

References

Abramowitz M and Stegun IA 1970 *Handbook of Mathematical Functions*. Dover Publications.

Ahyi AC, Pernod P, Gatti O, Lattard V, Merlen A, and Uberall H 1998 Experimental demonstration of the pseudo-Rayleigh wave. *Journal of Acoustical Society of America*, **104**, 2727–2732.

Augier B, Bot P, Auville F, and Durand M 2012 Dynamic behaviour of a flexible yatch sail plan. *Ocean Engineering*, **66**, 32–43.

Axisa F 2007 *Modelling of Mechanical Systems – Fluid-Structure Interaction*. Elsevier.

Bazilevs Y, Takizawa K, and Tezduyar TE 2013 *Computational Fluid-Structure Interaction: Methods and Applications*. John Wiley & Sons, Ltd.

Benson DJ and Souli M 2010 *Arbitrary Lagrangian Eulerian and Fluid-Structure Interaction: Numerical Simulation*. John Wiley & Sons, Ltd.

Bodnar T, Galdi GP, and Necasova S 2014 *Fluid-Structure Interaction and Biomedical Applications*. Springer-Verlag.

Bundgartz J and Schäffer M 2006 *Fluid-Structure Interaction: Modelling, Simulation, Optimization*. Springer-Verlag.

Casadëi F, alleux JP, Sala A, and Chille F. 2001 Transient fluid-structure interaction algorithms for large industrial applications. *Computer Methods in Applied Mechanics and Engineering*, **190**, 3081–3110.

Degroote J and Vierendeels J 2011 Multi-solver algorithms for the partitioned simulation of fluid-structure interaction. *Computer Methods in Applied Mechanics and Engineering*, **200**, 2195–2210.

De Langre E 2001 *Fluides et solides*. Editions de l'Ecole Polytechnique.

Derveau G, Chaigne A, Joly P, and Béchade E 2003 Time-domain simulation of a guitar: model and method. *Journal of the Acoustical Society of America*, **114**, 3368–3383.

Donea J, Guiliani S, and Halleux JP 1982 An Arbitrary-Lagrangian-Eulerian finite element method for transient dynamic fluid-structure interaction. *Computer Methods in Applied Mechanics and Engineering*, **33**, 689–723.

Ducoin A, Astolfi JA, Deniset F, and Sigrist JF 2009 Computational and experimental investigation of flow over a transient pitching hydrofoil. *European Journal of Mechanics - B/Fluids*, **28**, 728–743.

Ducoin A, Astolfi JA, and Sigrist JF 2012 An Experimental analysis of fluid-structure interaction on a flexible hydro-foil in various flow regimes including cavitating flow. *European Journal of Mechanics - B/Fluids*, **36**, 63–74.

Farhat C, Lesoinne M, and Le Tallec P 1998 Load and motion transfer algorithm for fluid-structure interaction problems with non-matching discrete interfaces: momentum and energy conservation, optimal discretisation and application to aeroelasticity. *Computer Methods in Applied Mechanics and Engineering*, **157**, 95–114.

Felippa C, Park KC, and Farhat C 2001 Partitioned analysis of coupled mechanical systems. *Computer Methods in Applied Mechanics and Engineering*, **190**, 3247–3270.

Fritz RJ 1972 The Effect of liquids on the dynamic motion of immersed solids. *Journal of Engineering for Industry*, **91**, 167–173.

Gaugain F 2013 Analyse expérimentale et simulations numériques de l'interaction fluide-structure d'un hydrofoil élastique en écoulement cavitant et subcavitant. PhD thesis, Ecole Nationale Supérieure des Arts & Métiers, Paris.

Gaugain F, Astolfi A, Sigrist JF, and Deniset F 2012 Numerical and experimental study of the hydroelastic behaviour of an hydrofoil. In *Proceedings of the 10th International Conference on Flow Induced Vibration (FIV 2012)*.

Guruswamy GP 1989 Integrated approach for active coupling of structures and fluid. *AIAA Journal*, **27**, 788–793.

Kiusalaas J 2005 *Numerical Methods in Engineering with MATLAB*. Cambridge University Press.

Leblond C, Iakovlev S, and Sigrist JF 2009. A fully elastic model for studying submerged circular cylindrical shells subjected to a weak shock wave. *Mécanique & Industries*, **10**, 275–284.

Leblond C, Sigrist JF, Auvity B, and Peerhossaini H 2009 A Semi-analytical approach for the study of an elastic circular cylinder in cylindrical fluid domain subjected to small amplitude transient motion. *Journal of Fluids and Structures*, **25**, 134–154.

Leroyer A and Visonneau M 2005 Numerical methods for RANSE simulations of a self-propelled fish-like body. *Journal of Fluids and Structures*, **20**, 975–991.

Le Tallec P and Mouro J 2001 Fluid-structure interaction with large structural displacements. *Computer Methods in Applied Mechanics and Engineering*, **190**, 3039–3067.

Maman N and Farhat C 1995 Matching fluid and structure meshes for aeroelastic computations: a parallel approach. *Computers & Structures*, **54**, 779–785.

Païdoussis MP 2004 *Fluid-Structure Interactions: Slender Structures and Axial Flow*. Academic Press.

Païdoussis MP, Price SJ, and De Langre E 2011 *Fluid-Structure Interactions – Cross-Flow-Induced Instabilities*. Cambridge University Press.

Piperno S 1997 Explicit-implicit fluid-structure staggered procedures with a structural predictor and fluid subcycling for 2D inviscid aeroelastic simulations. *International Journal for Numerical Methods in Fluids*, **25**, 1207–1226.

Piperno S and Farhat C 2001 Partitioned procedures for the transient solution of coupled aeroelastic problems. Part II: energy transfer analysis and three-dimensional application. *Computer Methods in Applied Mechanics and Engineering*, **190**, 3147–3170.

Piperno S, Farhat C, and Larrouturou B 1995 Partitioned procedures for the transient solution of coupled aeroelastic problems. Part I: model problem, theory and two-dimensional application. *Computer Methods in Applied Mechanical and Engineering*, **124**, 79–112.

Placzek A, Sigrist JF, and Hamdouni A 2009 Numerical simulation of an oscillating cylinder in a cross-flow at low Reynolds number: forced and free oscillations. *Computers & Fluids*, **38**, 80–100.

Schäffer M and Teschauer I 2001 Numerical simulation of coupled fluid-solid problems. *Computer Methods in Applied Mechanics and Engineering*, **190**, 3645–3667.

Schotté JS and Ohayon R 2009 Various modelling levels to represent internal liquid behaviour in the vibratory analysis of complex structures. *Computer Methods in Applied Mechanics and Engineering*, **198**, 1913–1925.

Shin Y 2004 Ship shock modelling and simulation of far-field underwater explosion. *Computers and Structures*, **82**, 2211–2219.

Souli M, Ouahsine A, and Lewin L 2000 ALE formulation for fluid-structure interaction problems. *Computer Methods in Applied Mechanics and Engineering*, **190**, 659–675.

Taylor GI 1941 *The Pressure and Impulse of Submarine Explosion Waves on Plate*. Cambridge University Press.

Yvin C 2014. Interaction fluide-structure pour des configurations multi-corps. Applications aux liaisons complexes, lois de commande d'actionneur et systèmes souples dans le domaine maritime. PhD thesis, Ecole Centrale, Nantes.

2

Structure Finite Elements

This chapter is concerned with the discretisation of the partial differential equations accounting for the linear vibrations of elastic structures by using the finite element method (FEM). Weighted integral formulation, geometry discretisation, polynomial approximation, definition of elementary and global matrices: the various steps of the FEM are recalled hereafter and are illustrated on simple cases.

Figure 2.1   Structural dynamics. The motion of complex structures, whether composed of continuous or discrete elements, of massive or slender subsystems, is governed by a set of partial differential equations. The objective of numerical methods which perform the discretisation of these equations is to reduce the problem to a finite number of degrees-of-freedom (DOF), while replacing the partial space and time derivative by linear algebraic equations, thus making the system amenable to automatic numerical computations. *Source*: © Jean-François Sigrist

Fluid–Structure Interaction: An Introduction to Finite Element Coupling, First Edition. Jean-François Sigrist.
© 2015 John Wiley & Sons, Ltd. Published 2015 by John Wiley & Sons, Ltd.
Companion Website: www.wiley.com/go/sigrist

2.1   Vibrations of an Elastic Structure

This section is concerned with the *linear vibrations* of an elastic structure, 'vibration' referring here to the small amplitude motion of a continuous system oscillating about a static equilibrium. To begin with, a review of the few basic notions of continuum mechanic theory, which are needed subsequently throughout this chapter, is proposed. Adopting a Lagrangian description of the problem, the strain and stress tensor are introduced. The stress–strain relation is defined by the so-called constitutive law, which is formulated for isotropic homogenous materials and discussed for other materials.

2.1.1   Modelling Assumptions

Structural dynamics is conveniently modelled within the framework of *continuum mechanics*[1] using the Lagrangian formulation. As depicted in Figure 2.2, C_0 is the *reference* configuration (associated with the un-deformed state of the mechanical system and attached to the coordinate system $\mathcal{R}_0$) and C_t is the *actual* configuration (associated with the deformed state and attached to the coordinate system $\mathcal{R}_t$). The motion of any point of the continuous system is accounted for with a mapping:

$$\chi : M_o(\mathbf{X}) \in C_o \hookrightarrow M_t(\mathbf{x}) \in C_t$$

$\mathbf{u} = \mathbf{M}_o\mathbf{M}_t$ is the *displacement field*, whose components in the Cartesian coordinate system are $(u_i)_{i \in [1,d]}$, where d denotes the space dimension of the problem.

The deformations which result from the structure displacements are quantified by the *strain tensor* $\varepsilon(\mathbf{u})$. The components of the strain tensor are denoted $\varepsilon_{ij}(\mathbf{u})$ with $(i,j) \in [1,d] \times [1,d]$; they are calculated from the displacement field according to:

$$\varepsilon_{ij}(\mathbf{u}) = \frac{1}{2} \left(\frac{\partial u_i}{\partial x_j} + \frac{\partial u_j}{\partial x_i} + \frac{\partial u_i}{\partial x_k} \frac{\partial u_j}{\partial x_k} \right)$$

Internal forces which are induced in the structure by its deformations are measured by the *stress tensor* $\sigma(\mathbf{u})$; the components of the stress tensor are denoted $\sigma_{ij}(\mathbf{u})$ with $(i,j) \in [1,d] \times [1,d]$. The stress–strain relation $\sigma(\mathbf{u}) = \Sigma(\varepsilon(\mathbf{u}))$ is modelled with the *constitutive equation*, which is dependent on the structure material properties.

The *vibrations* of a structure are described in the context of *small transformations*, that is, both the displacement and its gradient are supposed to be small compared to a characteristic length L of the problem and to unity ($\|\mathbf{u}\| \ll L$ and $\|\nabla\mathbf{u}\| \ll 1$); consequences of this hypothesis are twofold.

- The actual and the reference configurations of the continuous media are identical, and the equations of motion are therefore stated on a fixed domain Ω of boundary $\Gamma = \Gamma_o \cup \Gamma_\sigma$, with

---

[1] The formulation derived in what follows is carried out within the framework of tri-dimensional continuum mechanics. Similar derivations are available to deal with structural elements such as beams, membranes or shells, as further highlighted in the book by using a few specific examples. For the sake of simplicity, the mathematical model presented here is valid in the Euclidian framework, with an implicit hypothesis of orthogonality condition of distinct directions, so that no coupling occurs between two components i, j of the displacement field. An example described later in this chapter illustrates the coupling between DOF and some numerical problems which arise in such cases. Finally, it must be noted that hereafter, the equation makes use of the summation convention, that is, a repetition of the index indicates summation: $\bullet_i \bullet_{ij} = \Sigma_i \bullet_i \bullet_{ij}$.

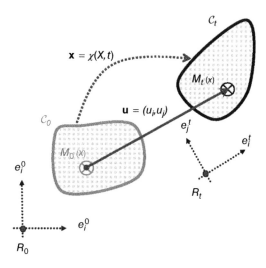

Figure 2.2   Lagrangian description in continuum mechanics: from reference configuration C_o to actual configuration C_t

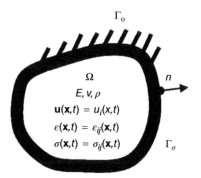

Figure 2.3   Definitions and notations of the structure problem

$\Gamma_o \cap \Gamma_\sigma = \emptyset$; on Γ, the outer unit normal vector is denoted **n**, according to the notations introduced in Figure 2.3.

- The second-order contribution of the displacement is discarded in the displacement–strain relation, so that the strain becomes:

$$\varepsilon_{ij}(\mathbf{u}) = \frac{1}{2}\left(\frac{\partial u_i}{\partial x_j} + \frac{\partial u_j}{\partial x_i}\right) \tag{2.1}$$

In addition, the structure constitutive material is supposed to be *homogeneous* and *isotropic* and have a *linear* behaviour, which means that the material properties are constant throughout the volume occupied by the structure, they do not depend on any particular direction of observation and $\Sigma(\bullet)$ is a linear function.

Linear elasticity is described by the Hooke law, which can be formulated analytically in various equivalent ways, all making use of two material parameters. A first formulation is as follows[2]:

$$\sigma_{ij}(\mathbf{u}) = \frac{E}{1+v}\left(\frac{v}{1-2v}\varepsilon_{kk}(\mathbf{u})\delta_{ij} + \varepsilon_{ij}(\mathbf{u})\right) \tag{2.2}$$

Using Equation (2.1), it can be equivalently formulated as follows:

$$\sigma_{ij}(\mathbf{u}) = \frac{E}{1+v}\left(\frac{v}{1-2v}\frac{\partial u_k}{\partial x_k}\delta_{ij} + \frac{1}{2}\left(\frac{\partial u_i}{\partial x_j} + \frac{\partial u_j}{\partial x_i}\right)\right)$$

E and v designate Young's modulus and Poisson's ratio, while the material density is denoted ρ. Two other formulations are worth quoted here.

- In the first one, Young's modulus and Poisson's ratio are replaced by the bulk modulus K and the shear modulus and G:

$$\sigma_{ij}(\mathbf{u}) = 3K(\varepsilon_{kk}(\mathbf{u})/3\delta_{ij}) + 2G(\varepsilon_{ij}(\mathbf{u}) - \varepsilon_{kk}(\mathbf{u})/3\delta_{ij})$$

with:

$$K = \frac{E}{3(1-2v)} \qquad G = \frac{E}{2(1+v)}$$

The bulk modulus and the shear modulus measure the resistance of the material, respectively, to compression and to shear.
- The second one is expressed in terms of the Lamé coefficients λ and μ:

$$\sigma_{ij}(\mathbf{u}) = \lambda\varepsilon_{kk}(\mathbf{u})\delta_{ij} + 2\mu\varepsilon_{ij}(\mathbf{u}) \tag{2.3}$$

μ is an alternative notation for the shear modulus; λ is related to the bulk and shear modulus according to $\lambda = K - 2/3G$ or also:

$$\lambda = \frac{Ev}{(1+v)(1-2v)}$$

Young's modulus measures the stiffness of an elastic material; it can be experimentally evidenced with a tensile test, as explained in Figure 2.4.

As evidenced in Table 2.1, Young's modulus varies over a wide extent, depending on the material under consideration. Steel, being a good representative for metallic materials, is used for most of the numerical applications throughout the book: the typical value of Young's modulus for steel is $E \sim 2.1 \times 10^{11}$ Pa.

v quantifies the Poisson effect: when stretched in a given direction, a material tends to contract in other directions. The Poisson's ratio varies between 0 (for materials which tend to be insensitive to the Poisson effect – such as cork) and 0.5 (in which case the material tends to be incompressible – such as rubber for instance). Among metallic materials, such as steel, the typical value of the Poisson's ratio is $v \sim 0.3$.

---

[2] δ_{ij} denotes the components of the identity tensor, more specifically: $\delta_{ii} = 1$ and $\delta_{ij} = 0$ for $i \neq j$.

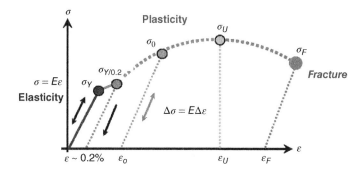

Figure 2.4 Stress–strain curve. A tensile test evidences the behaviour of a material with a sample being subjected to traction. In the present example, a typical stress–strain curve is presented for a metallic material. In the *elasticity* regime, the material returns to its initial (non-deformed) shape when the load is removed. The stress–strain relation is proportional $\sigma = E\varepsilon$ and the linear regime is valid in a rather limited range of small strains solely. The yield stress σ_Y identifies the upper bound of the elastic response. In the *plasticity* regime, a *ductile* material deforms under loading and exhibits a permanent deformation when the load is removed. This holds up to a certain limit, until the material breaks: the ultimate stress σ_U is the maximum strength the material can withstand, while ε_U denote the ultimate elongation. As the elasticity–plasticity transition is in general not sharply defined, the yield stress is usually defined as the stress which causes a permanent 0.2% plastic deformation. In the plasticity regime, the material gains permanent deformation and becomes harder, while more fragile – the process being referred to as *hardening*. The yield stress increases, and, as long as the plastic stress is lower than this threshold, a linear behaviour is observed from the unloaded state (with permanent strain ε_0) up to the yield stress σ_0. The plasticity regime is not observed – or it is significantly reduced – for a *brittle* material, such as glass or concrete

Table 2.1 Young's modulus for various materials

Material	E(GPa)
Aluminium	~75
Titane	~110
Bronze	~125
Granite	~50
Wood	~10
Carbon fibre	~250
Polyamid	~5
Silk	~25

Remark 2.1 Material constitutive properties *Without a significant loss of generality, considering a linear homogeneous and isotropic material is adapted to an introduction of structural elasticity and to a subsequent formulation of finite element techniques. However, many applications which can be run across in the industrial field involve other classes of material constitutive properties, among which the following are briefly discussed below.*

Time-/frequency-dependent properties *are encountered for instance in* viscoelastic materials[3], *which are often used to damp out structural vibrations, or in* elastomer materials, *which are commonly used for mechanical insulation or sealing purposes.*

Visco-elastic properties can be identified through various measuring techniques, for instance with frequency-based methods. With the dynamical mechanical analysis (DMA) test, for instance, a phase and amplitude lag between the applied strain and the stress state observed in the tested sample is evidenced; see Figure 2.5.

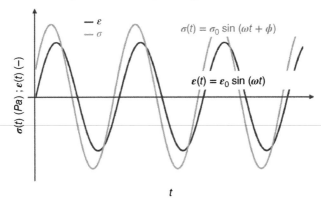

Figure 2.5 *Typical stress–strain response observed for a visco-elastic material with frequency-based methods*

The stress–strain relation is stated in the frequency-domain as follows:

$$\sigma^*(\omega) = E^*(\omega)\varepsilon^*(\omega)$$

where $E^*(\omega) = \|E(\omega)\| \exp(i\Phi(\omega)) = E'(\omega)(1 + i\eta(\omega))$ is a complex-valued scalar. The previous expression indicates that the visco-elastic material constitutive law is time dependent *and* dissipative.

- *$E'(\omega) = \Re(E^*(\omega))$ is the* storage modulus*: it accounts for energy storage in the material;*
- *$\eta(\omega)$ is the* loss factor *$\eta(\omega) = \Im(E^*(\omega))/\Re(E^*(\omega))$: it quantifies the energy dissipation.*

Several rheologic models are available to describe the visco-elastic behaviours of diverse complexity. All of them do comply with various physical constraints, such as the causality principle. In particular, the fractional derivative models[4] are of practical interest because they achieve a good compromise between consistency and complexity. With the fractional Zener model, the frequency-dependent Young's modulus is expressed as follows:

$$E^*(\omega) = \frac{E_o + E_\infty (i\omega\tau)^\alpha}{1 + (i\omega\tau)^\alpha}$$

---

[3] General considerations on such materials lay far beyond the scope of the present book: an introduction to viscoelasticity may be found, for instance, in Christensen (1982).
[4] See for instance Scheissel *et al.* (1995).

where E_o and E_∞ describe the material stiffness in the low- and high-frequency regimes, respectively, τ is the so-called relaxation time and α is the order of the fractional derivative. A typical master curve derived from the fractional model is plotted in Figure 2.6.

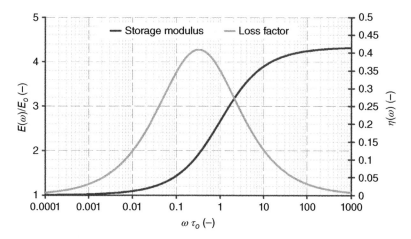

Figure 2.6 *Fractional derivative model: typical master curve with parameters $E_o = 7.5 \times 10^9$ Pa, $E_\infty = 32.5 \times 10^9$ Pa, $\tau = 5.0 \times 10^{-5}$ s and $\alpha = 0.65$*

Anisotropic properties *come upon for instance with* composite materials, *Figure 2.7, which are especially designed to various purposes – lightening structures being one of the most common, for instance, in civil construction and shipbuilding – and whose range of use is extended.*

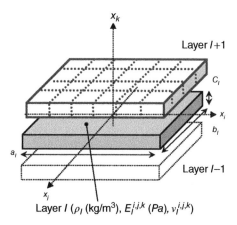

Layer I (ρ_I (kg/m³), $E_I^{j,j,k}$ (Pa), $v_I^{j,j,k}$)

Figure 2.7   Composite materials. *Composites are an assembly of multi-materials which possess different mechanical properties: the resulting material is expected to bear properties which combine the advantages of all of its constitutive sub-parts – for instance, in terms of weight, strength or versatility. In the present example, the composite is made of piled layers or weaved fibres, each layer or fibre being defined by its material properties, which depend on their geometrical orientation. In a numerical model, the anisotropy may be accounted for in a global manner, for instance, using a homogenised representation of traction-compression or shear stiffness, or with a detailed model of each sub-parts, when their individual material properties are well characterised*

The material properties of composite structures can vary within an extended range, depending, in particular, on their production process. Suffice it here to illustrate the anisotropy with a simple composite panel, as illustrated in Figure 2.8 and Table 2.2.

Specificities of finite element modelling for multi-material mechanical systems remain obviously beyond the object of this book[5]. It may be worth to emphasise that

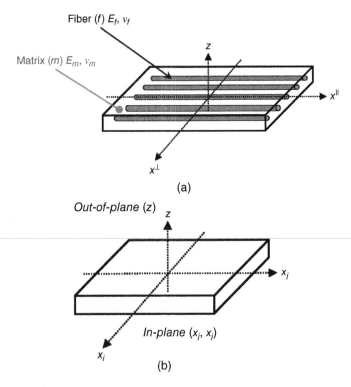

(a)

(b)

Figure 2.8   Anisotropy. *The anisotropic behaviour of composites is illustrated here with two simple examples. (a) In the first one, borrowed from Reddy (2004), a unidirectional fibre-reinforced composite lamina is considered. The lamina is composed of fibres oriented in a given direction and embedded within a matrix. The fibres and matrix are supposed to be isotropic and to comply with the Hooke law. Their individual material properties are denoted E_m, v_m and E_f, v_f respectively. Provided that the lamina behaves as a linear elastic material, the material properties of the composite are derived from a weighted average model: $E_\parallel = E_f \phi + E_m(1 - \phi)$, $E_\perp = \frac{E_f E_m}{E_f(1-\phi)+E_m\phi}$ and $v = v_f \phi + v_m(1 - \phi)$, where $E_\parallel$ and $E_\perp$ are the longitudinal and transverse Young's modulus, v is the Poisson coefficient and ϕ denotes the fibre volume fraction. (b) In the second example, the in-plane and out-of-plane directions are considered for a fibre-reinforced composite panel: as specified in Table 2.2, the material properties, derived here from laboratory tests, significantly differ depending on the in-plane and out-of-plane direction*

---

[5] An introduction on the topic for composites structures may be found in Matthews *et al.* (2000).

Table 2.2   *Material properties for a fibre-reinforced composite material (glass-roving/polyester with 70% reinforcement mass ratio)*

Material property	In-plane	Out-of-plane
E	~20 GPa	~10 GPa
σ_Y	~400 MPa	~20 MPa
ε_F	~2, 3%	~0, 2%

fluid–structure interaction (FSI) for composites is tackled by coupling structural finite ele-ment models to finite element or boundary element models of fluid spaces, in a similar way to what is presented in the present book for structures with isotropic and homogeneous mate-rials. On the other hand, rendering the geometrical complexity of a composite structure and representing each material layer with the appropriate property remains far from a trivial task in engineering finite element analysis.   ∎

2.1.2   Equations of Motion

Under the preceding assumptions, the motion of the structure is described by the momentum balance equation:

$$\rho\frac{\partial^2 u_i}{\partial t^2} - \frac{\partial \sigma_{ij}(\mathbf{u})}{\partial x_j} = 0$$

The above expression states that in the absence of external loads, the inertial forces are balanced by the deformation forces. In what follows, *harmonic* solutions of the equation of motion are of interest; the displacement field is expressed as $u_i(\mathbf{x}, t) = u_i(\mathbf{x})\exp(i\omega t)$, so that in the frequency-domain the former relation becomes

$$-\omega^2\,\rho u_i - \frac{\partial \sigma_{ij}(\mathbf{u})}{\partial x_j} = 0 \ \text{ in } \Omega \tag{2.4}$$

For the sake of clarity, it is assumed that the structure is free of external forces, either of volume nature in Ω or surface nature in Γ_σ; besides it is supposed to be clamped on boundary Γ_o: in the context of FSI, only fluid loading on the structure is accounted for. The corresponding boundary conditions are expressed as follows:

$$u_i = 0 \ \text{ on } \Gamma_o \tag{2.5}$$

$$\sigma_{ij}(\mathbf{u})n_j = 0 \ \text{ on } \Gamma_\sigma \tag{2.6}$$

Combining Equation (2.4) together with Equations (2.1) and (2.2) gives the *Navier equation*:

$$\omega^2 \, \rho u_i + \frac{E}{1+v}\left(\frac{1}{2(1-2v)}\frac{\partial^2 u_j}{\partial x_i \partial x_j} + \frac{1}{2}\frac{\partial^2 u_i}{\partial x_j \partial x_j}\right) = 0 \qquad (2.7)$$

In the general case, it is not possible to derive analytical solutions to the Navier equation, so that numerical resolution is usually resorted to. Among numerical techniques, the FEM is the most popular in structural dynamics. It is indeed adapted to various geometries, which can be meshed with elements of diverse sizes and shapes, as illustrated in Figure 2.9. Moreover, the method is conveniently versatile as the same computational model may serve to solve a large variety of distinct problems (involving linear or non-linear behaviours, static or dynamic

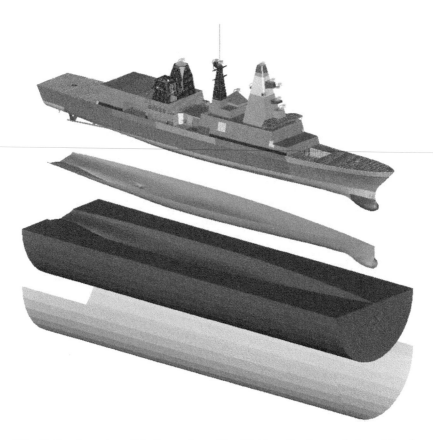

Figure 2.9   Finite element model of a surface ship. Highly refined numerical models of complex geometries are affordable thanks to the ever-increasing efficiency of FEM codes and the availability of larger storage resources. The constitutive part of mechanical systems is accounted for with mass, stiffness and damping characteristics, which may be either concentrated on specific locations or distributed throughout the structure. The choices made by the structural analyst to build a finite element model have a major influence as far as the computed dynamics of the system is concerned. To produce reliable simulations, an engineering expertise is therefore required to choose the best modelling assumptions among the various possible options. *Source*: DCNS Lorient, France, 2008. Reproduced with permission of DCNS

cases, time- or frequency-dependent formulations); see for instance Bathe (1982), Hughes (1987), MacNeal (1994) and Zienkiewicz and Taylor (2000).

2.2   Finite Element Method: Practical Implementation

2.2.1   Weighted Integral Formulation

As a first step in the procedure to build a finite element model of any continuum mechanical system set in motion, an equivalent formulation of the problem equations provided with appropriate boundary conditions is established. Referred to in what follows as the *weighted integral formulation* of the problem, it is stated according to an integral principle[6], and it is obtained with the so-called *test function method* applied to Equations (2.4)–(2.6). The equivalence between the original problem and its integral representation is understood with the following meaning:

$$-\omega^2 \rho u_i - \frac{\partial \sigma_{ij}(\mathbf{u})}{\partial x_j} = 0 \text{ in } \Omega$$

$$\Longleftrightarrow -\omega^2 \int_\Omega \rho u_i \, \delta u_i \, d\Omega - \int_\Omega \frac{\partial \sigma_{ij}(\mathbf{u})}{\partial x_j} \delta u_i \, d\Omega = 0 \qquad \forall \delta u_i \qquad (2.8)$$

$\delta \mathbf{u}$ stands for a test function, also referred to as a *virtual displacement field*, and it is assumed to comply with the boundary condition $\delta \mathbf{u}|_{\Gamma_o} = \mathbf{0}$.

Starting from Equation (2.8), calculations leading to a formulation which is suited to finite element approximation are detailed below, thereby illustrating some elementary mathematical manipulations with the index notation.

An integration by parts reduces the derivation order on the displacement field:

$$\int_\Omega \frac{\partial \sigma_{ij}(\mathbf{u})}{\partial x_j} \delta u_i \, d\Omega = -\int_\Omega \sigma_{ij}(\mathbf{u}) \frac{\partial \delta u_i}{\partial x_j} \, d\Omega + \int_\Gamma \sigma_{ij}(\mathbf{u}) n_j \delta u_i \, d\Gamma$$

Symmetry of the stress and strain tensors allows the following calculation sequence:

$$\sigma_{ij}(\mathbf{u}) \frac{\partial \delta u_i}{\partial x_j} = \frac{1}{2} \left(\sigma_{ij}(\mathbf{u}) \frac{\partial \delta u_i}{\partial x_j} + \sigma_{ij}(\mathbf{u}) \frac{\partial \delta u_i}{\partial x_j} \right)$$

$$= \frac{1}{2} \left(\sigma_{ij}(\mathbf{u}) \frac{\partial \delta u_i}{\partial x_j} + \sigma_{ji}(\mathbf{u}) \frac{\partial \delta u_j}{\partial x_i} \right)$$

$$= \frac{1}{2} \left(\sigma_{ij}(\mathbf{u}) \frac{\partial \delta u_i}{\partial x_j} + \sigma_{ij}(\mathbf{u}) \frac{\partial \delta u_j}{\partial x_i} \right)$$

Hence:

$$\int_\Omega \frac{\partial \sigma_{ij}(\mathbf{u})}{\partial x_j} \delta u_i \, d\Omega = \int_\Omega \sigma_{ij}(\mathbf{u}) \, \varepsilon_{ij}(\delta \mathbf{u}) \, d\Omega$$

---

[6] As discussed in Remark 2.3, the weighted integral formulation may be interpreted as the expression of the so-called *virtual work principle*.

On the one hand, the boundary condition (2.6) is satisfied by $\mathbf{u}$ on Γ_σ so that $\int_{\Gamma_\sigma} \sigma_{ij}(\mathbf{u})n_j \delta u_i\, d\Gamma_\sigma = 0$. On the other hand, $\delta\mathbf{u}$ complies with Equation (2.5) on Γ_o, which gives $\int_{\Gamma_o} \sigma_{ij}(\mathbf{u})n_j \delta u_i\, d\Gamma_o = 0$.

Thus:

$$\int_\Gamma \sigma_{ij}(\mathbf{u})n_j \delta u_i\, d\Gamma = 0$$

and the following expression is finally derived:

$$-\omega^2 \int_\Omega \rho u_i\, \delta u_i\, d\Omega + \int_\Omega \sigma_{ij}(\mathbf{u})\, \varepsilon_{ij}(\delta\mathbf{u})\, d\Omega = 0 \tag{2.9}$$

In order for this formulation to be meaningful, it is necessary that the integrals be finite, which is ensured when $\mathbf{u}$ and $\delta\mathbf{u}$ belong to the following functional space:

$$\mathcal{V} = \{\mathbf{u} \in H^1(\Omega), \mathbf{u}|_{\Gamma_o} = \mathbf{0}\} \tag{2.10}$$

$H^1(\Omega)$ is defined as:

$$H^1(\Omega) = \left\{ \psi \in L^2(\Omega), \frac{\partial\psi}{\partial x_i} \in L^2(\Omega), 1 \le i \le d \right\}$$

- ψ is a real-valued function of variables $\mathbf{x} = (x_i)_{1 \le i \le d}$ belonging in Ω whose first derivatives $\dfrac{\partial\psi}{\partial x_i}$ exist for all $1 \le i \le d$.
- $L^2(\Omega)$ is the space of squared-integrable functions over Ω:

$$L^2(\Omega) = \left\{ \psi, \int_\Omega |\psi|^2\, d\Omega < +\infty \right\}$$

$\mathcal{Y} = H^1(\Omega)$ is an *Hilbert space*. Hilbert spaces provide a convenient theoretical framework to deal with the numerical analysis of partial differential equations, such as the equations arising from the equilibrium of mechanical systems. Hilbert spaces are *vectorial* spaces endowed with a *scalar product* $\langle \bullet | \bullet \rangle$. The natural scalar product on $\mathcal{Y}$ is defined by

$$\langle \psi | \delta\psi \rangle = \int_\Omega \psi\, \delta\psi + \frac{\partial\psi}{\partial x_i} \frac{\partial\delta\psi}{\partial x_i}\, d\Omega$$

Any ψ belonging to $\mathcal{Y}$ may be expanded onto a *vectorial basis* $\Psi = \{\psi_m\}_{m \ge 1}$, according to:

$$\psi = \sum_{m \ge 1} \langle \psi | \psi_m \rangle \psi_m \tag{2.11}$$

where Ψ is *ortho-normal* with respect to the scalar product, meaning that $\langle \psi_m | \psi_{m'} \rangle = \delta_{m,m'}$ for all m, m'. $\langle \psi | \psi_m \rangle$ are the *modal coordinates* of ψ in the basis Ψ. Equation (6.11) is stated with the following meaning:

$$\lim_{M \to \infty} \left\| \psi - \sum_{m=1}^{m=M} \langle \psi | \psi_m \rangle \psi_m \right\| = 0,$$

where $\| \bullet \|$ is the natural norm associated with the scalar product $\langle \bullet | \bullet \rangle$:

$$\|\psi\| = \sqrt{\langle \psi | \psi \rangle} = \left(\int_\Omega |\psi|^2 + |\nabla \psi|^2 \, d\Omega \right)^{1/2}$$

This norm can be used to describe convergence,[7] as above with the modal projection; more generally, the mathematical results regarding the convergence of the FEM are stated using the natural norms of the Hilbert spaces, as evoked in Remark 2.4.

Equation (6.11) is the starting point of *modal expansion techniques*, which will be discussed and illustrated in Chapter 6, while the finite element computation of the modal basis for various mechanical systems will be illustrated throughout the other chapters.

The weighted integral formulation of the problem stated by Equations (2.4) to (2.6) is then stated as follows.

Find $\mathbf{u} \in \mathcal{V}$ and ω such that:

$$-\omega^2 \int_\Omega \rho u_i \, \delta u_i \, d\Omega + \int_\Omega \sigma_{ij}(\mathbf{u}) \, \varepsilon_{ij}(\delta \mathbf{u}) \, d\Omega = 0$$

with:

$$\varepsilon_{ij}(\mathbf{u}) = \frac{1}{2} \left(\frac{\partial u_i}{\partial x_j} + \frac{\partial u_j}{\partial x_i} \right)$$

and:

$$\sigma_{ij}(\mathbf{u}) = \frac{E}{1+v} \left(\frac{v}{1-2v} \varepsilon_{kk}(\mathbf{u}) \delta_{ij} + \varepsilon_{ij}(\mathbf{u}) \right)$$

for all $\delta \mathbf{u} \in \mathcal{V}$.

2.2.2 Finite Elements

Equation (2.9) defines *symmetric* and *bilinear* forms on $\mathcal{V}$,[8] which give rise to *symmetric matrices* after finite element discretisation.[9]

Ω is approximated by a set of polygonal elements $(\Omega_e)_{e \in [1,E]}$ according to $\Omega \simeq \bigcup_{e=1}^{e=E} \Omega_e$, which produces a *mesh* of the structure, as represented in Figure 2.9. Depending on the problem at hand, these elements can assume various shapes: for instance, triangle or quadrangle in 2D, tetrahedron or hexahedron in 3D, as illustrated in Figure 2.10.

---

[7] $\| \bullet \|$ involves the function and its derivatives, so that the convergence according to the H^1-norm is more restrictive than the convergence in the sense of the L^2-norm, which is defined as:

$$\|\psi\| = \left(\int_\Omega |\psi|^2 \, d\Omega \right)^{1/2}$$

for any square integrable function.

[8] $(\bullet, \bullet)$ is a bilinear form on $\mathcal{Y}$ if it defines a scalar-valued function, which is linear in each argument separately; it is symmetric if $(\psi, \delta\psi) = (\delta\psi, \psi)$ for all ψ and $\delta\psi$ belonging to $\mathcal{Y}$.

[9] In what follows, $[\bullet]$ denotes a matrix, $\langle \bullet \rangle$ and $\{ \bullet \}$ designate a line and a row vector, respectively, and $\bullet^\mathsf{T}$ stands for matrix/vector transposition. A matrix is *symmetric* when it is invariant with respect to transposition.

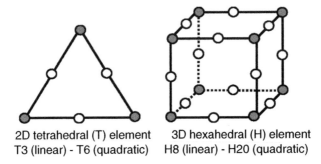

2D tetrahedral (T) element   3D hexahedral (H) element
T3 (linear) - T6 (quadratic)   H8 (linear) - H20 (quadratic)

Figure 2.10 Example of 2D and 3D finite elements. (a) 2D tetrahedral (T) element. (b) 3D hexahedral (H) element

The integrals in Equation (2.9) are evaluated according to:

$$\int_{\Omega} (\bullet)\, d\Omega \simeq \sum_{e=1}^{e=E} \int_{\Omega_e} (\bullet)\, d\Omega_e,$$

so that the following terms, which yield the *mass* and *stiffness* elementary matrices, are calculated on Ω_e:

$$\int_{\Omega_e} \rho u_i\, \delta u_i\, d\Omega_e \qquad \int_{\Omega_e} \sigma_{ij}(\mathbf{u})\, \varepsilon_{ij}(\delta \mathbf{u})\, d\Omega_e$$

On each element, the approximation of the displacement field is performed as follows:

$$u_i|_{\Omega_e}(\mathbf{x}) \simeq \mathbf{N}_i^e(\mathbf{x})\mathbf{U}_i^e$$

- $\mathbf{U}_i^e = \{u_i(\mathbf{x}_I)\}_{I \in [1, \mathcal{N}_e]}$ gathers the *nodal values* of u_i, which are determined at specific locations $\mathbf{x}_I$ on Ω_e; the approximation scheme defines the number of nodes $\mathcal{N}_e$ an element possesses.
- $\mathbf{N}_i^e = \langle N_I(\mathbf{x}) \rangle_{I \in [1, \mathcal{N}_e]}$ assembles the *shape functions* associated with each node of the element and performs the approximation of u_i with $\mathcal{N}_e$ nodal values. Each shape function $N_I(\mathbf{x})$ should be equal to one at its own node and vanish at all other nodes: this is written $N_I(\mathbf{x}_J) = \delta_{IJ}$ for any $\mathbf{x}_J$, with $\delta_{II} = 1$ and $\delta_{IJ} = 0$ for $I \neq J$. The Lagrange polynomial of degree $\mathcal{N}_e - 1$ is endowed with such properties. It is defined by:

$$N_I(\mathbf{x}) = \frac{\prod_{J=1, J\neq I}^{\mathcal{N}_e}(\mathbf{x} - \mathbf{x}_J)}{\prod_{J=1, J\neq I}^{\mathcal{N}_e}(\mathbf{x}_I - \mathbf{x}_J)}$$

Linear and quadratic shape functions for 1D approximation are illustrated in Figure 2.11.

2.2.3 Elementary Matrices

In what follows, let u_i and u_j denote two different components of the displacement field. According to the finite element approximation, they are calculated as follows:

$$\begin{Bmatrix} u_i \\ u_j \end{Bmatrix} = \begin{bmatrix} \mathbf{N}_i^e & 0 \\ 0 & \mathbf{N}_j^e \end{bmatrix} \begin{Bmatrix} \mathbf{U}_i^e \\ \mathbf{U}_j^e \end{Bmatrix}$$

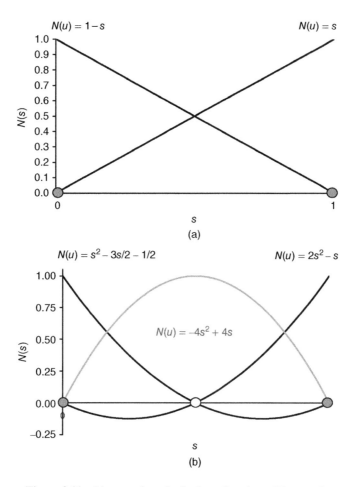

Figure 2.11   Linear and quadratic shape functions: 1D example

which is also written $\mathbf{u}|_{\Omega_e} = \mathbf{N}_e\mathbf{U}_e$, with:

$$\mathbf{N}^e = \begin{bmatrix} \mathbf{N}_i^e & \mathbf{0} \\ \mathbf{0} & \mathbf{N}_j^e \end{bmatrix}$$

The same procedure applies to the virtual field $\delta\mathbf{u}$, that is, $\mathbf{u}|_{\Omega_e} = \mathbf{N}_e\delta\mathbf{U}_e$. Such approximations of the displacement and virtual fields generate the discrete form of the weighted integral formulation of the problem, through the mass and stiffness matrices, which are obtained as follows.

The elementary mass matrix arises from the discretisation of the scalar product of vectors $\mathbf{u}$ and $\delta\mathbf{u}$. Using the vector representation in the base $(\mathbf{e}_i)$, it is $\mathbf{u} \cdot \delta\mathbf{u} = u_i\delta u_i + u_j\delta u_j$, when considering two components i,j without the summation convention. On Ω_e, the approximation of $\mathbf{u} \cdot \delta\mathbf{u}$ is $\delta\mathbf{U}_e^\mathsf{T}\mathbf{N}_e^\mathsf{T}\mathbf{N}_e\mathbf{U}_e$, so that:

$$\int_{\Omega_e} \rho\mathbf{u}\,\delta\mathbf{u}d\Omega_e = \delta\mathbf{U}_e^\mathsf{T}\mathbf{m}_S^e\mathbf{U}_e$$

with the elementary mass matrix:

$$\mathbf{m}_S^e = \int_{\Omega_e} \rho\mathbf{N}_e^\mathsf{T}\mathbf{N}_e\,d\Omega_e \tag{2.12}$$

The elementary stiffness matrix proceeds from the discretisation of the double contracted product of tensors $\sigma(\mathbf{u})$ and $\varepsilon(\delta\mathbf{u})$. Using the tensor representation in the base $(\mathbf{e}_i \otimes \mathbf{e}_j)$, it is $\sigma(\mathbf{u}) : \varepsilon(\delta\mathbf{u}) = \sigma_{ii}\varepsilon_{ii} + \sigma_{jj}\varepsilon_{jj} + 2\sigma_{ij}\varepsilon_{ij}$ with the two directions of interest i, j. It is equivalently formulated as $\varepsilon^\mathsf{T}\sigma$ with:

$$\varepsilon^\mathsf{T} = \langle \varepsilon_{ii}, \varepsilon_{jj}, 2\varepsilon_{ij} \rangle \qquad \sigma^\mathsf{T} = \langle \sigma_{ii}, \sigma_{jj}, \sigma_{ij} \rangle$$

ε and σ are derived from the stress–strain relation on the one hand and the constitutive material law on the other hand.

- Equation (2.1) is $\varepsilon = \mathbf{G}_S^e\mathbf{U}_e$ where $\mathbf{G}_S^e$ stands for the gradient operator:

$$\mathbf{G}_S^e = \begin{bmatrix} \dfrac{\partial N_i^e}{\partial x_i} & 0 \\[2mm] 0 & \dfrac{\partial N_j^e}{\partial x_j} \\[2mm] \dfrac{\partial N_i^e}{\partial x_j} & \dfrac{\partial N_j^e}{\partial x_i} \end{bmatrix}$$

- Equation (2.2) is $\sigma = \mathbf{H}\varepsilon$ with:

$$\mathbf{H} = \frac{E}{(1+v)(1-2v)} \begin{bmatrix} 1-v & v & 0 \\ v & 1-v & 0 \\ 0 & 0 & 1/2-v \end{bmatrix}$$

On Ω_e, the approximation of $\sigma(\mathbf{u}) : \varepsilon(\delta\mathbf{u})$ is $\delta\mathbf{U}_e^\mathsf{T}\mathbf{G}_S^{e\mathsf{T}}\mathbf{H}\mathbf{G}_S^e\mathbf{U}_e$, so that:

$$\int_{\Omega_e} \sigma(\mathbf{u}) : \varepsilon(\delta\mathbf{u})d\Omega_e = \delta\mathbf{U}_e^\mathsf{T}\mathbf{k}_S^e\mathbf{U}_e$$

with the elementary stiffness matrix:

$$\mathbf{k}_S^e = \int_{\Omega_e} \mathbf{G}_e^\mathsf{T}\mathbf{H}\mathbf{G}_e\,d\Omega_e \tag{2.13}$$

2.2.4 Mass and Stiffness Matrices

Taking into account the contribution of each element in the weighted integral formulation is stated as follows:

$$-\omega^2 \sum_{e=1}^{e=E} \delta\mathbf{U}_e^\mathsf{T}\mathbf{m}_S^e\mathbf{U}_e + \sum_{e=1}^{e=E} \delta\mathbf{U}_e^\mathsf{T}\mathbf{k}_S^e\mathbf{U}_e = 0 \qquad \forall\,\delta\mathbf{U}$$

where $\mathbf{U}_e$ stands for degrees-of-freedom (DOF) of each element. The DOF of the structure are subsequently assembled in a global vector $\mathbf{U}$. The former is extracted from the latter by a matrix operation:

$$\mathbf{U}_e = \mathbf{\Lambda}_e \mathbf{U}$$

$\mathbf{\Lambda}_e$ is the *localisation matrix*: it relates the nodal information from the global mesh to the local element. In the same manner, $\delta\mathbf{U}_e = \mathbf{\Lambda}_e \delta\mathbf{U}$, so that the summation displayed above is also:

$$-\omega^2 \delta\mathbf{U}^\mathsf{T}\mathbf{M}_S\mathbf{U} + \delta\mathbf{U}^\mathsf{T}\mathbf{K}_S\mathbf{U} = 0 \qquad \forall\, \delta\mathbf{U}$$

Since the former expression holds for any $\delta\mathbf{U}$, the following relation is arrived at:

$$(-\omega^2 \mathbf{M}_S + \mathbf{K}_S)\mathbf{U} = \mathbf{0} \tag{2.14}$$

Equation (2.14) defines a generalised linear eigenvalue problem, where

- $\mathbf{M}_S = \sum\limits_{e=1}^{e=E} \mathbf{\Lambda}_e{}^\mathsf{T} \mathbf{m}_S^e \mathbf{\Lambda}_e$ is the *mass matrix* of the structure;
- $\mathbf{K}_S = \sum\limits_{e=1}^{e=E} \mathbf{\Lambda}_e{}^\mathsf{T} \mathbf{k}_S^e \mathbf{\Lambda}_e$ is the *stiffness matrix* of the fluid.

$\mathbf{M}_S$ and $\mathbf{K}_S$ are *symmetric* and, respectively, *positive definite* and *non-negative definite* matrices[10]. The physical interpretation of these properties is as follows. Let $\mathcal{E}_K^S$ and $\mathcal{E}_P^S$ the kinetic and potential elastic energy of the structure; they are calculated as follows:

$$\mathcal{E}_K^S = \frac{1}{2}\int_\Omega \rho\left(\frac{\partial\mathbf{u}}{\partial t}\right)^2 d\Omega \qquad \mathcal{E}_P^S = \frac{1}{2}\int_\Omega \varepsilon(\mathbf{u})\cdot\cdot\,\sigma(\mathbf{u})\, d\Omega$$

For an harmonic vibration, the displacement is $\mathbf{u}(t) = \mathbf{U}\exp(i\omega t)$, so that the kinetic energy is proportional to

$$-\frac{\omega^2}{2}\int_\Omega \rho u_i u_i d\Omega$$

Hence:

$$\mathcal{E}_K^S \propto -\frac{\omega^2}{2}\mathbf{U}^\mathsf{T}\mathbf{M}_S\mathbf{U} \tag{2.15}$$

The potential energy is proportional to:

$$\frac{1}{2}\int_\Omega \varepsilon_{ij}(\mathbf{u})\sigma_{ij}(\mathbf{u})\, d\Omega$$

Hence:

$$\mathcal{E}_P^S \propto \frac{1}{2}\mathbf{U}^\mathsf{T}\mathbf{K}_S\mathbf{U} \tag{2.16}$$

Symmetry of $\mathbf{M}_S$ and $\mathbf{K}_S$ refers to the *reversibility* of mechanical systems: the energy associated with $\mathbf{U}$ is the same as the energy associated with $-\mathbf{U}$. The positive definite character of $\mathbf{M}_S$ expresses that the kinetic energy is always a positive quantity. The non-negative definite character of $\mathbf{K}_S$ stems from the fact that small motions of a mechanical system around a stable

---

[10] A positive definite matrix $\mathbf{M}$ satisfies $\mathbf{X}^\mathsf{T}\mathbf{M}\mathbf{X} \geq 0, \forall\mathbf{X}$ with $\mathbf{X}^\mathsf{T}\mathbf{M}\mathbf{X} = 0$ only for $\mathbf{X} = 0$; a non-negative definite matrix verifies $\mathbf{X}^\mathsf{T}\mathbf{K}\mathbf{X} \geq 0, \forall\mathbf{X}$ but it exists $\mathbf{X} \neq 0$ such that $\mathbf{X}^\mathsf{T}\mathbf{K}\mathbf{X} = 0$.

equilibrium position involve positive stiffness. A rigid body displacement of the structure entails no potential energy, hence complies with $\mathbf{U}^T\mathbf{K}\mathbf{U} = \mathbf{0}$, even if $\mathbf{U}$ differs from a non-null vector.

The total mass of the structure ensues from $m_S = \int_\Omega \rho\, \mathbf{d}^2\, d\Omega$, where $\mathbf{d}$ is a unit vector in a given direction. The discretisation of the latter expression according to the definition of the mass matrix is:

$$\int_\Omega \rho\, \mathbf{d} \cdot \mathbf{d}\, d\Omega = \mathbf{\Delta}^T \mathbf{M}_S \mathbf{\Delta}$$

Hence, the system mass m_S may be retrieved from the mass matrix according to:

$$\mathbf{\Delta}^T \mathbf{M}_S \mathbf{\Delta} = m_S \tag{2.17}$$

with $\mathbf{\Delta}$ a vector of unit component in a given direction.

Equation (2.17) is, for instance, of practical interest in a modal analysis to indicate modes with preponderant influence on the system dynamics, as will be illustrated in Chapters 5 and 6.

Remark 2.2 Structures with frequency-dependent materials *As evoked in Remark 2.1, the linear vibrations of structures composed of frequency-dependent materials are accounted for with appropriate constitutive laws. For visco-elastic materials, the stress–strain relation is, for instance, stated as follows:*

$$\sigma_{ij}(\mathbf{u}, \omega) = \frac{E^*(\omega)}{1+v}\left(\varepsilon_{ij}(\mathbf{u}) + \frac{v}{1-2v}\varepsilon_{kk}(\mathbf{u})\delta_{ij}\right) \tag{2.18}$$

where (i, j) denote the Cartesian coordinates. Young's modulus is complex:

$$E^*(\omega) = E'(\omega)(1 + i\eta(\omega))$$

Substituting Equation (2.18) into the weighted integral formulation of the problem gives:

$$-\omega^2 \int_\Omega \rho u_i\, \delta u_i\, d\Omega +$$

$$\int_\Omega E'(\omega)(1 + i\eta(\omega))(\varepsilon_{ij}(\mathbf{u})\, \varepsilon_{ij}(\delta\mathbf{u}) + \frac{v}{1-2v}\varepsilon_{ii}(\mathbf{u})\, \varepsilon_{jj}(\delta\mathbf{u}))\, d\Omega = 0$$

The finite element discretisation of this expression yields a non-linear eigenvalue problem:

$$(-\omega^2\mathbf{M}_S + \mathbf{K}_S(\omega))\mathbf{U} = \mathbf{0} \tag{2.19}$$

The stiffness matrix $\mathbf{K}_S(\omega)$ bears the frequency dependency of the material. It is assembled from the elementary matrices $\mathbf{k}_S^e(\omega)$ which are:

$$\mathbf{k}_S^e(\omega) = \int_{\Omega_e} E'(\omega)(1 + i\eta(\omega))\left(\mathbf{G}_S^{e\,T}\mathbf{L}\mathbf{G}_S^e + \frac{v}{1-2v}\mathbf{D}_S^{e\,T}\mathbf{D}_S^e\right)\, d\Omega$$

$\mathbf{G}_S^e$ is the gradient operator, introduced in the previous subsection, and $\mathbf{L}$ is:

$$\mathbf{L} = \begin{bmatrix} 1 & 0 & 0 \\ 0 & 1 & 0 \\ 0 & 0 & 1/2 \end{bmatrix}$$

$\mathbf{D}_S^e$ *is the divergence operator; for two directions of interest i and j, it reads as follows:*

$$\mathbf{D}_S^e = \left\langle \frac{\partial \mathbf{N}_i^e}{\partial x_i}, \frac{\partial \mathbf{N}_j^e}{\partial x_j} \right\rangle$$

Solving the non-linear eigenvalue problem (2.19) requires specific methods, some of which will be briefly dealt with in Chapter 6. More on the finite element modelling of structures with visco-elastic materials is proposed, for instance, in Soize and Ohayon (1998). ∎

Remark 2.3 Integral principles and constrained relations *The vibrations of an elastic structure are described here by a set of equations of the type (2.4)–(2.6) completed with relations of the type (2.1)–(2.3): obviously, the problem involves partial differential equations.*

With the test function method, an equivalent formulation of the problem may be proposed: the weighted integral formulation, as per Equation (2.9), is an expression of the principle of virtual work. It is an integral principle which postulates that the internal and external works $\mathcal{W}_i$ and $\mathcal{W}_e$, associated with a virtual displacement $\delta\mathbf{u}$, is stationary. It is stated as follows:

$$\delta \mathcal{W}_i + \delta \mathcal{W}_e = 0,$$

where $\delta(\bullet)$ is standing for the variation operator.[11]

The work of the internal stress force in the virtual displacement is calculated as follows:

$$\delta \mathcal{W}_i = -\int_\Omega \sigma(\mathbf{u}) : \varepsilon(\delta\mathbf{u}) \, d\Omega$$

The external forces reduce to the inertial force $-\rho\omega^2\mathbf{u}$ in the absence of volume and surface forces; the virtual work associated with $\delta\mathbf{u}$ is calculated as follows:

$$\delta \mathcal{W}_e = -\int_\Omega (-\rho\omega^2\mathbf{u}) \cdot \delta\mathbf{u} \, d\Omega$$

The principle of virtual work is expressed as follows:

$$-\int_\Omega \sigma(\mathbf{u}) : \varepsilon(\delta\mathbf{u}) \, d\Omega + \int_\Omega \rho\omega^2\mathbf{u} \cdot \delta\mathbf{u} \, d\Omega = 0$$

It is shown to be equivalent to the equations of equilibrium: the solution to the problem is $\mathbf{u}$ belonging in $\mathcal{U}$ and it is such that $\delta\mathcal{W}(\mathbf{u}, \delta\mathbf{u}) = 0$ for all $\delta\mathbf{u}$ belonging in $\mathcal{U}$. $\delta\mathbf{u}$ stands for any virtual motion: by 'virtual', it is meant that it does not modify the forces applied on the system, while the actual displacement $\mathbf{u}$ does.

---

[11] For a scalar valued-function ψ of the variable $\mathbf{u}$, $\delta\psi$ is calculated with the directional derivative according to:

$$\delta\psi = \frac{\partial\psi}{\partial\varepsilon}(\mathbf{u} + \varepsilon\delta\mathbf{u})\Big|_{\varepsilon\to 0} \quad \text{for all } \delta\mathbf{u}$$

ψ is stationary when $\delta\psi = 0$. For instance, $\psi(\mathbf{u}) = \frac{\lambda}{2}\mathbf{u}^2$ is stationary for $\mathbf{u} = \mathbf{0}$. Indeed:

$$\frac{\partial\psi}{\partial\varepsilon}(\mathbf{u} + \varepsilon\delta\mathbf{u}) = \lambda\varepsilon\delta u_i\delta u_i + \lambda u_i\delta u_i$$

so that $\delta\psi = \lambda u_i\delta u_i$ for all δu_i which is equivalent to $u_i = 0$ for all i; it is noted that, depending on the sign of λ, ψ may be maximum or minimum at the point of stationarity.

In general manner, the functional space $\mathcal{V}$ may incorporate various constraining relations: as prescribed in Equation (2.10) for instance, $\mathcal{V}$ includes the clamping condition of the structure on Γ_o. In the matrix form of the integral formulation, such a boundary condition is enforced by extracting lines and rows in $\mathbf{M}_S$ and $\mathbf{K}_S$, which correspond to the DOF of the constrained nodes. As a result, sub-matrices are handled in the eigenvalue problem (2.14).

The implementation of other types of constrained relations – such as per Equation (4.2), the signification of which is discussed in Chapter 4 – might become cumbersome, if not impracticable, with the preceding approach. The Lagrange formulation offers an alternative framework which is suited for the representation of various constraining relations.

The energy function of the system, referred to as the Lagrangian, is denoted $\mathcal{L}$, and it may be calculated with a set of generalised coordinates, denoted $\mathbf{q}(t)$. The time integral of this quantity defines a mapping $\bullet \mapsto \int_t^{t'} \bullet(\dot{\mathbf{q}}, \mathbf{q}, \tau) \, d\tau$ which is referred to as the action: its value depends on the considered quantity for all times $\tau \in [t, t']$.

The principle of least action states that the dynamics of a mechanical system is described by the stationarity of the action of the Lagrangian between t and t'. It is stated as follows:

$$\delta \left(\int_t^{t'} \mathcal{L}(\dot{\mathbf{q}}, \mathbf{q}, \tau) \, d\tau \right) = 0$$

- For a non-constrained mechanical system, $\mathcal{L}$ is calculated with the kinetic energy $\mathcal{K}(\dot{\mathbf{q}}, \mathbf{q}, t)$ and the potential energy $\mathcal{P}(\dot{\mathbf{q}}, \mathbf{q}, t)$ of the system:

$$\mathcal{L}(\dot{\mathbf{q}}, \mathbf{q}, t) = \mathcal{K}(\dot{\mathbf{q}}, \mathbf{q}, t) - \mathcal{P}(\dot{\mathbf{q}}, \mathbf{q}, t)$$

The Lagrange equations are then derived from the stationarity principle applied to the action of $\mathcal{L}$; they are formulated as follows:

$$\frac{d}{dt}\left(\frac{\partial \mathcal{L}}{\partial \dot{\mathbf{q}}} \right) - \frac{\partial \mathcal{L}}{\partial \mathbf{q}} = \mathbf{0}$$

When the kinetic and potential energies are expressed according to $\mathcal{K}(\dot{\mathbf{q}}, \mathbf{q}, t) = \dot{\mathbf{q}}^T \mathbf{M} \dot{\mathbf{q}}/2$ and $\mathcal{P}(\dot{\mathbf{q}}, \mathbf{q}, t) = \mathbf{q}^T \mathbf{K} \mathbf{q}/2$, where $\mathbf{M}$ and $\mathbf{K}$ are the mass and stiffness operators of the system, the Lagrange equations yield the dynamic equation:

$$\mathbf{M}\ddot{\mathbf{q}}(t) + \mathbf{K}\mathbf{q}(t) = \mathbf{0}$$

- For a constrained mechanical system, the Lagrangian of the system $\mathcal{L}'$ is expressed with the Lagrangian of the non-constrained system $\mathcal{L}$, to which is added the work of the external forces $\mathcal{W}(\dot{\mathbf{q}}, \mathbf{q}, t)$:

$$\mathcal{L}'(\dot{\mathbf{q}}, \mathbf{q}, t) = \mathcal{K}(\dot{\mathbf{q}}, \mathbf{q}, t) - \mathcal{P}(\dot{\mathbf{q}}, \mathbf{q}, t) + \mathcal{W}(\dot{\mathbf{q}}, \mathbf{q}, t)$$

Assuming that $\mathcal{W}$ can be written as a linear combination of L relations, $\psi_l(\dot{\mathbf{q}}, \mathbf{q}) = 0$ for $l \in [1, L]$, $\mathcal{L}'$ is calculated as follows:

$$\mathcal{L}'(\dot{\mathbf{q}}, \mathbf{q}, \lambda) = \mathcal{K}(\dot{\mathbf{q}}, \mathbf{q}, t) - \mathcal{P}(\dot{\mathbf{q}}, \mathbf{q}, t) + \sum_{l=1}^{l=L} \lambda_l \psi_l(\dot{\mathbf{q}}, \mathbf{q})$$

$(\lambda_l)_{l \in [1,L]}$ *are the* Lagrange multipliers *associated with the constraining relations* $\psi_l(\dot{\mathbf{q}}, \mathbf{q})_{l \in [1,L]}$.

The stationarity principle applied to the action of $\mathcal{L}'$ *yields the Lagrange equations for the constrained system:*

$$\begin{cases} \dfrac{d}{dt}\left(\dfrac{\partial \mathcal{L}'}{\partial \dot{\mathbf{q}}}\right) - \dfrac{\partial \mathcal{L}'}{\partial \mathbf{q}} = \mathbf{0} \\[2mm] \psi_l(\dot{\mathbf{q}}, \mathbf{q}) = 0 \qquad \forall l \in [0, L] \end{cases}$$

When the constraining relations are expressed as $\mathbf{Lq} = \mathbf{0}$*, where* $\mathbf{L}$ *is a linear operator, and when the kinetic and potential energies are calculated with linear operators* $\mathbf{M}$ *and* $\mathbf{K}$*, the Lagrangian is*

$$\mathcal{L}'(\dot{\mathbf{q}}, \mathbf{q}, t, \lambda) = \dot{\mathbf{q}}^T \mathbf{M}\dot{\mathbf{q}}/2 - \mathbf{q}^T \mathbf{K}\mathbf{q}/2 + \mathbf{q}^T \mathbf{L}^T \lambda$$

The following matrix system is then derived from the Lagrange equations:

$$\begin{bmatrix} \mathbf{M} & \mathbf{0} \\ \mathbf{0} & \mathbf{0} \end{bmatrix} \begin{Bmatrix} \ddot{\mathbf{q}}(t) \\ \ddot{\lambda}(t) \end{Bmatrix} + \begin{bmatrix} \mathbf{K} & -\mathbf{L}^T \\ -\mathbf{L} & \mathbf{0} \end{bmatrix} \begin{Bmatrix} \mathbf{q}(t) \\ \lambda(t) \end{Bmatrix} = \begin{Bmatrix} \mathbf{0} \\ \mathbf{0} \end{Bmatrix} \tag{2.20}$$

With the Lagrange formulation, the constraining relations are associated with Lagrange multipliers, which become additional DOFs of the system. The Lagrange multipliers also have a physical meaning, which seems to be of practical interest for the analyst. For instance, using a displacement-based formulation of structural dynamics, the Lagrange multipliers associated with Equation (2.5) represent the reaction forces on interfaces where the displacement is constrained, as further illustrated in Remark 2.6. More on the theoretical background of the Lagrangian formulation may be found in Axisa (2005) and references herein. ∎

2.2.5   Calculating and Assembling Matrices

2.2.5.1   Calculating Matrices: Example of Gaussian Quadrature

The computation of the elementary matrices is usually carried out on a *reference element* using *numerical integration*. The reference element is a simple geometric dimensionless domain, endowed with a coordinate system $\xi = (\xi_i, \xi_j)$ (reference frame). It is related to each element in the physical frame (with the coordinate system $\mathbf{x} = (x_i, x_j)$) through a mapping $\Xi : \xi \hookrightarrow \mathbf{x}$, as represented in Figure 2.12.

Changing from the physical element to the reference element, the expression of the integrals involved in the definition of the elementary matrices is expressed as follows:

$$\int_{\Omega_e} \psi\left(\mathbf{u}\left(\mathbf{x}, \dfrac{\partial \mathbf{u}}{\partial \mathbf{x}}\right)\right) dx = \int_{-1}^{+1} \int_{-1}^{+1} \psi(\mathbf{u}(\xi, \dfrac{\partial \mathbf{u}}{\partial \xi}))|\mathbf{J}(\xi)| \, d\xi$$

$|\mathbf{J}(\xi)|$ is the determinant of the Jacobian $\mathbf{J}$, whose components are $J_{ij} = \dfrac{\partial x_i}{\partial \xi_j}$.

Various methods are available to perform the integration of ψ over $[-1, +1]$. The Gaussian quadrature rule is a numerical scheme which yields an exact result for polynomial of degrees $2L - 1$. It reads as follows:

$$\int_{-1}^{+1} \psi(\xi)d\xi \simeq \sum_{l=0}^{l=L} w_l \psi(\xi_l) \tag{2.21}$$

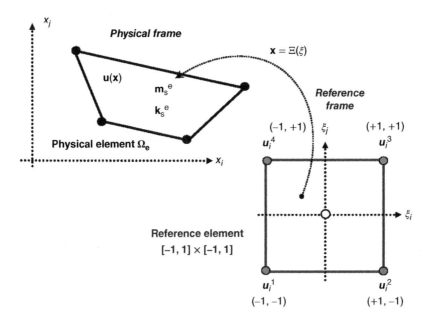

Figure 2.12 Finite element Ω_e, reference element and associated coordinate system

Table 2.3 Integration points ξ_l and weighting coefficients w_l for the Gauss integration scheme

L	ξ_l	w_l	Integration order
1	0	2	1
2	± 0.577350269189626	1	3
3	0	0.888888888888889	5
	± 0.774596669241483	0.55555555555555556	
4	± 0.339981043584856	0.652145154862546	7
	± 0.861136311594053	0.347854845137454	
5	0	0.56888 88888 88889	9
	± 0.538469310105683	0.478628670499366	
	± 0.906179845938664	0.236926885056189	
6	± 0.238619186083197	0.467913934572691	11
	± 0.661209386466265	0.360761573048139	
	± 0.932469514203152	0.171324493279170	
7	0	0.41795 91836 73469	13
	± 0.405845151377397	0.381830050505119	
	± 0.741531185599394	0.279705391489277	
	± 0.949107912342759	0.129484966168870	

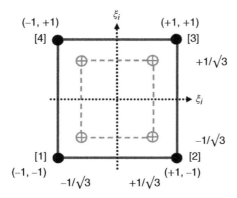

Figure 2.13   Example of a quadrangular element associated with a third-order integration scheme using four integration points

with an appropriate set of L integration points ξ_l and associated weighting coefficients w_l. For the example of a quadrangular element, see Table 2.3 and Figure 2.13.

2.2.5.2   Assembling Matrices: Example of the Cuthill–MacKee Algorithm

The assembling process is described according to the following formulation:

$$[\bullet] = \sum_{e=1}^{e=E} \Lambda_e^T [\bullet]_e \Lambda_e$$

Although represented in a straightforward manner, it is one of the most complicated stages in a finite element program. In particular, it necessitates a connectivity table which identifies the relation between nodes and elements, and it requires that the material properties be properly associated with the corresponding elements.

Matrices assembled from finite elements are *sparse* (i.e. they have a few non-zero terms): it is therefore convenient to store the non-zero values solely. Assembling matrices is performed with dedicated algorithms, and in order to achieve numerical efficiency (for instance in terms of matrices manipulation), it is also desirable to produce matrices with limited bandwidth: to that end, the Cuthill–MacKee algorithm (Cuthill and MacKee, 1969) proves optimal; see Figure 2.14.

Remark 2.4 Convergence of the FEM *As numerical techniques become part of the design process of many mechanical systems, issues regarding the reliability of the FEM are of major concern for the engineer: convergence results are therefore associated with the accuracy of the FEM. It is of course not the purpose of this book to dwell in the realm of complex mathematics, which are required to establish convergence results for FEM. Suffice it here to grasp within few lines the essence of these issues – more on this aspect is examined for instance in Babuska and Strouboulis (2001).*

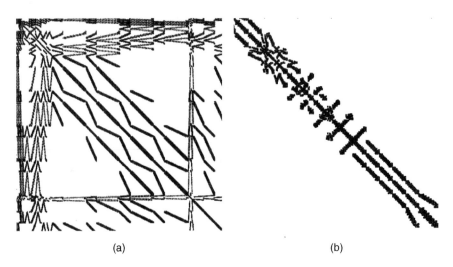

(a)   (b)

Figure 2.14   Achieving optimal bandwidth with the Cuthill–MacKee algorithm. (a) Non-optimal bandwidth. (b) Optimal bandwidth

*As mentioned above, the weighted integral formulation is an equivalent statement of the
initial problem which involves partial differential equations. As far as eigenvalue problems are
concerned, an abstract version of the weighted integral formulation may be stated as follows.*

(Π) *Find* $(v, \lambda) \in \mathcal{V} \times \Lambda$ *such that* $a(v, \delta v) = \lambda(v, \delta v)$, $\qquad \forall \delta v \in \mathcal{V}$

$\mathcal{V}$ *is an Hilbert space endowed with the scalar product* $(\bullet, \bullet)$ *and the associated norm* $\| \bullet \|$.
$a(\bullet, \bullet)$ *is a bilinear form, that is, it is linear with respect to each variable.*

When $a(\bullet, \bullet)$ *is symmetric, the eigenvalue problem has solutions* $(v_n, \lambda_n)_{n \geq 1}$ *such that* $\mathbf{v} =$
$(v_n)_{n \geq 1}$ *forms an orthogonal basis of* $\mathcal{V}$ – *orthogonality being expressed with the following
meaning:* $a(v_n, v_{n'}) = \delta_{nn'} \lambda_n$ *with* $\|v_n\| = \sqrt{(v_n, v_n)} = 1$.

The eigenvalues form a non-decreasing sequence $(\lambda_{n+1} \geq \lambda_n)$, *and they are characterised
by the so-called* Min–Max *principle:*

$$\lambda_n = \min_{\mathcal{V}_n} \max_{w_n \neq 0, \ w_n \in \mathcal{V}_n} \frac{a(w_n, w_n)}{\|w_n\|} \tag{2.22}$$

where $\mathcal{V}_n$ *is a n–dimensional subspace of* $\mathcal{V}$.

The numerical discretisation aims at producing a finite dimension problem (Π_h), *where* $\mathcal{V}_h$
is an approximation of $\mathcal{V}$; *for the eigenvalue problem* (Π), *it is formulated as follows:*

(Π_h) *Find* $(v_h, \lambda_h) \in \mathcal{V}_h \times \Lambda$ *such that* $a(v_h, \delta v_h) = \lambda_h(v_h, \delta v_h)$, $\forall \delta v_h \in \mathcal{V}_h$

The numerical scheme is said to be consistent *if the problem* (Π_h) *is an approximation of*
(Π). v_h, λ_h *are an approximation of* v, λ, *which is convergent when:*

$$\|v - v_h\| \to 0 \quad \text{and} \quad \|\lambda - \lambda_h\| \quad \text{for} \quad h \to 0$$

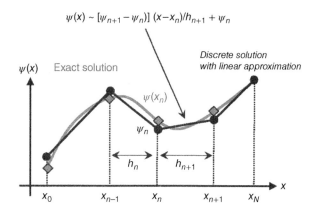

$$\psi(x) \sim [\psi_{n+1} - \psi_n)] \, (x-x_n)/h_{n+1} + \psi_n$$

Figure 2.15 Errors in a finite element computation *originate from the approximation of the solution at nodal values (the computed value of ψ at node x_n is denoted ψ_n and may differ from the exact value $\psi(x_n)$) and from the polynomial interpolation (the value of ψ within interval $[x_n, x_{n+1}]$ is computed from the nodal values according to a polynomial shape, which may not fit the exact solution). The convergence of the finite element approximation is shown to be governed by the spatial discretisation and the order of polynomial approximation*

With the finite element technique, which is shown to be consistent, an approximation of (Π) takes the form:

(Π)   *Find* $\mathbf{V} \in V, \lambda \in \Lambda$ *such that* $\delta \mathbf{V}^T \mathbf{A} \mathbf{V} = \lambda \delta \mathbf{V}^T \mathbf{V}, \; \forall \delta \mathbf{V} \in V$

V is a scalar-valued vectorial space of dimension $\mathcal{N}$ and $\mathbf{A}$ is a symmetric scalar-valued matrix of size $\mathcal{N} \times \mathcal{N}$, where $\mathcal{N}$ is the number of DOF of the discrete problem. (Π) yields the classical eigenvalue problem $\mathbf{A}\mathbf{V} = \lambda \mathbf{V}$.

Errors in the FEM stem from the discretisation procedure, (Π_h) being solved in lieu of (Π) and from the polynomial approximation, as illustrated in a straightforward manner in Figure 2.15.

Error estimates give general results of convergence stated as[12]:

$$\|v - v_h\| \le Ch^p \tag{2.23}$$

Equation (2.23) indicates that the convergence of the finite element approximation is governed by

- *the spatial discretisation (h−convergence, h referring to a spatial measure of the finite element size). In the h−version of the finite element, the mesh is therefore refined so as to achieve the desired accuracy. Classical finite element programs are based on the FEM h−version;*

---

[12] For eigenvalue problems, mathematical results such as expressed by Equation (2.23) are well established; see for instance Boffi (2010).

- *the order of approximation (p–convergence, p referring to the degree of the polynomial approximation). In the p–version of the FEM, the mesh is fixed and the accuracy is obtained by adaptively increasing the degrees of elements;*
- *a combination of spatial discretisation and order of approximation (h, p–convergence).*

Basic features of the h, p and h − p versions of the FEM are discussed and illustrated by Babuska and Guo (1992), which develops some theoretical aspects of these convergence issues.   ∎

2.2.6   Modal Analysis

2.2.6.1   Eigenvalue Problem

Performing the *modal analysis* of mechanical systems consists of solving a generalised eigenvalue problem of the form $(-\omega^2 \mathbf{M} + \mathbf{K})\mathbf{X} = \mathbf{0}$, as written in Equation (2.14). It yields the *natural modes* of vibration of the structure. The latter are described by various quantities: some are *vectorial* in nature (such as the mode shape in terms of displacement – as in the example of Figure 2.16 – or in terms of strain or stress), others are *scalar* in nature (such as the frequency, modal mass or stiffness, participation factor, the effective mass – see the next subsection).

As emphasised in Chapter 6, the eigenmodes also provide the analyst with useful information concerning the intrinsic properties of the system response to any external load.

2.2.6.2   Eigenvectors and Eigenvalues

Let $(\mathbf{X}_n)_{n \geq 1}$ denote the eigenvectors in terms of displacement and let $(\omega_n)_{n \geq 1}$ be the associated eigenpulsations in ascending order ($\omega_{n+1} \geq \omega_n$), that is the solutions of the generalised eigenvalue problem $(\mathbf{K} - \omega^2 \mathbf{M})\mathbf{X} = \mathbf{0}$, formulated with *symmetric* definite positive or non-negative

Figure 2.16   Example of modal analysis for a propeller: finite element discretisation of the geometry and first natural mode of vibration. *Source*: DCNS Nantes, France, 2005. Reproduced with permission of DCNS

matrices $\mathbf{M}$ and $\mathbf{K}$. As evoked before, the set of eigenvectors $(\mathbf{X}_n)_{n\geq1}$ forms an *orthogonal basis*, with the following meaning.

- The eigenmodes comply with the orthogonality condition:

$$\mathbf{X}_{n'}^{\mathsf{T}}\mathbf{M}\mathbf{X}_n = \delta_{n,n'}m_n \qquad \mathbf{X}_{n'}^{\mathsf{T}}\mathbf{K}\mathbf{X}_n = \delta_{n,n'}k_n \qquad \forall\,(n,n') \tag{2.24}$$

where $m_n = \mathbf{X}_n^{\mathsf{T}}\mathbf{M}\mathbf{X}_n$ and $k_n = \mathbf{X}_n^{\mathsf{T}}\mathbf{K}\mathbf{X}_n$ are the *modal mass* and the *modal stiffness*. The eigenpulsation ω_n is:

$$\omega_n = \sqrt{\frac{k_n}{m_n}} \qquad \forall\,n \tag{2.25}$$

- A space-time separated formulation can be written for the solution $\mathbf{X}(\mathbf{x},t)$ of the time evolution problem $\mathbf{M}\ddot{\mathbf{X}}(t) + \mathbf{K}\mathbf{X}(t) = \mathbf{F}(t)$ (with given $\mathbf{F}(t)$ and initial conditions $\dot{\mathbf{X}}(t=0)$ and $\mathbf{X}(t=0)$), using the eigenmode basis. Namely:

$$\mathbf{X}(\mathbf{x},t) = \sum_{n\geq1} \xi_n(t)\mathbf{X}_n(\mathbf{x}) \tag{2.26}$$

ξ_n is the *generalised coordinate* of $\mathbf{X}$ in the basis $(\mathbf{X}_n)_{n\geq1}$. Equation (2.26) indicates that the dynamic behaviour of the system is the *linear* superposition of the responses of each individual mode. The contribution of eigenmode $\mathbf{X}_n$ is quantified by the participation factor q_n; for an imposed force, the latter is:

$$q_n = \frac{\mathbf{X}_n^{\mathsf{T}}\mathbf{F}}{\mathbf{X}_n^{\mathsf{T}}\mathbf{M}\mathbf{X}_n}$$

As will be made more explicit in Chapter 6, for seismic-type analysis, the individual contribution of any eigenmode of displacement shape $\mathbf{X}_n$ can be determined by its *effective mass* μ_n. The latter is given by

$$\mu_n = \frac{(\mathbf{X}_n^{\mathsf{T}}\mathbf{M}\boldsymbol{\Delta})^2}{\mathbf{X}_n^{\mathsf{T}}\mathbf{M}\mathbf{X}_n} \tag{2.27}$$

μ_n is a fraction of the system total mass m, with the following meaning:

$$\sum_{n\geq1} \mu_n = \boldsymbol{\Delta}^{\mathsf{T}}\mathbf{M}\boldsymbol{\Delta} = m \tag{2.28}$$

Remark 2.5 Orthogonality of eigenmodes *Proof of properties (2.24) and (2.28) is worthy of interest as an example of calculation with scalar products involving eigenmodes. Let $\mathbf{X}_n$ and $\mathbf{X}_{n'}$ be two different eigenvectors associated with eigenpulsations ω_n and $\omega_{n'}$, hence:*

$$\mathbf{K}\mathbf{X}_n = \omega_n^2\mathbf{M}\mathbf{X}_n \qquad \mathbf{K}\mathbf{X}_{n'} = \omega_{n'}^2\mathbf{M}\mathbf{X}_{n'}$$

A left-hand side multiplication of the first expression by $\mathbf{X}_{n'}^{\mathsf{T}}$ and a right-hand side of the transpose of the second by $\mathbf{X}_n^{\mathsf{T}}$ produces the following:

$$\mathbf{X}_{n'}^{\mathsf{T}}\mathbf{K}\mathbf{X}_n = \omega_n^2\mathbf{X}_{n'}^{\mathsf{T}}\mathbf{M}\mathbf{X}_n \qquad \mathbf{X}_{n'}^{\mathsf{T}}\mathbf{K}^{\mathsf{T}}\mathbf{X}_n = \omega_{n'}^2\mathbf{X}_{n'}^{\mathsf{T}}\mathbf{M}^{\mathsf{T}}\mathbf{X}_n$$

Taking into account the symmetry of matrices $\mathbf{M}$ *and* $\mathbf{K}$ *and comparing the latter expressions gives*

$$(\omega_n^2 - \omega_{n'}^2)\mathbf{X}_{n'}^T\mathbf{M}\mathbf{X}_n = 0$$

From $n \neq n'$, *or* $\omega_n - \omega_{n'} \neq 0$,[13] *the following relations ensue:*

$$\mathbf{X}_{n'}^T\mathbf{M}\mathbf{X}_n = 0 \qquad \mathbf{X}_{n'}^T\mathbf{K}\mathbf{X}_n = 0$$

The scalar product of $\mathbf{K}\mathbf{X}_n = \omega_n^2\mathbf{M}\mathbf{X}_n$ *and* $\mathbf{X}_n$ *yields:*

$$k_n = \omega_n^2 m_n$$

retrieving so Equation (2.25).

Using the modal expansion of a given vector $\mathbf{X}$, projecting onto eigenmode $\mathbf{X}_n$ and taking into account the orthogonality properties provides:

$$\mathbf{X}_n^T\mathbf{M}\mathbf{X} = \mathbf{X}_n^T\mathbf{M}\sum_{n'\geq 1}\xi_n\mathbf{X}_{n'} = \xi_n\mathbf{X}_n^T\mathbf{M}\mathbf{X}_n$$

The modal expansion of $\mathbf{X}$ *onto the modal basis* $(\mathbf{X}_n)_{n\geq 1}$ *is also stated as follows:*

$$\mathbf{X} = \sum_{n\geq 1}\frac{\mathbf{X}_n^T\mathbf{M}\mathbf{X}}{\mathbf{X}_n^T\mathbf{M}\mathbf{X}_n}\mathbf{X}_n$$

The mass-weighted scalar product $\mathbf{X}^T\mathbf{M}\mathbf{X}'$ *of vectors* $\mathbf{X}$ *and* $\mathbf{X}'$ *is therefore termed as*

$$\mathbf{X}^T\mathbf{M}\mathbf{X}' = \sum_{n\geq 1}\frac{\mathbf{X}_n^T\mathbf{M}\mathbf{X}\ \mathbf{X}_n^T\mathbf{M}\mathbf{X}'}{\mathbf{X}_n^T\mathbf{M}\mathbf{X}_n}$$

With $\mathbf{X} = \mathbf{X}' = \Delta$, *Equation (2.28) is arrived at.*   ∎

2.3   Example: Bending Modes

2.3.1   *Bending Motion of a Straight Elastic Beam*

The problem under concern is described in Figure 2.17: the bending modes of a straight beam of circular section with mixed clamped/free boundary conditions are considered. The beam geometry is defined by its section radius R, thickness h, length L; relevant material properties are the density ρ and Young's modulus E.

The Bernoulli–Euler model accounts for the deflection of cross sections, which are supposed to remain plane and rotate about the bending axis, although rotation inertia is neglected. The model is shown to be valid for *slender beams*, that is for $2R \ll L$.

Denoting $u(z)$ the longitudinal displacement of the tube cross section at z, the equation of motion is

$$-\omega^2\rho Su + EI\frac{\partial^4 u}{\partial z^4} = 0 \ \ \text{for } z \in [0, L]$$

---

[13] The proposed demonstration holds only for eigenmodes with different eigenpulsation; however, the orthogonality condition remains valid for modes with identical eigenpulsation; see for instance Axisa (2001).

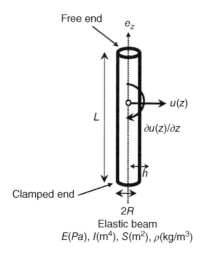

Figure 2.17   Bending of an elastic straight beam

The displacement and the rotation at the clamped end are null, as well as the shear moment and the shear force at the free end:

$$u|_{z=0} = 0 \qquad \frac{\partial u}{\partial z}\bigg|_{z=0} = 0 \qquad +EI\frac{\partial^2 u}{\partial z^2}\bigg|_{z=L} = 0 \qquad -EI\frac{\partial^3 u}{\partial z^3}\bigg|_{z=L} = 0$$

The bending modes are found to be a linear superposition of sine/cosine and hyperbolic sine/cosine functions:

$$u_n(z) = \cos(\Omega_n z/L) - \cosh(\Omega_n z/L) + \lambda_n[\sin(\Omega_n z/L) - \sinh(\Omega_n z/L)] \qquad (2.29)$$

with:

$$\lambda_n = \frac{\sin(\Omega_n) - \sinh(\Omega_n)}{\cos(\Omega_n) + \cosh(\Omega_n)}$$

The corresponding pulsation is

$$\omega_n = \frac{\Omega_n^2}{L^2}\sqrt{\frac{EI}{\rho S}}$$

$(\Omega_n)_{n\geq 1}$ is the nth zero of function $D(x) = \cosh(x)\cos(x) + 1$, see Table 2.4.[14]

Table 2.4   Zeros of function $D(x) = \cosh(x)\cos(x) + 1$

n	1	2	3	4	≥ 5
Ω_n	1.875	4.694	7.855	10.996	$\sim (2n-1)\pi/2$

[14] For other combinations of boundary conditions, see, for instance, Blevins (1979).

Table 2.5 Effective mass of bending modes

n	1	2	3	4	5
$\mu_n/\rho SL$	61.31%	18.83%	6.47%	3.31%	2.00%
n	6	7	8	9	10
$\mu_n/\rho SL$	1.34%	0.96%	0.72%	0.56%	0.45%

The effective mass μ_n of mode u_n, with the meaning of Equation (2.27), is calculated as follows:

$$\frac{\mu_n}{\rho SL} = \frac{\left(\int_0^L u_n(z)dz\right)^2}{\int_0^L u_n^2(z)dz} = \frac{4}{\Omega_n^2} \frac{(\cosh(\Omega_n) + \cos(\Omega_n))^2}{(\sinh(\Omega_n) + \sin(\Omega_n))^2}$$

The corresponding values are given in Table 2.5, for the first 10 modes. The accumulated effective masses yield the total mass of the system, according to Equation (2.28):

$$\sum_{n \geq 1} \frac{\mu_n}{\rho SL} = 1$$

2.3.2 Bernoulli Beam Elements: 1D Element

The weighted integral formulation of the problem reads as follows:

$$-\omega^2 \int_0^L \rho S u\, \delta u\, dz + \int_0^L EI \frac{\partial^2 u}{\partial z^2} \frac{\partial^2 \delta u}{\partial z^2}\, dz = 0$$

for all virtual displacements δu, which comply with the boundary conditions at $z = 0$. The finite element discretisation is performed with I unidimensional elements $[z_i, z_{i+1}]$, such that

$$\int_0^L (\bullet)\, dz = \sum_{i=1}^{i=I} \int_{z_i}^{z_{i+1}} (\bullet)\, dz$$

Finite elements with two nodes and two DOFs per node, as depicted in Figure 2.18, are used so that on $[z_i, z_{i+1}]$, the approximation of the displacement field is:

$$u|_{[z_i, z_{i+1}]} = \mathbf{N}_u^i \mathbf{U}_i$$

- $\mathbf{N}_u^i$ is the vector of shape functions. In the weighted integral formulation of the problem, second-order derivatives of the displacement field have to be calculated: a polynomial approximation of the second order is at least required. Using cubic shape functions ensures that the third-order derivative of the displacement field, namely the shear force, is constant on the element. A third-order polynomial function is entirely defined by a set of four scalar coefficients. Using an element of two nodes and two DOFs per nodes yields four DOF per element. As illustrated in Figure 2.19, the shape function associated with any

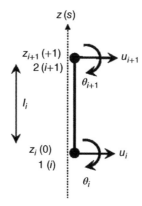

z (s)

$z_{i+1}(+1)$
2 (i+1)

u_{i+1}

θ_{i+1}

l_i

$z_i(0)$
1 (i)

u_i

θ_i

Figure 2.18   Straight beam element with two nodes and two degrees-of-freedom per node

degree-of-freedom should be equal to one at this DOF and to zero at other DOFs: for the element represented in Figure 2.18, a set of four relations is stated for each shape function, uniquely determining the cubic polynomial. These shape functions are

$$\mathbf{N}_u^i(s) = \langle 1 - 3s^2 + 2s^3, l_i(s - 2s^2 + s^3), 3s^2 - 2s^3, l_i(-s^2 + s^3) \rangle$$

with $z = z_i + sl_i$ for $s \in [0, 1]$, l_i being the length of the element $l_i = z_{i+1} - z_i$; Figure 2.19 sketches these functions with $l_i = 1$.

- $\mathbf{U}_i$ is the vector of the displacement unknowns of the element:

$$\mathbf{U}_i = \langle u_i, \theta_i, u_{i+1}, \theta_{i+1} \rangle^{\mathsf{T}}$$

where $\theta = \dfrac{\partial u}{\partial z}$ is standing for the *rotation* of the beam section.

The elementary mass and stiffness matrices are calculated as follows:

$$\mathbf{m}_u^i = \int_{z_i}^{z_{i+1}} \rho S \mathbf{N}_u^{i\,\mathsf{T}} \mathbf{N}_u^i \, dz \qquad \mathbf{k}_u^i = \int_{z_i}^{z_{i+1}} \frac{d^2 \mathbf{N}_u^i}{dz^2}^{\mathsf{T}} \frac{d^2 \mathbf{N}_u^i}{dz^2} \, dz$$

This example is simple enough for an analytical calculation of the matrices to be preferred to a numerical integration procedure; the analytical expressions of $\mathbf{m}_u^i$ and $\mathbf{k}_u^i$ are

$$\mathbf{m}_u^i = \frac{\rho S l_i}{420} \begin{bmatrix} 156 & 22l_i & 54 & -13l_i \\ 22l_i & 4l_i^2 & 13l_i & -3l_i^2 \\ 54 & 13l_i & 156 & -22l_i \\ -13l_i & -3l_i^2 & -22l_i & 4l_i^2 \end{bmatrix}$$

$$\mathbf{k}_u^i = \frac{EI}{l_i^3} \begin{bmatrix} 12 & 6l_i & -12 & 6l_i \\ 6l_i & 4l_i^2 & -6l_i & 2l_i^2 \\ -12 & -6l_i & 12 & -6l_i \\ 6l_i & 2l_i^2 & -6l_i & 4l_i^2 \end{bmatrix}$$

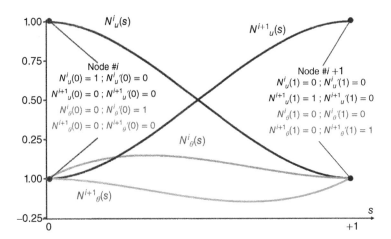

Figure 2.19   Cubic shape functions. The Bernoulli beam element featured in Figure 2.18 has two nodes and two degrees-of-freedom per node, standing, respectively, for the displacement and its derivative. On the reference element, the approximation of the displacement is performed with cubic shape functions. The local variable is s and varies from 0 (for node i) to 1 (for node $i + 1$). The shape functions associated with node i are denoted N_u^i (for the displacement) and N_θ^i (for the displacement derivative – which represents the rotation of the beam section). In order for the approximation to be uniquely determined from the values of the element degree-of-freedom, the shape function should comply with the so-called *unisolvance* property, which is illustrated in the figure. For instance, $N_u^i(s)$ verifies $N_u^i(0) = 1$, $N'_u{}^i(0) = 0$, $N_u^i(1) = 0$ and $N'_u{}^i(1) = 0$. For a cubic shape function, $N_u(s) = a_3 s^3 + a_2 s^2 + a_1 s + a_0$, the former relations yield: $a_0 = 1$, $a_1 = 0$, $a_3 + a_2 + a_1 + a_0 = 0$ and $3a_2 + 2a_2 + a_1 = 0$; hence: $a_0 = 1$, $a_1 = 0$, $a_2 = -3$ and $a_3 = 2$. Shape functions associated with the other degrees-of-freedom of nodes i and $i + 1$ are obtained in a similar manner

The global mass and stiffness matrices are assembled according to

$$[\bullet] = \sum_{i=1}^{i=I} \Lambda_i{}^T [\bullet]_u^i \Lambda_i$$

where Λ_i is the connectivity matrix for the beam element i. It is a $4 \times 2(I + 1)$ matrix which relates the element nodes to the mesh nodes, as detailed in Table 2.6.

The discrete weighted integral formulation yields the eigenvalue problem (2.14); the boundary conditions at $z = 0$ are accounted for by eliminating the displacement u_1 and rotation θ_1

Table 2.6   Connectivity matrix of 1D beam element i

Element			Mesh		
Node	Degree-of-freedom	Number	Node	Degree-of-freedom	Number
1	u_1	1	i	u_i	$2i - 1$
1	θ_1	2	i	θ_i	$2i$
2	u_2	3	$i+1$	u_{i+1}	$2i + 1$
2	θ_2	4	$i+1$	θ_{i+1}	$2i + 2$

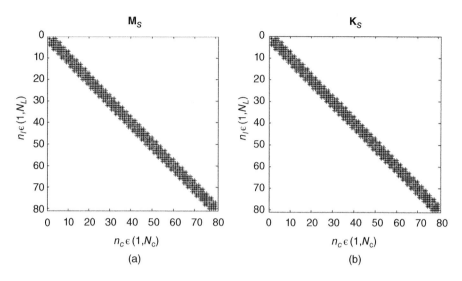

Figure 2.20 (a) Mass and (b) stiffness matrices for beam bending modes

of the first node, that is, by retaining the first two lines and rows of the assembled matrices $\mathbf{M}_S$ and $\mathbf{K}_S$.

For 1D problem such as considered here for beam bending modes, assembling the matrices according to the connectivity matrix proposed in Table 2.6 is straightforward; furthermore, it yields sparse matrices with limited bandwidth, as illustrated in Figure 2.20.

2.3.3 Bending Modes

Geometrical and physical parameters are set as follows: $L = 1$ m, $R = 0.1$ m, $h = 0.01$ m, $\rho = 7,800$ kg/ m^3 and $E = 2.1 \cdot 10^{11}$ Pa. The finite element computation of the eigenfrequencies with $I = 20$ elements compares exactly with the analytical solution, at least for the first four modes, as evidenced in Table 2.7. The convergence of the numerical computation is highlighted by Figure 2.21 in terms of frequency and in Figure 2.22 in terms of effective mass.

Figure 2.23 represents the first four bending modes: in order to obtain a mode shape in good qualitative agreement with the analytical expression, a fine finite element mesh is used (in this case, with $I = 40$ elements).

Remark 2.6 Boundary conditions with Lagrange multipliers *As explained in Remark 2.3, the boundary conditions may be viewed as constraining relations which can be dealt with using Lagrange multipliers. The weighted integral formulation of the problem using the boundary condition at $z = 0$ is written as follows:*

$$-\omega^2 \int_0^L \rho S u \delta u \, dz + \int_0^L EI \frac{\partial^2 u}{\partial z^2} \frac{\partial^2 \delta u}{\partial z^2} \, dz - \lambda \delta u|_{z=0} - \lambda' \frac{\partial \delta u}{\partial z}\bigg|_{z=0} = 0$$

for all δu.

Table 2.7 Eigenfrequencies of the elastic beam: analytical solution
and numerical computation

Frequency	Analytical (Hz)	Numerical (Hz)
f_1	195	195
f_2	1,224	1,224
f_3	3,427	3,427
f_4	6,716	6,716

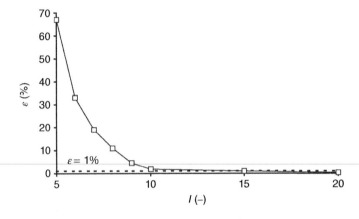

Figure 2.21 Convergence of the finite element computation: relative error (numerical vs analytical) on the eigenfrequency of the 10th mode

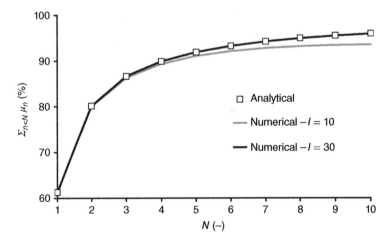

Figure 2.22 Accumulated effective mass of bending modes

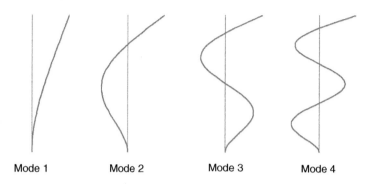

Mode 1   Mode 2   Mode 3   Mode 4

Figure 2.23 Mode shape of bending modes 1–4 in terms of displacement

$T(z)$ and $M(z)$ are the shear moment and the shear force exerted by the beam part $[z, L]$ onto the beam part $[0, z]$; using notations of Figure 2.17, they are calculated as follows:

$$T(z) = -EI\frac{\partial^3 u}{\partial z^3} \qquad M(z) = +EI\frac{\partial^2 u}{\partial z^2}$$

Defining $\lambda = -T(0) = +EI\frac{\partial^3 u}{\partial z^3}\Big|_{z=0}$ and $\lambda' = -M(0) = -EI\frac{\partial^2 u}{\partial z^2}\Big|_{z=0}$, the weighted integral formulation of the problem is also termed as:

$$-\omega^2 \int_0^L \rho S u \delta u \, dz + \int_0^L EI\frac{\partial^2 u}{\partial z^2}\frac{\partial^2 \delta u}{\partial z^2} \, dz - \lambda \delta u|_{z=0} - \lambda'\frac{\partial \delta u}{\partial z}\Big|_{z=0} = 0$$

λ and λ' are the Lagrange multipliers associated with constraints $u|_{z=0} = 0$ and $\frac{\partial u}{\partial z}\Big|_{z=0} = 0$, they are also the reaction force and moment at the constrained cross section $z = 0$.
The weighted formulation of the constraint condition is:

$$\delta\lambda u|_{z=0} + \delta\lambda'\frac{\partial u}{\partial z}\Big|_{z=0} = 0$$

for all $\delta\lambda$ and $\delta\lambda'$ and its finite element form reads as:

$$\delta\mathbf{\Lambda}^T\mathbf{LU} = 0$$

with $\mathbf{\Lambda}^T = \langle \lambda\ \lambda'\rangle$ and where $\mathbf{L}$ is a $2 \times 2(I + 2)$ matrix:

$$\mathbf{L} = \begin{bmatrix} 1 & 0 & \mathbf{0} \\ 0 & 1 & \mathbf{0} \end{bmatrix}$$

The corresponding eigenvalue problem is derived from Equation (2.20).
The modal analysis with the Lagrange multipliers gives frequencies, mode shapes and reaction forces/moments, providing thus useful information in a straightforward way.

The reaction force T_o and moment M_o on the beam at $z = 0$ are $-T(0)$ and $-M(0)$, so that:

$$T_{o,n} = +\frac{T_o}{EI/L^3/u_n(L)} = +\Omega_n^3 \frac{\sin(\Omega_n) - \sinh(\Omega_n)}{\sin(\Omega_n)\sinh(\Omega_n)}$$

$$M_{o,n} = -\frac{M_o}{EI/L^2/u_n(L)} = -\Omega_n^2 \frac{\cos(\Omega_n) + \cosh(\Omega_n)}{\sin(\Omega_n)\sinh(\Omega_n)}$$

Using the finite element approximation, the shear force T_i and shear moment M_i at z_i belonging to the finite element $[z_i, z_{i+1}]$ are calculated as follows:

$$T_i = -EI\frac{d^3 \mathbf{N}_u^i}{dz^3}\mathbf{U}_i \qquad M_i = +EI\frac{d^2 \mathbf{N}_u^i}{dz^2}\mathbf{U}_i$$

where the derivatives of the shape functions at $z = z_i$ are expressed as follows:

$$\frac{d^2 \mathbf{N}_u^i}{dz^2}\bigg|_{z=z_i} = \langle -6/l_i^2, -4/l_i, 6/l_i^2, -2/l_i \rangle \qquad \frac{d^3 \mathbf{N}_u^i}{dz^3}\bigg|_{z=z_i} = \langle 12/l_i^3, 6/l_i^2, -12/l_i^3, 6/l_i^2 \rangle$$

Table 2.8 gives the analytical values of $T_{o,n}$ and $M_{o,n}$ for the first four bending modes and the numerical computation of these quantities:

- from the Lagrange multipliers $\lambda_n/EIL^3 u_n(L)$ and $\lambda'_n/EIL^2 u_n(L)$;
- from a finite element computation of $T(0)$ and $M(0)$.

Table 2.8 Reaction force and moment of bending modes: analytical solution and numerical computation with Lagrange multipliers and shape functions

n	1	2	3	4
	Analytical values			
$T_{o,n}$	$-4.84 \cdot 10^0$	$+1.05 \cdot 10^2$	$-4.84 \cdot 10^2$	$+1.33 \cdot 10^3$
$M_{o,n}$	$-3.52 \cdot 10^0$	$+2.20 \cdot 10^1$	$-6.17 \cdot 10^2$	$+1.21 \cdot 10^2$
	Numerical values from Lagrange multipliers			
$\lambda_n/EIL^3 u_n(L)$	$-4.839 \cdot 10^0$	$+1.053 \cdot 10^2$	$-4.842 \cdot 10^2$	$+1.329 \cdot 10^3$
$\lambda'_n/EIL^2 u_n(L)$	$-3.516 \cdot 10^0$	$+2.203 \cdot 10^1$	$-6.690 \cdot 10^2$	$+1.209 \cdot 10^2$
	Numerical values from shape functions			
$T(0)/EIL^3 u_n(L)$	$+4,838 \cdot 10^0$	$-1.050 \cdot 10^2$	$+4.775 \cdot 10^2$	$-1.282 \cdot 10^3$
$M(0)/EIL^2 u_n(L)$	$+3,516 \cdot 10^0$	$-2.202 \cdot 10^1$	$+6.530 \cdot 10^2$	$-1.197 \cdot 10^2$

Both approaches yield similar yet different results with $I = 10$ finite elements, as presented in the table. However, the discrepancies which are observed between the computed and analytical values vanish when the mesh is further refined.

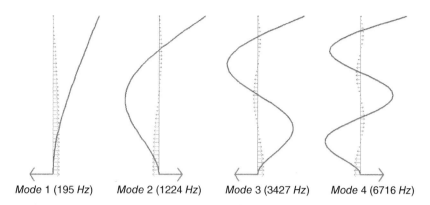

Mode 1 (195 Hz)   Mode 2 (1224 Hz)   Mode 3 (3427 Hz)   Mode 4 (6716 Hz)

Figure 2.24   *Mode shape of beam bending with shear force and reaction force (the full line stands for the transverse displacement, arrows represent the shear stress along the tube and the shear reaction at the support)*

Figure 2.24 sketches the bending mode shapes together with the shear force calculated at various nodes of the finite element mesh and the Lagrange multiplier associated with the constraint $u|_{z=0} = 0$ (once again, a refined mesh with $I = 40$ elements is used here). ∎

2.4   Example: Coupled Bending/Membrane Modes

2.4.1   Bending and Membrane Motion of a Circular Elastic Ring

Figure 2.25 depicts the problem studied in this section: the coupled bending/membrane (or flexion/traction-compression) modes of a circular ring, represented here as a section of a tri-dimensional thin elastic shell, are considered.

Let R and h be the ring radius and thickness, let E and ρ be the ring material properties and let u_r and u_θ be the radial and ortho-radial displacement of the ring. The bending/membrane

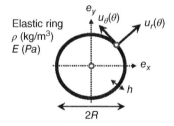

Figure 2.25   Bending/membrane modes of an elastic circular ring

modes of the ring are described by the Love–Kirchhoff model[15]:

$$-\omega^2 \rho h u_r + \frac{Eh}{R^2}\left[u_r + \frac{\partial u_\theta}{\partial \theta} + \frac{h^2}{12R^2}\left(\frac{\partial^4 u_r}{\partial \theta^4} - \frac{\partial^3 u_\theta}{\partial \theta^3}\right)\right] = 0 \tag{2.30}$$

$$-\omega^2 \rho h u_\theta - \frac{Eh}{R^2}\left[\frac{\partial^2 u_\theta}{\partial \theta^2} + \frac{\partial u_r}{\partial \theta} + \frac{h^2}{12R^2}\left(\frac{\partial^2 u_\theta}{\partial \theta^2} - \frac{\partial^3 u_r}{\partial \theta^3}\right)\right] = 0 \tag{2.31}$$

2.4.2   Fourier Component Representation: 0D Element

Given the periodicity of the problem, it is convenient to use a series expansion in Fourier space harmonics of the displacement fields:

$$u_r(\theta) = u_r^0 + \sum_{m \geq 1} u_r^m \cos(m\theta) + \sum_{m' \geq 1} u_r^{m'} \sin(m'\theta)$$

$$u_\theta(\theta) = u_\theta^0 + \sum_{m \geq 1} u_\theta^m \sin(m\theta) + \sum_{m' \geq 1} u_\theta^{m'} \cos(m'\theta)$$

In the above series, it is found convenient to individualise the contribution of three distinct types of components, namely, the *axisymmetric* harmonic ($m = 0$), and the *symmetric* (even) and *anti-symmetric* (odd) harmonics ($m > 1$ and $m' > 1$, respectively); see Figure 2.26.

The projection of Equations (2.30) and (2.31) on the Fourier component of order m (or m') yields:

$$-\omega^2 \rho h u_r^m + \frac{Eh}{R^2}\left[u_r^m + m u_\theta^m + \frac{h^2}{12R^2}\left(m^4 u_r^m + m^3 u_\theta^m\right)\right] = 0$$

---

[15] The bi-dimensional model adopted here is derived from the Love–Kirchhoff tri-dimensional shell model (see for instance Axisa (2005)), which is the equivalent of the Euler–Bernoulli model for thin shells. Let $\mathbf{u} = (u_r, u_\theta, u_z)$ be the displacement in a cylindrical coordinate system. The equations of motion are stated as follows:

$$-\omega^2 \rho h u_r + \frac{Eh}{(1-v^2)R^2}\left[u_r + \frac{\partial u_\theta}{\partial \theta} + vR\frac{\partial u_z}{\partial z} + \right.$$

$$\left. \frac{h^2}{12R^2}\left(\frac{\partial^4 u_r}{\partial \theta^4} + 2R^2\frac{\partial^4 u_r}{\partial \theta^2 \partial z^2} + R^4\frac{\partial^4 u_r}{\partial z^4} - \frac{\partial^3 u_\theta}{\partial \theta^3} - R^2\frac{\partial^3 u_\theta}{\partial \theta \partial z^2}\right)\right] = 0$$

for the radial displacement,

$$-\omega^2 \rho h u_\theta - \frac{Eh}{R^2}\left[\frac{\partial^2 u_\theta}{\partial \theta^2} + \frac{\partial u_r}{\partial \theta} + \frac{(1-v)R^2}{2}\frac{\partial^2 u_\theta}{\partial z^2} + \frac{(1+v)R}{2}\frac{\partial^2 u_z}{\partial \theta \partial z} + \right.$$

$$\left. \frac{h^2}{12R^2}\left(\frac{\partial^2 u_\theta}{\partial \theta^2} - \frac{\partial^3 u_r}{\partial \theta^3} - R^2\frac{\partial^3 u_r}{\partial \theta \partial z^2} + \frac{(1-v)R^2}{2}\frac{\partial^2 u_z}{\partial z^2}\right)\right] = 0$$

for the ortho-radial displacement and:

$$-\omega^2 \rho h u_z - \frac{Eh}{(1-v^2)R^2}\left[vR\frac{\partial u_r}{\partial z} + \frac{(1+v)R}{2}\frac{\partial^2 u_\theta}{\partial \theta \partial z} + \frac{(1-v)}{2}\frac{\partial^2 u_z}{\partial \theta^2} + R^2\frac{\partial^2 u_z}{\partial u}\right] = 0$$

for the axial displacement. Retaining the in-plane motion solely $\left(\frac{\partial(\bullet)}{\partial z} = 0\right)$ and neglecting the Poisson effect ($v = 0$) produces the 2D model of interest here. As discussed in Remark 2.8, the model is shown to be valid for *thin shell*, that is for $h \ll R$.

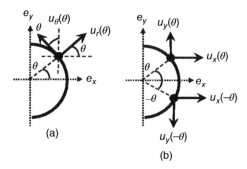

(a)

(b)

Figure 2.26   Fourier component. (a) $u_r(\theta)$ and $u_\theta(\theta)$ are the radial and ortho-radial displacements; they are defined in the coordinate system $\mathbf{e}_r, \mathbf{e}_\theta$. In the Cartesian coordinate system, the displacement field is $\mathbf{u}(\theta) = u_x(\theta)\mathbf{e}_x + u_y(\theta)\mathbf{e}_y$, with $u_x(\theta) = u_r(\theta)\cos\theta - u_\theta(\theta)\sin\theta$ and $u_y(\theta) = u_r(\theta)\sin\theta + u_\theta(\theta)\cos\theta$. $u_r(\theta)$ and $u_\theta(\theta)$ are 2π–periodic functions which can be expanded in Fourier series. (b) The *symmetric* component produces a displacement field which is symmetric with respect to the $(O, \mathbf{e}_x)$ axis so that $u_x(\theta) = u_x(-\theta)$ and $u_y(\theta) = -u_y(-\theta)$. $u_r(\theta)$ and $u_\theta(\theta)$ are consequently even and odd functions of θ. The Fourier series of symmetric components is expressed in terms of cosine functions for the radial displacement $(u_r(\theta) = \sum_m u_r^m \cos(m\theta))$ and sine functions for the ortho-radial displacement $(u_\theta(\theta) = \sum_m u_\theta^m \sin(m\theta))$

and:

$$-\omega^2 \rho h u_\theta^m + \frac{Eh}{R^2}\left[m^2 u_\theta^m + m u_r^m + \frac{h^2}{12R^2}\left(m^2 u_\theta^m + m^3 u_r^m\right)\right] = 0$$

These equations are written in a concise form as $(-\omega^2 \mathbf{M}_S^m + \mathbf{K}_S^m)\mathbf{U}_m = \mathbf{0}$, with $\mathbf{U}_m = \begin{Bmatrix} u_r^m \\ u_\theta^m \end{Bmatrix}$

and:

$$\mathbf{M}_S^m = \rho h \begin{bmatrix} 1 & 0 \\ 0 & 1 \end{bmatrix} \qquad \mathbf{K}_S^m = \frac{Eh}{R^2}\begin{bmatrix} \left(1 + m^4 \dfrac{h^2}{12R^2}\right) & m\left(1 + m^2 \dfrac{h^2}{12R^2}\right) \\ m\left(1 + m^2 \dfrac{h^2}{12R^2}\right) & m^2\left(1 + \dfrac{h^2}{12R^2}\right) \end{bmatrix}$$

From the finite element standpoint, this is equivalent to using a 0D element for each Fourier component, with a single node and two DOFs, as sketched in Figure 2.27.

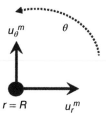

Figure 2.27   0D element with one node and two degrees-of-freedom

2.4.3 Bending/Membrane Modes

The geometrical and physical parameters are set as follows: $R = 0.5$ m, $h = 0.05$ m, $\rho = 7,800$ kg/ m^3 and $E = 2.1 \cdot 10^{11}$ Pa.

Two eigenmodes are calculated for each m or m'; they will be referred to as the first and the second branches of the symmetric or the anti-symmetric Fourier components (in what follows, only the symmetric Fourier components are considered, but similar observations can be drawn for the anti-symmetric components). The displacement shapes of a few eigenmodes are depicted as polar plots in Figure 2.28.

- For $m = 0$, the first branch is the rotation mode ($u_r^1 = 0$, $u_\theta^1 = 1$), which is the first rigid mode at zero frequency, while the second branch is the breathing mode ($u_r^1 = 1$, $u_\theta^1 = 0$) at frequency $1,652$ Hz.

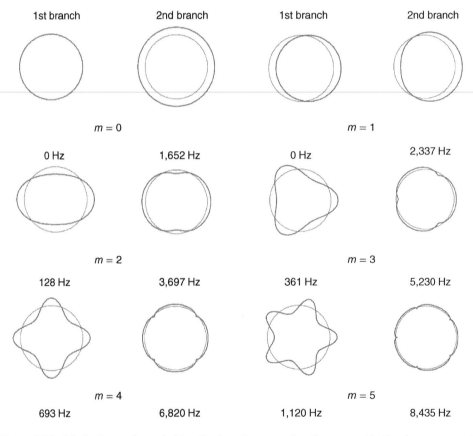

Figure 2.28 Mode shape of coupled bending/membrane modes, for symmetric Fourier components $m = 0$ to $m = 5$ (note that the mode shapes are symmetric with respect to the horizontal axis)

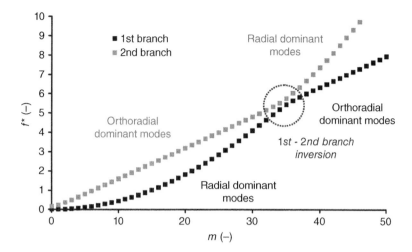

Figure 2.29 Bending or membrane dominated modes, first and second branches ($f^* = fR/\sqrt{E/\rho}$ is a non-dimensional frequency)

- For $m = 1$, the first branch is the translation mode ($u_r^1 = 1$, $u_\theta^1 = -1$), which is the second rigid mode at zero frequency, and the second branch is an elastic mode at frequency $2,337$ Hz, with $u_r^1 = 1$, $u_\theta^1 = 1$.
- For each $m > 1$, the first and second branch are elastic modes, which are either radial dominant (when $|u_r^1| > |u_\theta^1|$) or ortho-radial dominant (when $|u_r^1| < |u_\theta^1|$).

Figure 2.29 identifies which branch corresponds to a predominantly bending (or flexion) motion and which stands for a predominantly membrane (or traction/compression) motion. An inversion of the dominated modes between the two branches is observed, for $m = 35$ in the present case.

It should be noted that the two branches never cross each other: this denotes a *conservative* mechanical coupling, matrices $\mathbf{M}_S^m$ and $\mathbf{K}_S^m$ being symmetric[16].

Remark 2.7 0D or 1D elements? *The use of Fourier series in the case of periodic geometries is a quite convenient technique as a mean to reduce the number of DOFs of the problem, and it is of particular interest in the case of tri-dimensional problems with axial periodicity. Although full 3D model are of common use in the industrial field, 2D axisymmetric representations remain pertinent in many cases.*

A bi-dimensional finite element resolution of the problem can be obtained using 1D elements. In the present case, it seems mathematically consistent to derive an equivalent

---

[16] Non-conservative coupling mechanism usually arises from non-symmetric energy operators: this encompasses a vast domain of theoretical as well as practical issues, which lay far beyond the scope of the present book. As far as FSI finite element formulation is concerned, it should nonetheless be stressed that conservative mechanisms can be represented with non-symmetric matrices, as will be detailed in Chapter 5 in the context of vibro-acoustics.

formulation of the problem starting directly from Equations (2.30) and (2.31). The corresponding weighted integral formulation is termed as follows:

- *Radial displacement:*

$$-\omega^2 \int_0^{2\pi} \rho h \, u_r \delta u_r d\theta + \frac{Eh}{R^2} \left(\int_0^{2\pi} u_r \delta u_r d\theta + \int_0^{2\pi} \frac{\partial u_\theta}{\partial\theta} \delta u_r d\theta \right.$$

$$\left. + \frac{h^2}{12R^2} \int_0^{2\pi} \frac{\partial^4 u_r}{\partial\theta^4} \delta u_r d\theta - \frac{h^2}{12R^2} \int_0^{2\pi} \frac{\partial^3 u_\theta}{\partial\theta^3} \delta u_r d\theta \right) = 0 \qquad \forall \delta u_r$$

- *Ortho-radial displacement:*

$$-\omega^2 \int_0^{2\pi} \rho h \, u_\theta \delta u_\theta d\theta - \frac{Eh}{R^2} \left(\int_0^{2\pi} \frac{\partial^2 u_\theta}{\partial\theta^2} \delta u_\theta d\theta + \int_0^{2\pi} \frac{\partial u_r}{\partial\theta} \delta u_\theta d\theta \right.$$

$$\left. + \frac{h^2}{12R^2} \int_0^{2\pi} \frac{\partial^2 u_\theta}{\partial\theta^2} \delta u_\theta d\theta - \frac{h^2}{12R^2} \int_0^{2\pi} \frac{\partial^3 u_r}{\partial\theta^3} \delta u_\theta d\theta \right) = 0 \qquad \forall \delta u_\theta$$

Integrating by parts the above expressions leads to a symmetric formulation of the problem, which was not apparent in the original equations. It also reduces the derivation order in some terms. After a few mathematical manipulations, the following expressions are arrived at:

$$-\omega^2 \int_0^{2\pi} \rho h \, u_r \delta u_r d\theta + \frac{Eh}{R^2} \left(\int_0^{2\pi} u_r \delta u_r d\theta - \int_0^{2\pi} u_\theta \frac{\partial \delta u_r}{\partial\theta} d\theta \right.$$

$$\left. + \frac{h^2}{12R^2} \int_0^{2\pi} \frac{\partial^2 u_r}{\partial\theta^2} \frac{\partial^2 \delta u_r}{\partial\theta^2} d\theta - \frac{h^2}{12R^2} \int_0^{2\pi} \frac{\partial u_\theta}{\partial\theta} \frac{\partial^2 \delta u_r}{\partial\theta^2} d\theta \right) = 0$$

and:

$$-\omega^2 \int_0^{2\pi} \rho h \, u_\theta \delta u_\theta d\theta + \frac{Eh}{R^2} \left(\int_0^{2\pi} \frac{\partial u_\theta}{\partial\theta} \frac{\partial \delta u_\theta}{\partial\theta} d\theta - \int_0^{2\pi} \frac{\partial u_r}{\partial\theta} \delta u_\theta d\theta \right.$$

$$\left. + \frac{h^2}{12R^2} \int_0^{2\pi} \frac{\partial u_\theta}{\partial\theta} \frac{\partial \delta u_\theta}{\partial\theta} d\theta - \frac{h^2}{12R^2} \int_0^{2\pi} \frac{\partial^2 u_r}{\partial\theta^2} \frac{\partial \delta u_\theta}{\partial\theta} d\theta \right) = 0$$

Using ring elements with two nodes and three degrees-of-freedom per nodes, as depicted in Figure 2.30, seems suitable to perform the discretisation of the above expression. With $[0, 2\pi]$ split in I intervals $[\theta_i, \theta_{i+1}]$, u_r and u_θ may be approximated as:

$$u_r|_{[\theta_i, \theta_{i+1}]} = \mathbf{N}_r^i \mathbf{U}_r^i \qquad u_\theta|_{[\theta_i, \theta_{i+1}]} = \mathbf{N}_\theta^i \mathbf{U}_\theta^i$$

- $\mathbf{N}_r^i$ *and* $\mathbf{N}_\theta^i$ *denote the vectors of shape functions. Since the weighted integral formulation of the problem involves derivatives up to the second order for u_r and up to the first order of u_θ, a logical choice of shape functions would be of cubic type for the former and linear type for the latter, namely:*

$$\mathbf{N}_r^i = \langle 1 - 3s^2 + 2s^3, \phi_i(s - 2s^2 + s^3), 3s^2 - 2s^3, \phi_i(-s^2 + s^3) \rangle$$

$$\mathbf{N}_\theta^i = \langle 1 - s, s \rangle$$

with $\theta = \theta_i + s\phi_i$, $s \in [0, 1]$ and $\phi_i = \theta_{i+1} - \theta_i$.

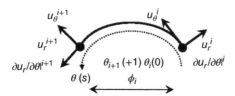

Figure 2.30 *Curved ring element with two nodes and three degrees-of-freedom per node*

- U_r^i and U_θ^i are the nodal values of the radial and ortho-radial displacements of element i:

$$U_r^i = \left\langle u_r^i, \frac{\partial u_r}{\partial \theta}^i, u_r^{i+1}, \frac{\partial u_r}{\partial \theta}^{i+1} \right\rangle^T \qquad U_\theta^i = \langle u_\theta^i, u_\theta^{i+1} \rangle^T$$

The finite element discretisation leads to the following matrix system:[17]

$$\left(-\omega^2 \begin{bmatrix} M_{rr} & 0 \\ 0 & M_{\theta\theta} \end{bmatrix} + \begin{bmatrix} K_{rr} & K_{r\theta} \\ K_{\theta r} & K_{\theta\theta} \end{bmatrix} \right) \begin{Bmatrix} U_r \\ U_\theta \end{Bmatrix} = \begin{Bmatrix} 0 \\ 0 \end{Bmatrix}$$

Figure 2.31 gives the mode shapes of the first elastic modes, that is, non-rigid modes. Double modes correspond to the symmetric and anti-symmetric components of non-null order; the corresponding frequencies (128 Hz, 362 Hz) compare well with the Fourier model. The convergence of the finite element model is illustrated in Figure 2.32: with $I = 100$ elements, the relative difference between the 1D and 0D finite element discretisations is less than 1%.

The convergence of the ring finite element model is rather slow, which is a practical limitation on the use of such an element in the present case. Another limitation arises when dealing

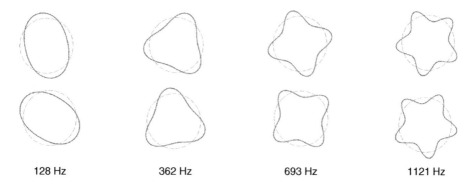

| 128 Hz | 362 Hz | 693 Hz | 1121 Hz |

Figure 2.31 First modes of the elastic ring. *The modes displayed here are computed using the 1D curved elements depicted in Figure* 2.30. *The DOFs are* U_r *and* U_θ. *It is noted that (i) the modes may be ordered in pair, except for the modes whose shape is independent on the angular position; (ii) the axis of symmetry of the modes are not aligned with the horizontal and vertical axes in the Cartesian coordinate systems, as is inherently the case if the Fourier series is used – see Figure* 2.28. *With an appropriate linear combination of paired modes, the symmetry features of the Fourier components may however be retrieved*

---

[17] In the present case, elementary matrices are computed with a numerical procedure, as discussed in Section 2.2.5.

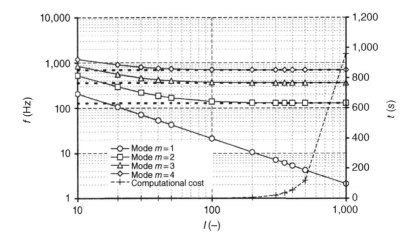

Figure 2.32   *Convergence on frequencies with ring elements*

with the rigid modes. Though the rotation is accurately described, in terms of both mode shape and frequency, see Figure 2.33, a non-null frequency is calculated for the translation modes (5.9 Hz with I = 360 elements). As the mesh is further refined, a convergence towards the appropriate zero value is observed, as evidenced in Figure 2.32, though in practice, the computed value is always positive.

0.0 Hz   5.9 Hz

Figure 2.33   *First ring rigid modes computed with 1D elements*

In the present case, the use of 1D ring elements as formulated here is limited by both the relative slow convergence and the impossibility to account for the translation modes of the ring: from the numerical and practical standpoints, 0D element based on Fourier series expansion is an efficient alternative to circumvent this shortcoming of the 1D ring element. Another alternative is to use Bernoulli straight beam elements, as represented in Figure 2.34. Within each element, the bending and membrane motions are decoupled.

- The radial displacement is described by the following bending equation:

$$-\omega^2 \rho h u_r + \frac{Eh^3}{12R^2} \frac{\partial^4 u_r}{\partial \theta^4} = 0$$

Cubic shape functions are used, with nodal DOF u_r^i, u_r^{i+1} and $\left. \frac{\partial u_r}{\partial \theta}\right|^i, \left. \frac{\partial u_r}{\partial \theta}\right|^{i+1}$; matrices describing bending are given in Section 2.3.2, with $I = \frac{h^3}{12R^4}$, $S = h$ and $l_i = \theta_{i+1} - \theta_i$.

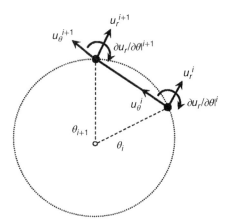

Figure 2.34 *Beam element with decoupled bending and membrane motions for the discretisation of an elastic circular ring*

- *The ortho-radial displacement is described by the following membrane equation:*

$$-\omega^2 \rho h u_\theta - \frac{Eh}{R^2}\frac{\partial^2 u_\theta}{\partial \theta^2} = 0$$

Linear shape functions $\mathbf{N}_u^i = \langle 1 - s, s \rangle$ with associated nodal DOF $u_\theta^i, u_\theta^{i+1}$ are used; the elementary matrices are calculated as follows:

$$\mathbf{m}_u^i = \int_{\theta_i}^{\theta_{i+1}} \rho h \mathbf{N}_u^{i\,T} \mathbf{N}_u^i \, d\theta \qquad \mathbf{k}_u^i = \int_{\theta_i}^{\theta_{i+1}} Eh \frac{d\mathbf{N}_u^i}{d\theta}^T \frac{d\mathbf{N}_u^i}{d\theta} \, d\theta$$

Hence:

$$\mathbf{m}_u^i = \frac{\rho h l_i}{6}\begin{bmatrix} 2 & 1 \\ 1 & 2 \end{bmatrix} \qquad \mathbf{k}_u^i = \frac{Eh}{l_i^2}\begin{bmatrix} 1 & -1 \\ -1 & 1 \end{bmatrix}$$

As made conspicuous in Tables 2.9, such an approach yields better convergence and allows for the description of the rigid modes.   ∎

Remark 2.8 Validity of the Bernoulli–Euler and the Love–Kirchhoff models *The Bernoulli–Euler model describes the bending modes of vibrations of a slender straight beam, neglecting both elastic stresses in the beam cross section and rotational inertia of the cross section. Shear induces an additional elastic potential energy, while rotational inertia induces an additional term in the kinetic energy. The increased mechanical energy is shown to be proportional to the square of the length to radius ratio L/R[18].*

The Love–Kirchhoff model is based on similar simplifying hypothesis to describe the motion of thin shells, and it is valid for small thickness-to-curvature ratios h/R. The Reissner–Mindlin model, schematically represented in Figure 2.35, is of common use to deal with thick shells.

---

[18] In this book, the numerical applications are proposed for beam and shell with aspect ratio R/L and h/R about 10%, which may be considered as the validity limit for the Bernoulli–Euler and Love–Kirchhoff models. The applications remain worthy of interest as the purpose here is mainly to illustrate the application of the FEM with some of the simplest available analytical models.

Table 2.9   *Comparison of curved and beam elements discretisation*

Frequency (Fourier component)	I = 36 elements		I = 180 elements	
	Curved elements (Hz)	Beam elements (Hz)	Curved elements (Hz)	Beam elements (Hz)
f_1 $(m = 0)$	0	0	0	0
f_2, f_3 $(m = 1)$	59	0	12	0
f_4, f_5 $(m = 2)$	197	129	131	128
f_6, f_7 $(m = 3)$	433	364	365	362
f_8, f_9 $(m = 4)$	766	698	696	692

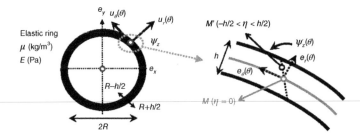

Figure 2.35   *Elastic ring: bending and membrane modes with the thick shell model*

The displacement of a point M at mid-thickness is denoted $\mathbf{u}_M(\theta) = u_r(\theta)\mathbf{e}_r + u_\theta(\theta)\mathbf{e}_\theta$, *while the displacement of a current point M′ within the shell thickness is calculated as* $\mathbf{u}_{M'}(\theta) = u_r(\theta)\mathbf{e}_r + (u_\theta(\theta) + \eta(\theta)\psi_z(\theta))\mathbf{e}_\theta$, *with* $\eta \in [-h/2, +h/2]$ *and with* $\psi_z(\theta)$ *the small angle of rotation of the shell about the axial direction.*

The equation of motion is expressed in terms of radial displacement u_r, *ortho-radial displacement* u_θ *and rotation of section* ψ_z. *The following representation is adopted here in order to highlight the differences between the thin and thick shell models[19].*

- *Radial displacement:*

$$-\omega^2 \rho h u_r + \frac{E}{R}\left[\lambda\left(u_r + \frac{\partial u_\theta}{\partial \theta}\right) + (h - \lambda R)\frac{\partial \psi_z}{\partial \theta}\right]$$

$$+\frac{\lambda G}{R}\left[\frac{\partial u_\theta}{\partial \theta} - \frac{\partial^2 u_r}{\partial \theta^2} - R\frac{\partial \psi_z}{\partial \theta}\right] = 0$$

---

[19] The equation of motion for the thick shell model with the Reissner–Mindlin description is taken here for granted; more on the mathematical formulation of shell dynamics is proposed for instance in Axisa (2005). More on the issues raised in the previous and the present remarks regarding the finite element formulation of curved shells may be found in Chapelle and Bathe (2010).

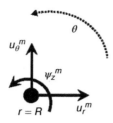

Figure 2.36 *0D element with one node and three degrees-of-freedom*

- *Ortho-radial displacement:*

$$-\omega^2 \rho h u_\theta - \omega^2 \mu \frac{h^3}{12R} \psi_z - \frac{E}{R} \left[\lambda \left(\frac{\partial u_r}{\partial \theta} + \frac{\partial^2 u_\theta}{\partial \theta^2} \right) + (h - \lambda R) \frac{\partial^2 \psi_z}{\partial \theta^2} \right]$$

$$+ \frac{\lambda G}{R} \left[u_\theta - \frac{\partial u_r}{\partial \theta} - R \psi_z \right] = 0$$

- *Rotation:*

$$-\omega^2 \rho \frac{h^3}{12R} u_\theta + \mu \frac{h^3}{12} \psi_z - \frac{E}{R} \left[(h - \lambda R) \left(\frac{\partial u_r}{\partial \theta} + \frac{\partial^2 u_\theta}{\partial \theta^2} \right) + R(h - \lambda R) \frac{\partial^2 \psi_z}{\partial \theta^2} \right]$$

$$+ \frac{\lambda G}{R} \left[\frac{\partial u_r}{\partial \theta} - u_\theta + R \psi_z \right] = 0$$

with $\lambda = \ln \left(\dfrac{1 + h/2R}{1 - h/2R} \right)$ *and* $G = \dfrac{E}{2(1 + v)}$ *the shear modulus.*

u_r, u_θ *and* ψ_z *are expanded as Fourier series. The expressions retained for* u_r *and* u_θ *are given in Section 2.4.2, while the following expansion is used for* ψ_z:

$$\psi_z = \psi_z^o + \sum_{m \geq 1} \psi_z^m \sin(m\theta) + \sum_{m' \geq 1} \psi_z^{m'} \cos(m'\theta)$$

with m and m' standing for the symmetric and anti-symmetric Fourier components.

   The projection of the equation of motion on each Fourier component is in this case equivalent to describing the structure dynamics using a 0D finite element with three DOFs, as depicted in Figure 2.36.

   The matrices associated with the element are given as follows:

$$\mathbf{M}_m = \rho h \begin{bmatrix} 1 & 0 & 0 \\ 0 & 1 & \dfrac{h^2}{12R} \\ 0 & \dfrac{h^2}{12R} & \dfrac{h^2}{12} \end{bmatrix}$$

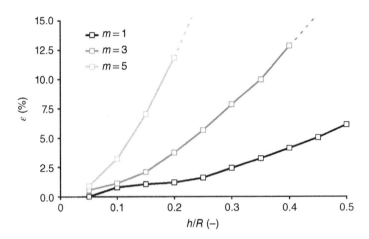

Figure 2.37 *Comparison of the thick and thin shell models in terms of computed eigenfrequencies versus the thickness-to-curvature ratio for various Fourier component*

and:

$$\mathbf{K}_m = \frac{E}{R} \begin{bmatrix} \lambda & \lambda m & (h - \lambda R)m \\ \lambda m & \lambda m^2 & (h - \lambda R)m^2 \\ (h - \lambda R)m & (h - \lambda R)m^2 & -R(h - \lambda R)m^2 \end{bmatrix}$$

$$+ \frac{G}{R} \begin{bmatrix} \lambda m^2 & \lambda m & -R\lambda m \\ \lambda m & \lambda & -\lambda R \\ -R\lambda m & -\lambda R & \lambda R^2 \end{bmatrix}$$

The frequencies of the elastic ring calculated with thin or thick shell models are denoted f_{thin} and f_{thick}, respectively. $\varepsilon = \frac{f_{\text{thin}}}{f_{\text{thick}}} - 1$ is represented in Figure 2.37 as a function of the thickness-to-curvature ratio. As conveyed by the plots in the figure, $h/R \approx 10\%$ may be assumed to be a reasonable upper bound for the validity of the thin shell model – in most applications at least. ∎

References

Axisa F 2001 *Modelling of Mechanical Systems – Discrete Systems*. Kogan Page Science.

Axisa F 2005 *Modelling of Mechanical Systems – Structural Elements*. Elsevier.

Babuska I and Guo BQ 1992 The *h, p* and *h − p* versions of the finite element method: basis theory and applications. *Advances in Engineering Software*, **15**, 159–174.

Babuska I and Strouboulis T 2001 *The Finite Element Method and its Reliability*. Oxford University Press.

Bathe KJ 1982 *Finite Element Procedures in Engineering Analysis*. Prentice-Hall.

Blevins RD 1979 *Formulas for Natural Frequency and Mode Shape*. Krieger Publishing Company.

Boffi D 2010 Finite element approximation of eigenvalue problems. *Acta Numerica*, **10**, 1–20.

Chapelle D and Bathe KJ 2010 *The Finite Element Analysis of Shells*. Springer-Verlag.

Christensen RM 1982 *Theory of Viscoelasticity: An Introduction*. Academic Press.

Cuthill E and MacKee J 1969 Reducing the bandwidth of sparse symmetric matrices. In *Proceedings of the 24th National Conference of the Association for Computing Machinery (ACM 1969)*.

Hughes T 1987 *The Finite Element Method: Linear Static and Dynamic Finite Element Analysis*. Prentice-Hall.

MacNeal RH 1994 *Finite Elements: Their Design and Performance*. Marcel Dekker.

Matthews FL, Davies GA, Hitchings D, and Soutis C 2000 *Finite Element Modeling of Composite Materials and Structures*. Woodhead Publishing.

Reddy JN 2004 *Mechanics of Laminated Composite Plates and Shells: Theory and Analysis*, CRC Press LLC.

Scheissel HV, Metzleri R, Blumen'f A, and Nonnenmacheri TF 1995 Generalized visco-elastic models: their fraction equations with solutions. *Journal of Physics*, **28**, 6567–6584.

Soize C and Ohayon R 1998 *Structural Acoustics and Vibrations: Mechanical Models, Variational Formulations and Discretisation*. Academic Press.

Zienkiewicz OC and Taylor RL 2000 *The Finite Element Method – Volume 1: The Basis, Volume 2: Solid Mechanics, Volume 3: Fluid Dynamics*. Butterworth-Heinemann.

3

Fluid Finite Elements

Wave equations arising for fluids on account of compressibility or gravity effects are considered in this chapter. They derive from the mass and momentum equations, which constitute the Navier–Stokes model, when considering the small transformations of perfect fluids initially at rest. Pressure and displacement-based formulations of the wave equations are investigated. The associated boundary conditions, holding for either the standing waves or the propagating waves, are also discussed. The finite element and the boundary element discretisations of the wave equations are detailed and illustrated on simple examples.

Figure 3.1   From quiescent fluid to flowing fluid. Based on the space and time evolutions of the velocity and pressure fields, the Navier–Stokes model allows for the description of viscous fluid flows. For a quiescent and inviscid fluid, the small variations of p and $\mathbf{v}$ about a steady state are considered: they are described using the Euler model. In both models, free surface effects may be accounted for. *Source*: © Jean-François Sigrist

Fluid–Structure Interaction: An Introduction to Finite Element Coupling, First Edition. Jean-François Sigrist.
© 2015 John Wiley & Sons, Ltd. Published 2015 by John Wiley & Sons, Ltd.
Companion Website: www.wiley.com/go/sigrist

3.1 Fluid Flow Equations

As the shape of a flowing fluid continuously changes over time, the description of the flow may conveniently be achieved within the Eulerian framework. While the Lagrangian formulation tracks the evolution of a defined set of matter, the Eulerian formulation focusses on the rate of change of the quantity of interest. The mass and momentum balancing of a fluid are established at time t in an elementary volume C, the so-called *control volume* as represented in Figure 3.2, when the volume vanishes to zero. The following set of equations, written here with an external volume force, is obtained; they hold for a given point $M \in C$.

- Mass equation:

$$\frac{\partial \rho}{\partial t} + \frac{\partial (\rho v_j)}{\partial x_j} = 0 \tag{3.1}$$

- Momentum equation:

$$\frac{\partial (\rho v_i)}{\partial t} + \frac{\partial (\rho v_i v_j)}{\partial x_j} = \rho f_i + \frac{\partial \sigma_{ij}(\mathbf{v})}{\partial x_j} \tag{3.2}$$

$p(\mathbf{x}, t)$ is the pressure field; the components of the velocity field $\mathbf{v}(\mathbf{x}, t)$ in a Cartesian coordinate system are denoted $v_i(\mathbf{x}, t)$ for $i \in [1, d]$, where d is the dimension of the problem. $\sigma(\mathbf{x}, t)$ and $\rho(\mathbf{x}, t)$ denote the stress and density fields at a given point. The external volume force is denoted $\mathbf{f}$, and its components are f_i: in the following, it stands for gravity. The stress tensor is expressed as $\sigma_{ij}(\mathbf{v}) = -p\delta_{ij} + \tau_{ij}(\mathbf{v})$ with $\tau(\mathbf{v})$ the shear stress tensor.

Equations (3.1) and (3.2) are complemented with the state equation of the fluid $\rho(p)$, which relates the density to the pressure field[1].

For *Newtonian fluids*, the shear stress–strain relation is linear, as illustrated in Figure 3.3; it is written as follows:

$$\tau_{ij} = \mu \left(\frac{\partial v_i}{\partial x_j} + \frac{\partial v_j}{\partial x_i} \right) + \mu' \frac{\partial v_k}{\partial x_k} \delta_{ij} \tag{3.3}$$

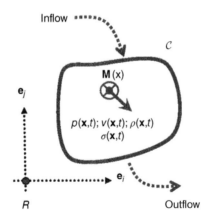

Figure 3.2 Eulerian description in continuum mechanics: fluid flow through a control volume C

---

[1] In the context of fluid–structure interaction (FSI), thermal energy exchanges are accounted for in a few and very specific cases solely. Hence, in this book, the fluid properties are assumed to be independent of the temperature.

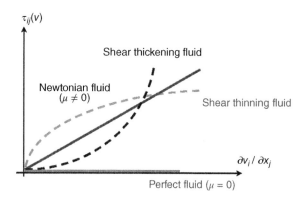

Figure 3.3   Newtonian and non-Newtonian fluids. For Newtonian fluids, the viscous stresses τ_{ij} are proportional to the shear strain $\partial v_i / \partial x_j$. As conveyed by Equation (3.3), the viscosity tensor reduces to two coefficients; under the Stokes hypothesis, they are linearly dependent. Most common liquids and gases – water, air, thin oil or solvent being some obvious examples – are assumed to be Newtonian. For non-Newtonian fluids, the viscosity depends on the shear strain or the shear strain rate. The graph above features the behaviour of some non-Newtonian fluids with a stress–strain relation, which is invariant in time. The viscosity of shear thickening fluids increases with increased stress, whereas it decreases with increased stress for shear thinning fluids. In some cases, the shear stress–shear strain relation may be time dependant, so that the viscosity varies with the duration of the applied stresses. Biological, natural or edible substances – such as blood, lava or yoghurt to name a very few – are typical examples of non-Newtonian fluids (Chassaing, 2000).

μ and μ' are the fluid viscosity coefficients: they respectively account for the fluid resistance to shear deformation and compression. Under the Stokes hypothesis[2], the viscosity coefficients verify $2\mu + 3\mu' = 0$.

For flows of Newtonian fluids, the mass and momentum equations read as follows:

$$\frac{\partial \rho}{\partial t} + \frac{\partial (\rho v_j)}{\partial x_j} = 0 \tag{3.4}$$

$$\frac{\partial (\rho v_i)}{\partial t} + \frac{\partial (\rho v_i v_j)}{\partial x_j} = \rho f_i - \frac{\partial p}{\partial x_i} + \mu \frac{\partial}{\partial x_j} \left(\frac{\partial v_i}{\partial x_j} + \frac{\partial v_j}{\partial x_i} - \frac{2}{3} \frac{\partial v_k}{\partial x_k} \delta_{ij} \right) \tag{3.5}$$

The general form of the balancing equation may be expressed as follows:

$$\frac{\partial (\rho \psi)}{\partial t} + \frac{\partial (\rho \psi v_j)}{\partial x_j} = \Sigma_\psi + \frac{\partial}{\partial x_j} \left(\Gamma_\psi \frac{\partial \psi}{\partial x_i} \right) \tag{3.6}$$

---

[2] The Stokes hypothesis stipulates that the thermodynamic pressure p is equivalent to the mechanical pressure π, the latter being defined as the trace of the stress tensor:

$$\pi = \sigma_{kk}(\mathbf{v}) = p - \frac{2\mu + 3\mu'}{3} \frac{\partial v_k}{\partial x_k}$$

$\pi \equiv p$ implies $2\mu + 3\mu' = 0$.

for a given quantity ψ (for instance, the mass equation is of the type [3.6] with $\psi = 1$ and $\Sigma_\psi = 0$, $\Gamma_\psi = 0$).

According to Equation (3.6), the space and time evolutions of ψ result from various processes: the creation or the destruction of ψ is described by the source term Σ_ψ, while the time fluctuations of ψ are represented by the non-stationary term $\dfrac{\partial(\rho\psi)}{\partial t}$; the convection of ψ by the fluid, with or without bulk motion, is accounted for by the advection and diffusion terms, $\dfrac{\partial(\rho\psi v_j)}{\partial x_j}$ and $\dfrac{\partial}{\partial x_j}\left(\Gamma_\psi \dfrac{\partial\psi}{\partial x_i}\right)$, respectively. Γ_ψ is the so-called *diffusion coefficient* associated with the scalar quantity ψ.

As stated by Equation (3.4), in the absence of source terms, the balance of mass results only from the advection and from the time fluctuation of density. Equation (3.5), on the other hand, conveys that the balance of momentum stems from the advection and from the time evolution of velocity, balanced by diffusion associated with viscosity, while the pressure gradient and the external forces generate fluid motion.

The mass equation and the equation of state $\rho(p)$ are of scalar nature, while the momentum conservation equation is vectorial. These three equations constitute the Navier–Stokes model[3]; it involves three unknowns which are scalar (p, ρ) or vectorial $(\mathbf{v})$, so that the problem is well posed.

Various numerical methods may be resorted to for solving the flow equations, some of which are briefly exposed in Remark 3.1. In the context of small perturbations of the fluid flow about a steady state, the fluid equations may be simplified in various ways to make them tractable for discretisation according to the finite element method, as detailed in Section 3.3.

For *perfect fluids*, the viscosity effect is discarded so that no shear stress is observed in the fluid flow; setting $\mu = 0$ in the Navier–Stokes equations yields the Euler equations:

$$\frac{\partial\rho}{\partial t} + \frac{\partial(\rho v_j)}{\partial x_j} = 0 \tag{3.7}$$

$$\frac{\partial(\rho v_i)}{\partial t} + \frac{\partial(\rho v_i v_j)}{\partial x_j} = \rho f_i - \frac{\partial p}{\partial x_i} \tag{3.8}$$

Wave propagation equations in a quiescent and non-viscous fluid are derived from the Euler equations. They are formulated in the context of *small transformations* about a steady state (fluid at rest or in permanent flow), that is, on a fixed domain. In order to describe the fluid flow, it is convenient to separate any field $\psi(\mathbf{x}, t)$ (standing for pressure, density or any velocity component) as follows:

$$\psi(\mathbf{x}, t) = \psi_o(\mathbf{x}) + \psi'(\mathbf{x}, t) \tag{3.9}$$

where $\psi_o(\mathbf{x})$ and $\psi'(\mathbf{x}, t)$ are the *steady* and *fluctuating* parts.

The perturbation of the fluid flow is supposed to be small with respect to the steady state, so that linearisation of the Euler equations is made possible.

---

[3] Who would be charmed by stories embedded in mathematical models, such as those usually told about the Navier–Stokes equations, might turn to Stewart (2013) who offers a delightful introduction on the way equations and mathematicians shape elegant representations of our world.

Substituting these expressions into the mass and momentum equations and retaining terms of the same order yield the balancing equations for the steady state (zero-order) and fluctuating flows (first-order).

- The mass and momentum equations for the *steady state* flow are given as follows:

$$\frac{\partial(\rho_o v_{oi})}{\partial x_i} = 0$$

$$\rho_o \frac{\partial(v_{oi} v_{oj})}{\partial x_j} = \rho_o f_i - \frac{\partial p_o}{\partial x_i}$$

- The mass and momentum equations for the *fluctuating* flow are given as follows:

$$\frac{\partial \rho'}{\partial t} + \frac{\partial(\rho_o v_i')}{\partial x_i} + \frac{\partial(\rho' v_{oi})}{\partial x_i} = 0$$

$$\rho_o \frac{\partial v_i'}{\partial t} + v_{oi} \frac{\partial \rho'}{\partial t} + \frac{\partial(\rho_o v_{oi} v_j')}{\partial x_j} + \frac{\partial(\rho_o v_i' v_{oj})}{\partial x_j} = \rho' f_i - \frac{\partial p'}{\partial x_i}$$

Remark 3.1 Computational Fluid Dynamics *Computational Fluid Dynamics (CFD) is concerned with the numerical techniques aimed at computing approximate solutions of the flow equations. Among various methods, whether* mesh-based *(e.g. finite element or finite difference methods) or based on* meshless *approaches (e.g. lattice Boltzmann or particle element methods), two of the most common methods for engineering applications are introduced in what follows.*

Finite volume method (FVM) The finite volume method (FVM) is adapted to an extended class of problems and it is one of the most popular methods for engineering purposes. Involving reliable computational coding for complex problems, the FVM is easily formulated on tri-dimensional geometries. It allows for meshes of complex shape to be handled, see for instance Figure 3.4, while achieving the level of accuracy and robustness required in most industrial applications, see for instance Figure 3.5.

The FVM is an Eulerian-formulated and mesh-based method, which emanates from an integral formulation of Equation (3.6) over the control volume Ω bounded by the surface Γ. It yields the following form of the balance equation for the fluid variable ψ[4]:

$$\frac{\partial}{\partial t} \int_\Omega \rho \psi \, d\Omega + \int_\Gamma \rho v_i n_i d\Gamma = \int_\Omega \Sigma_\psi d\Omega + \int_\Gamma \Gamma_\psi \frac{\partial \psi}{\partial x_i} n_i \tag{3.10}$$

---

[4] In Equation (3.10), the volume integrals of the convective and diffusive terms are transformed into surface integrals by using the Gauss theorem:

$$\int_\Omega \nabla \cdot \boldsymbol{\psi} \, d\Omega = \int_\Gamma \boldsymbol{\psi} \cdot \mathbf{n} \, d\Gamma$$

which holds for any vectorial function $\boldsymbol{\psi}$ or:

$$\int_\Omega \nabla \psi \, d\Omega = \int_\Gamma \psi \cdot \mathbf{n} \, d\Gamma$$

which holds for any scalar function ψ. In the above expressions, $\mathbf{n}$ denote the unit normal vector on Γ, pointing outwards.

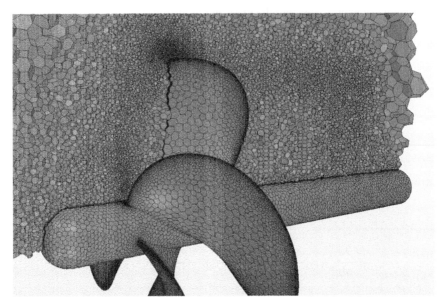

Figure 3.4   Finite volume mesh. *For the simulation of the fluid flow around a propeller, the finite volume mesh is composed of polyhedral cells, which proves in some instances suited to tackle the geometrical complexity of the problem under concern while ensuring a satisfactory level of accuracy.* Source: *Julien* MANERA, *CD-adapco, Bobigny, France, 2014. Reproduced with permission of CD-adapco*

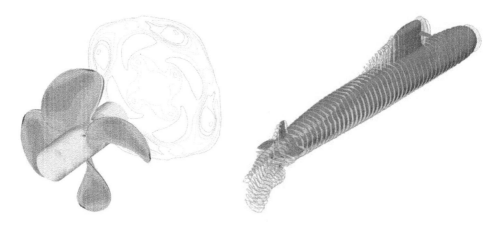

Figure 3.5   Hydrodynamics with the finite volume method. *The finite volume method is of broad use in various engineering applications of CFD, for either internal or external flows, as in these examples. The simulations of the fluid flow around the propeller or the submarine are performed with an FVM-based code and allow for the assessment of their hydrodynamic performance.* Source: *Jean-François* SIGRIST *and Jean-Jacques* MAISONNEUVE, *DCNS research, Nantes, France, 2012 and 2014. Reproduced with permission of DCNS*

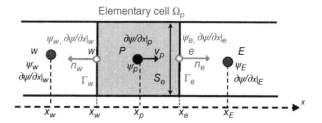

Figure 3.6   Finite volume method for a unidimensional problem. Ω_P is an elementary cell centred on node P (of abscissae x_P), on which Equation (3.10) is stated. The neighbouring cells are, respectively, Ω_W (on the left-hand, or west, side) and Ω_E (on the right-hand, or east, side). They are centred on nodes W and E (of abscissae x_W and x_E). w and e, of abscissae x_w and x_w stand for the nodes centred at boundaries Γ_e and Γ_w, with unit normal vector $\mathbf{n}_e$ and $\mathbf{n}_w$ pointing outwards

The numerical approximation of the different terms in the preceding equation is conducted with various schemes, as discussed thoroughly by Ferziger and Peric (2002) for instance. However, in order to grasp the essential ideas of the method, a brief presentation of the discretisation principles is proposed here for a unidimensional problem, as depicted in Figure 3.6, and using first-order schemes.

- A first-order approximation (in time) of the non-stationary term reads

$$\frac{\partial}{\partial t} \int_{\Omega} \rho \psi \, d\Omega \approx \frac{\rho(\psi_P^{n+1} - \psi_P^n)|\Omega_P|}{\delta t}$$

where $|\Omega_P|$ is the volume of the elementary cell centred on P, $\psi_P = \psi(x_P)$ and superscripts n and n + 1 designate the value of ψ at time t_n and t_{n+1}.

- A first-order approximation (in space) of the source term at time t_{n+1} within Ω is

$$\int_{\Omega} \Sigma_\psi \, d\Omega \approx \Sigma_{\psi_P}^{n+1} |\Omega_P|$$

- A first-order approximation (in space) of the fluxes through the control surface Γ is

$$\int_{\Gamma} \psi v_j n_j \, d\Gamma \approx \psi_e(v_j n_j)_e |\Gamma_e| + \psi_w(v_j n_j)_w |\Gamma_w|$$

for the convective flux, and

$$\int_{\Gamma} \Lambda_\psi \frac{\partial \psi}{\partial x_j} n_j \, d\Gamma \approx \Lambda_\psi^e \left(\frac{\partial \psi}{\partial x_j} n_j \right)_e |\Gamma_e| + \Lambda_\psi^w \left(\frac{\partial \psi}{\partial x_j} n_j \right)_w |\Gamma_w|$$

for the diffusive flux.

The former expressions require the evaluation of ψ at different nodes of the cell surface: ψ_e, ψ_w and their derivatives are calculated from the corresponding values at the cell centre ψ_E and ψ_W using finite difference schemes. For instance, with a centred-differencing scheme, $\psi_e = \psi_E \lambda_e + \psi_P(1 - \lambda_e)$, where λ_e is the so-called linear approximation factor:

$$\lambda_e = \frac{x_{j_e} - x_{j_P}}{x_{j_E} - x_{j_P}}$$

Various approximations of the derivatives $\frac{\partial \psi}{\partial x_j}$ at node P and E may be used, for instance:

$$\left.\frac{\partial \psi}{\partial x_j}\right|_E = \frac{\psi_E - \psi_P}{x_{jE} - x_{jP}} \qquad \left.\frac{\partial \psi}{\partial x_j}\right|_P = \frac{1}{2}\left(\frac{\psi_P - \psi_W}{x_{jP} - x_{jW}} + \frac{\psi_P - \psi_E}{x_{jP} - x_{jE}}\right)$$

With the approximation schemes of the integrals and of the nodal quantities, a linear relation is stated between the value of ψ at the cell centre (ψ_P) and its value at the centre of the adjacent cells (ψ_N):

$$A_P\,\psi_P^{n+1} + \sum_N A_N\,\psi_N^{n+1} = b_P^n$$

A_P denotes the contribution of the volume integral for cell Ω_P at time t_{n+1}, A_N contains the contribution of the surrounding cells to the surface integrals at time t_{n+1} and b_P gathers the contribution of all terms at time t_n. Assembling the contribution of each cell, a matrix system is constructed:

$$\mathbf{A}(\psi)\,\mathbf{\Psi} = \mathbf{b}_\psi \tag{3.11}$$

$\mathbf{\Psi}$ gathers the nodal values of ψ, the matrix $\mathbf{A}(\psi)$ and vector $\mathbf{b}_\psi$ are composed of coefficients A_P and b_P, respectively. Equation (3.11) is solved with dedicated numerical procedures, as discussed, for instance, in Ferziger and Peric (2002).

Smoothed particle hydrodynamics (SPH) *is a meshless, Langrangian-based method offering a convenient alternative to Eulerian-based techniques, which often find their limits when the flow involves large deformation, separation or recirculation[5].*

The SPH approximation of the flow field $\psi(\mathbf{x})$ originates from its integral representation with the Dirac distribution, namely:

$$\psi(\mathbf{x}) = \int_\Omega \delta(\mathbf{x} - \mathbf{x}')\psi(\mathbf{x}')\,d\Omega(\mathbf{x}')$$

which holds for any function with sufficient regularity. With a smoothing function $W_h(\mathbf{x} - \mathbf{x}')$ instead of the Dirac function[6], the former relation yields the so-called kernel approximation *of ψ:*

$$\langle\psi(\mathbf{x})\rangle = \int_\Omega W_h(\mathbf{x} - \mathbf{x}')\psi(\mathbf{x}')\,d\Omega(\mathbf{x}')$$

A kernel approximation of the derivative of ψ is obtained using the divergence theorem; it reads as follows:

$$\left\langle\frac{\partial \psi}{\partial x_j}(\mathbf{x})\right\rangle = -\int_\Omega \psi(\mathbf{x}')\frac{\partial W_h}{\partial x_j}(\mathbf{x} - \mathbf{x}')\,d\Omega(\mathbf{x}')$$

The spatial integration in the kernel approximation is converted into a summation over all the particles contained within a support domain, as described in Figure 3.7. The SPH

---

[5] SPH has been originally introduced in astrophysics, see, for instance, Gingold and Monaghan (1977) and Lucy (1977), and has been applied since to various physics, among which fluid dynamics. SPH techniques are also used in real-time animation: in such cases, the level of accuracy of an SPH-based simulation is often degraded in order to achieve numerical efficiency.

[6] W_h is the so-called *kernel function* and may be interpreted as an approximation of the Dirac distribution. It is defined by a smoothing length such as (i) $W_h(\mathbf{x} - \mathbf{x}') = 0$ when $\|\mathbf{x} - \mathbf{x}'\| > \kappa h$; (ii) it tends to the Dirac as the length h vanishes, $\lim_{h \to 0} W_h(\mathbf{x} - \mathbf{x}') = \delta(\mathbf{x} - \mathbf{x}')$; (iii) it satisfies a normalisation condition $\int_\Omega W_h(\mathbf{x} - \mathbf{x}')\,d\Omega(\mathbf{x}') = 1$.

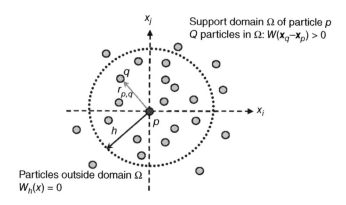

Figure 3.7   Kernel approximation. *The kernel function is a function which vanishes outside a support domain Ω; for 2D problem Ω is included in a circle of radius κh. The kernel approximation of ψ at each point P is obtained by summing up the weighted values of the function at the position of the particles in the support domain*

representation for $\psi(\mathbf{x}_p)$ is written using the particle approximation according to

$$\langle \psi(\mathbf{x}_p) \rangle = \sum_{q=1}^{q=Q} \frac{m_q}{\rho_q} \psi(\mathbf{x}_q) W_h(\mathbf{x}_p - \mathbf{x}_q)$$

m_q and ρ_q denote the mass of particle q and the density of the fluid at point $\mathbf{x}_q$: m_q/ρ_q may be interpreted as the elementary volume $\Delta\Omega_q$ associated with particle q, which gives ground to the approximation $\int_\Omega \bullet \, d\Omega \simeq \sum_q \bullet (\mathbf{x}_q)\Delta\Omega_q$.

The SPH approximation of the time derivative, gradient and divergence is also derived; for instance, a particle approximation of $\frac{\partial \psi}{\partial x_j}(\mathbf{x}_p)$ may be written as follows:

$$\left\langle \frac{\partial \psi}{\partial x_j}(\mathbf{x}_p) \right\rangle = -\sum_{q=1}^{q=Q} \frac{m_q}{\rho_q} \psi(\mathbf{x}_q) \frac{\partial W_h}{\partial x_j}(\mathbf{x}_p - \mathbf{x}_q)$$

The SPH approximation is subsequently written for Equation (3.6); it produces a set of ordinary differential equations with respect to time:

$$\frac{\partial \mathbf{\Psi}}{\partial t} + \mathbf{A}(\psi)\mathbf{\Psi} = \mathbf{b}_\psi$$

where $\mathbf{\Psi}$ gathers the values of ψ at the particle location. The above equations are commonly solved using an explicit numerical time integration algorithm. 'Highly non-linear' problems can be tackled with SPH simulations, achieving an accuracy which is acceptable for engineering purposes, such as in the example provided in Figure 3.8.

The convergence of the SPH method being controlled by the length h associated with the kernel function, as well as by the number of particles P in the fluid domain, SPH simulations habitually involve a large number of particles and are therefore computationally expensive. As with other particle-based methods, efficient parallel computing sequences are developed

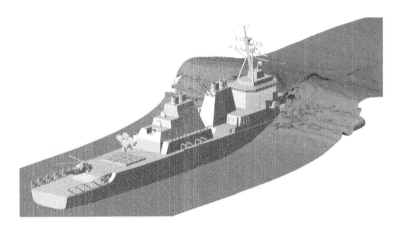

Figure 3.8 Rough wave impacting a surface ship. *The 20-m high wave breaks onto the 200-m long frigate: 12 millions of particles are used to describe the fluid flow. Smoothed Particle Hydrodynamics allows for the description of 'highly non-linear' fluid flows, which is in general not affordable with mesh-based techniques; from the computational viewpoint, SPH-based simulations remain however extremely demanding, even with high parallel computing techniques. Source: Erwan* JACQUIN, *HydrOcéan, Nantes, France, 2012. Reproduced with permission of HydrOcéan*

in SPH codes in order to enhance the computational performance, while ensuring a high level of physical accuracy.

SPH is also considered for the simulation of wave propagation equations[7]: this raises numerous issues (ranging from numerical accuracy to physical consistency) regarding the ability of the SPH method to become an alternative to the finite element method (FEM) and boundary element method (BEM), which remain the reference techniques for solving most of the wave problems in acoustics and vibro-acoustics. The basics of the FEM and BEM are discussed in the following sections.   ■

3.2 Compressibility Waves

3.2.1 Wave Equation

Acoustic waves propagate in a compressible fluid initially at rest in which the influence of gravity is discarded. In such cases, separating the steady state and the fluctuating components of pressure, velocity or density according to Equation (3.9) reads as follows:

$$p(\mathbf{x}, t) = p_o(\mathbf{x}) + p'(\mathbf{x}, t) \qquad \mathbf{v}(\mathbf{x}, t) = \mathbf{v}'(\mathbf{x}, t) \qquad \rho(\mathbf{x}, t) = \rho_o + \rho'(\mathbf{x}, t)$$

while the source term is null $\mathbf{f}(\mathbf{x}, t) = \mathbf{0}$.

The steady state pressure verifies $\dfrac{\partial p_o}{\partial x_i} = 0$; hence, it is constant throughout the fluid domain $p_o = P_o$.

The linearisation of the fluid equation of state about the static equilibrium with constant pressure is

$$p' = \left.\frac{\partial p}{\partial \rho}\right|_{(P_o, \rho_o)} \rho' \tag{3.12}$$

---
[7] See, for instance, Zhang *et al.* (2014).

which is also written as follows:

$$p' = c_o^2 \rho'$$

where c_o is the speed of sound in the fluid.

The fluctuations of the velocity and density about the steady state are governed by the momentum equation:

$$\frac{\partial p'}{\partial x_i} = -\rho_o \frac{\partial v_i'}{\partial t} \qquad (3.13)$$

together with the mass equation:

$$\frac{\partial \rho'}{\partial t} + \rho_o \frac{\partial v_j'}{\partial x_j} = 0$$

Taking the divergence of the momentum equations gives the following:

$$\rho_o \frac{\partial^2 v_i'}{\partial t \partial x_i} + \frac{\partial^2 p'}{\partial x_i \partial x_i} = 0$$

while a time derivation of the mass equation provides

$$\frac{\partial^2 \rho'}{\partial t^2} + \rho_o \frac{\partial^2 v_i'}{\partial t \partial x_i} = 0$$

Eliminating the velocity out of these equations yields the wave equation formulated in terms of the pressure solely. It is of second order with respect to time and space coordinates, and it is written as follows:

$$\frac{1}{c_o^2} \frac{\partial^2 p'}{\partial t^2} - \frac{\partial^2 p'}{\partial x_i \partial x_i} = 0$$

Boundary conditions associated with the wave equation are discussed in the following subsection.

Remark 3.2 Gravity waves *While the presentation focusses here on acoustic waves, gravity waves, which are observed at the free level of liquids, are also of interest in many instances. They are briefly discussed in this book as a complement to compressibility waves. The presentation is restricted to surface waves in closed rigid or deformable reservoirs. Various aspects of the physical modelling as well as the mathematical or numerical representations of gravity waves are therefore not considered here and may be found for instance in Axisa and Antunes (2007) or in Ibrahim (2005).*

Sloshing waves, also broadly known as water waves, involve inertia of the liquid and the stiffness of the free level, the latter being pre-stressed by a permanent field of gravity. To derive the governing equations of wave propagation, the compressibility of the liquid may be discarded so that $\rho'(\mathbf{x}, t) = 0$, while the steady velocity field is $\mathbf{v}_o(\mathbf{x}, t) = \mathbf{0}$. Writing Equation (3.9) for the pressure, velocity, density and external force gives

$$p(\mathbf{x}, t) = p_o(\mathbf{x}) + p'(\mathbf{x}, t) \qquad \mathbf{v}(\mathbf{x}, t) = \mathbf{v}'(\mathbf{x}, t) \qquad \rho(\mathbf{x}, t) = \rho_o \qquad \mathbf{f}(\mathbf{x}, t) = \rho_o \mathbf{g}$$

p_o *verifies* $\dfrac{\partial p}{\partial x_i} = \rho_o g_i$*: the steady state of the liquid is described using the hydrostatic pressure* $p_o(\mathbf{x}) = \rho_o g_i x_i + P_o$.

The mass conservation equation reads $\dfrac{\partial v_i'}{\partial x_i} = 0$ *(continuity equation), while the momentum equation is given by Equation (3.13).*

Within the fluid domain, the pressure fluctuations are thus governed by the Laplace equation:

$$\frac{\partial^2 p'}{\partial x_i \partial x_i} = 0$$

which results from time derivation of the continuity equation together with the pressure/acceleration relation, Equation (3.13). In addition to its hydrostatic component, gravity influences the pressure fluctuations at the free surface, according to the following equation:

$$p' = \rho_o g \eta \qquad\qquad (3.14)$$

η is the elevation of the free surface, that is, the displacement of the fluid particle in the direction of gravity **n**. $\eta = \xi_i n_i$, where ξ is the displacement field, as discussed later in Remark 3.4.

The former relation stipulates that the pressure fluctuation (p') emanates from the weight of the free level ($\rho_o g \eta$ is the gravitational force of a unit surface of fluid). It is assumed that Equation (3.14) holds whatever the sign of the elevation may be, in order to preserve the linearity of the formulation.

Using the relations $\dfrac{\partial p'}{\partial x_i} = -\rho_o \dfrac{\partial v_i'}{\partial t} = -\rho_o \dfrac{\partial^2 \xi_i}{\partial t^2}$ in direction **n**, the gravity waves equation is arrived at:

$$\frac{\partial^2 p'}{\partial t^2} + g\frac{\partial p'}{\partial n} = 0$$

■

Remark 3.3 Speed of sound in linear acoustics

The speed of sound in the fluid is expressed as follows:

$$c_o = \sqrt{\frac{\partial p}{\partial \rho}}_{(p_o, \rho_o)} \qquad\qquad (3.15)$$

For single phase fluids, the speed of sound is obtained from the thermodynamic properties of liquids or gases, as specified for instance in Table 3.1[8] and Figure 3.9.

For two-phase fluids, the calculation of the speed of sound according to this relation is not straightforward; with a two-phase homogenous model, as developed for instance by Axisa and Antunes (2007), an equivalent speed of sound c_H, and density ρ_H may be derived from the characteristics of each individual phase c_L, ρ_L (liquid phase) and c_V, ρ_V (vapour phase):

$$\rho_H = \rho_V \alpha_H + \rho_L (1 - \alpha_H)$$

$$c_H = 1 \Big/ \sqrt{\left(\frac{1}{c_V^2}\left(\alpha_H^2 + \alpha_H(1 - \alpha_H)\right)\frac{\rho_L}{\rho_V} + \frac{1}{c_L^2}\left((1 - \alpha_H)^2 + \alpha_H(1 - \alpha_H)\right)\frac{\rho_V}{\rho_L}\right)}$$

α_H is the void fraction: it is defined as the ratio of the volume of the vapour phase over the total volume.

---

[8] As put forward by Table 3.1 and more generally by the data compiled in Selfridge (1985), there is no obvious correlation between the speed of sound and density of many fluids of industrial significance – in particular, low or high speeds of sound do not necessarily correspond to low or high densities of the fluid!

Table 3.1   *Density and speed of sound for various liquids and gas*

Substance	ρ (kg/m^3)	c (m/s)
Gas		
Oxygen @ 20 °C	1.308	320
Hydrogen @ 20 °C	0.082	930
Air @ 20 °C	1.270	330
Liquid		
Mercury @ 20 °C	13,595	1,450
Water @ 20 °C (natural water)	998	1,450
Water @ 20 °C (sea water)	1,025	1,520
Hydrogen @ −250 °C	71	1,185
Oxygen @ −185 °C	1,155	1,010
Oil (Diesel) @ 20 °C	800	1,250

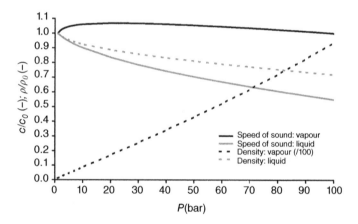

Figure 3.9   Density and speed of sound in water. *The density and the speed of sound in water are plotted versus the static pressure, both for the liquid and vapour phases at the point of saturation. The reference values are given for $P = 1$ bar; the corresponding values for saturated vapour and liquid are, respectively, $c_o = 472$ m/s, $\rho_o = 0.59$ kg/m^3 and $c_o = 1,545$ m/s, $\rho_o = 958$ kg/m^3*

In Figure 3.10, the equivalent density and speed of sound for an air/water mixture are plotted versus the homogenised void fraction. It is worth noticing that the speed of sound is minimum at $\alpha_H \sim 0.5$, being there much lower than the value in pure liquid – and even in pure vapour ($c_H \sim 20$ m/s). In terms of order of magnitude indeed, at $\alpha_H \sim 0.5$, the inertia of the two-phase fluid is that of the liquid ($\rho_H \sim 500$ kg/m^3), while the elasticity of the two-phase fluid is that of the vapour phase.

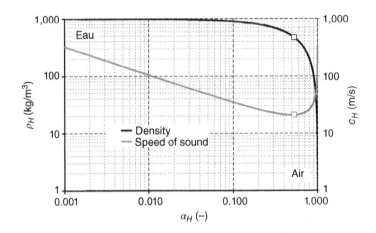

Figure 3.10   *Density and speed of sound of a two-phase homogenous fluid*

Hereafter, to alleviate the notations, p and $\mathbf{v}$ denote the fluctuations of the pressure and velocity fields, while the fluid properties are designated as ρ and c, without any further reference to the equilibrium state.

3.2.2   Boundary Conditions

The boundary conditions associated with the wave equation are discussed below, considering first the case of a finite volume of fluid limited by solid walls and by free levels. Provided such boundaries are conservative in nature, they do not absorb any amount of acoustic energy of the incident waves, which experience thus total reflection when impinging on them. The case of an infinite extension of fluid is discussed later.

When rejected to infinity, a boundary, whatever its physical nature, cannot induce any reflected wave inside the fluid. Modelling non-reflecting boundaries in a finite element model, which necessarily deals with a finite volume of fluid, still reveals as a challenging problem. Several distinct approaches have been proposed in the open literature, however, the presentation of this topical subject is restricted here to one of them, which proves reasonably simple and efficient from the computational viewpoint. It makes use of the so-called *Sommerfeld condition* as formulated in terms of pressure solely.

3.2.2.1   Acoustic Impedance

Using an harmonic representation of the pressure field $p(\mathbf{x}, t) = p(\mathbf{x}) \exp(i\omega t)$, the acoustic wave propagation is formulated throughout the fluid domain as follows:

$$-\frac{\omega^2}{c^2}p + \frac{\partial^2 p}{\partial x_i \partial x_i} = 0$$

The condition associated with the compressibility wave propagation equation in a bounded domain is stated as follows:

$$\frac{\partial p}{\partial n} + \frac{i\omega}{c}\rho c Z(\omega)p = 0 \tag{3.16}$$

where $Z(\omega)$ is the *acoustic impedance* of the considered boundary. Defined as the ratio of the pressure p and the normal velocity $\mathbf{v} \cdot \mathbf{n}$, the acoustic impedance quantifies the resistance of a medium towards sound propagation[9]. Using $\dfrac{\partial p}{\partial n} = -\rho i \omega \mathbf{v} \cdot \mathbf{n}$ yields:

$$\frac{1}{Z(\omega)} = -\frac{1}{\rho c} \frac{1}{i\omega/c} \frac{\partial p}{\partial n} \frac{1}{p}$$

The boundary condition (3.16) may also account for acoustic damping, which is associated with the non-reflection of waves propagating in an infinite domain, as detailed in the following subsection.

- The coupling condition with a moving wall is as follows: for non-viscous flow, the fluid particles are free to move along the wall while adhering to it. At the interface, the displacement of the fluid, denoted ξ, and of the wall, denoted $\mathbf{u}$, verify:

$$\xi \cdot \mathbf{n} = \mathbf{u} \cdot \mathbf{n}$$

with $\mathbf{n}$ the unit normal vector on the wall. The latter relation may be expressed in terms of acceleration by a twofold derivation in time, as discussed in Remark 3.4:

$$\frac{\partial p}{\partial n} = -\rho \frac{\partial^2 \mathbf{u}}{\partial t^2} \cdot \mathbf{n} \tag{3.17}$$

For a fixed wall ($\mathbf{u} = \mathbf{0}$), this condition yields a Neumann-type boundary condition:

$$\frac{\partial p}{\partial n} = 0 \Longleftrightarrow Z(\omega) = \infty, \forall \omega$$

The acoustic impedance of a fixed wall is infinite, which means total reflection of the wave without change in sign for the pressure and change in sign of the velocity. The latter condition also stands for a *symmetry plane* for the fluid problem.
- The boundary condition with an acoustic-free surface, or isobar, states that the pressure fluctuation is null (Dirichlet-type boundary condition):

$$p = 0 \Longleftrightarrow Z(\omega) = 0, \forall \omega$$

The acoustic impedance is null, which means total reflection of the wave without change in sign for the velocity and with change in sign of the pressure. This condition also represents an *antisymmetry plane* for the fluid problem.

3.2.2.2 Radiation Condition

A wave propagated in an infinite extent of fluid must comply with the *radiation condition*, also known as the Sommerfeld condition, which stipulates that waves are not reflected at infinity.

---

[9] For instance, under normal conditions ($P_o = 1 \times 10^{+5}$ Pa, $T_o = 20\,°C$), the acoustic impedance of air is $Z_{air} = 0.415 \times 10^{+9}$ Pa.s/m while it is $Z_{water} = 1.45 \times 10^{+9}$ Pa.s/m for water, highlighting the difference in nature for such fluids, when it comes to sound propagation – a fact well known to and experienced by any scuba diver!

It is formulated as follows:

$$\lim_{r \to \infty} r^{(d-1)/2} \left(\frac{\partial p}{\partial r} + \frac{i\omega}{c} p \right) = 0$$

where d is the dimension of the problem.

The radiation condition is stated here as an asymptotic condition: its numerical implementation with finite element–based techniques is not straightforward. In the general case, an *approximation* of this condition is proposed, which allows for a finite element representation.

As evidenced below, the radiation condition may be formulated in an exact manner for plane, cylindrical and spherical waves solely (Bettess, 1992).

Plane wave In the harmonic regime, the pressure field is $p(r, t) = p(r, \omega) \exp(i\omega t)$, so that the wave equation for a unidimensional pressure field reads as follows:

$$\frac{\partial^2 p}{\partial r^2} + \frac{\omega^2}{c^2} p = 0$$

Harmonic solutions are $p(r, \omega) = \exp(\pm i\omega r/c)$ so that the propagative solution is a linear combination of functions $\exp(i\omega(t \pm r/c))$: the component $\exp(i\omega(t - r/c))$ corresponds to a wave travelling *to infinity*, while the component $\exp(i\omega(t + r/c))$ stands for a wave propagating *from infinity*. The reflection condition is obtained by discarding the contribution of the incoming wave, hence: $p(r, t) = \Pi \exp(i\omega(t - r/c))$ and it verifies:

$$\frac{\partial p}{\partial r} = -\frac{i\omega}{c} \Pi \exp(i\omega(t - r/c)) = -\frac{i\omega}{c} p$$

The following relation is thereby the exact radiation condition for plane wave propagation:

$$\frac{\partial p}{\partial r} + \frac{i\omega}{c} p = 0 \tag{3.18}$$

Cylindrical wave In the same manner, the harmonic wave equation for an axisymmetric pressure field in cylindrical coordinate system is stated as follows:

$$\frac{\partial^2 p}{\partial r^2} + \frac{1}{r} \frac{\partial p}{\partial r} + \frac{\omega^2}{c^2} p = 0$$

The propagative solutions are $\frac{1}{\sqrt{r}} \exp(i\omega(t \pm r/c))$ and the wave travelling to infinity is $\frac{\Pi}{\sqrt{r}} \exp(i\omega(t - r/c))$, so that:

$$\frac{\partial p}{\partial r} = -\frac{\Pi}{2r\sqrt{r}} \exp(i\omega(t - r/c)) - \frac{i\omega}{c} \frac{\Pi}{\sqrt{r}} \exp(i\omega(t - r/c))$$

Hence, the exact radiation condition:

$$\frac{\partial p}{\partial r} + \frac{p}{2r} + \frac{i\omega}{c} p = 0 \tag{3.19}$$

Table 3.2 Radiation conditions for particular wave propagation modes

Plane wave	Cylindrical wave	Spherical wave
	Propagation equation	
$\dfrac{\partial^2 p}{\partial r^2} + \dfrac{\omega^2}{c^2} p = 0$	$\dfrac{\partial^2 p}{\partial r^2} + \dfrac{1}{r}\dfrac{\partial p}{\partial r} + \dfrac{\omega^2}{c^2} p = 0$	$\dfrac{\partial^2 p}{\partial r^2} + \dfrac{2}{r}\dfrac{\partial p}{\partial r} + \dfrac{\omega^2}{c^2} p = 0$
	Radiation condition	
$\dfrac{\partial p}{\partial r} + \dfrac{i\omega}{c} p = 0$	$\dfrac{\partial p}{\partial r} + \left(\dfrac{1}{2r} + \dfrac{i\omega}{c} \right) p = 0$	$\dfrac{\partial p}{\partial r} + \left(\dfrac{1}{r} + \dfrac{i\omega}{c} \right) p = 0$

Spherical wave The harmonic wave equation for a spherical field in spherical coordinate system is expressed as follows:

$$\frac{\partial^2 p}{\partial r^2} + \frac{2}{r}\frac{\partial p}{\partial r} + \frac{\omega^2}{c^2} p = 0$$

Propagative solutions are $\frac{1}{r}\exp(i\omega(t \pm r/c))$. With the wave travelling to infinity $\frac{\Pi}{r}\exp(i\omega(t - r/c))$, it is stated as follows:

$$\frac{\partial p}{\partial r} = -\frac{\Pi}{r^2}\exp(i\omega(t - r/c)) - \frac{i\omega}{c}\frac{\Pi}{r}\exp(i\omega(t - r/c))$$

The exact radiation condition is as follows:

$$\frac{\partial p}{\partial r} + \frac{p}{r} + \frac{i\omega}{c} p = 0 \tag{3.20}$$

Table 3.2 presents these particular propagation modes and the associated exact radiation conditions.

In order to deal with other cases, various approximations of the Sommerfeld condition have been proposed, among which the so-called plane wave approximation (PWA), discussed above, the doubly asymptotic approximation (DAA) or the m-order Boundary Conditions. The PWA has first been proposed as one of the simplest approximation of the radiation condition (Mindlin and Bleich, 1953). It is valid for high values of the *wave number* $k = \omega/c$, while the DAA is suited to very high and very low wave numbers and, in advanced versions, at selected intermediate values of k (Geers and Zhang, 1994a; 1994b).

Approximations of high order allow a larger frequency span to be considered: following, for instance, Bayliss *et al.* (1982), an anechoic condition is formulated in terms of an infinite series of operators. To that end, let the harmonic wave equation be described using the spherical coordinate system (r, θ, φ):

$$\frac{1}{r^2}\frac{\partial}{\partial r}\left(r^2 \frac{\partial p}{\partial r} \right) + \frac{1}{r^2 \sin\theta}\frac{\partial}{\partial \theta}\left(\frac{\partial p}{\partial \theta}\sin\theta \right) + \frac{1}{r^2\sin^2\theta}\frac{\partial^2 p}{\partial \varphi^2} + \frac{\omega^2}{c^2} p = 0$$

The general solution is readily found in terms of the following series:

$$p(r, \theta, \varphi, \omega) = \frac{\exp(i\omega r/c)}{\omega r/c}\sum_{n=1}^{n=\infty}\frac{\Pi_n(\theta, \varphi)}{(\omega r/c)^n}$$

Table 3.3 Radiation conditions of order 0, 1 and 2

Zero-order condition	First-order condition
$\dfrac{\partial p}{\partial r} = -\dfrac{i\omega}{c}p$	$\dfrac{\partial p}{\partial r} = -\left(\dfrac{1}{r} + \dfrac{i\omega}{c}\right)p$

Second-order condition
$\dfrac{\partial p}{\partial r} = -\left(\dfrac{1}{r} + \dfrac{i\omega}{c}\right)p + \dfrac{1}{2r(1 + i\omega r/c)}\left(\dfrac{1}{\sin\theta}\dfrac{\partial}{\partial\theta}\left(\dfrac{\partial p}{\partial\theta}\sin\theta\right) + \dfrac{1}{\sin^2\theta}\dfrac{\partial^2 p}{\partial\varphi^2}\right)$

A condition which ensures the term-to-term absorption of the pressure field is of the generic type $B_m p = 0$, where B_m is a differential operator of order m. For tri-dimensional harmonic problems[10], it is expressed as follows:

$$B_m = \prod_{\mu=1}^{\mu=m}\left(\frac{\partial}{\partial r} + \frac{i\omega}{c} + \frac{2\mu - 1}{r}\right) \tag{3.21}$$

Named after its authors, this kind of approximation is referred to as the *BGT Boundary Conditions* of order m. The zero, first- and second-order conditions are specified in Table 3.3.

- The zero-order radiation condition (BGT-0) corresponds to the direct application of the Sommerfeld at the boundary of the truncated domain:

$$B_0 p = \frac{\partial p}{\partial r} + \frac{i\omega}{c}p = 0$$

It is also equivalent to the plane wave non-reflexion condition, Table 3.2; the corresponding impedance is:

$$Z(\omega) = \rho c$$

- The first-order condition (BGT-1) is similar to the non-reflection condition for a spherical wave:

$$B_1 p = \frac{\partial p}{\partial r} + \frac{i\omega}{c}p + \frac{p}{r} = 0$$

When imposed at distance R away from the acoustic source, the first-order condition corresponds to the impedance:

$$Z(\omega) = \frac{\rho c}{1 + \dfrac{1}{i\omega R/c}}$$

For high wave numbers, $\omega R/c \gg 1$, the conditions of order 0 and 1 are therefore equivalent: the propagation of an acoustic spherical wave at low speed and high pulsation may safely be described using the PWA.

---

[10] A similar boundary condition for bi- or tri-dimensional frequency-dependent as well as time-dependent problems may be found in Chapter 7 of Bettess (1992).

Figure 3.11 Pulsating sphere in an infinite acoustic medium. The pressure field which results from the sphere motion $\mathbf{u}(\omega) = u(\omega)\mathbf{e}_r$ at distance r away from its centre is $p(r, \omega) = -\rho\mu(\omega r/c)\omega^2 u(\omega) + \rho c \zeta(\omega) i \omega u(\omega)$, with $\mu(\omega r/c) = \frac{1}{1+(\omega r/c)^2}$ and $\zeta(\omega) = \frac{(\omega r/c)^2}{1+(\omega r/c)^2}$. In order to validate a finite element modelling with BGT conditions, the numerical computation of coefficients μ and ζ is performed in a bounded finite element domain of increasing size. *Source*: Cédric LEBLOND, DCNS Research/IRT Jules VERNE, Nantes, France, 2014. Reproduced with permission of IRT Jules VERNE

- The second-order condition (BGT-2) is stated as follows:

$$B_2 p = \frac{\partial^2 p}{\partial r^2} + \left(\frac{4}{r} + \frac{2i\omega}{c} \right) \frac{\partial p}{\partial r} + \left(\frac{2}{r} + \frac{4i\omega}{c} \right) \frac{p}{r} - \frac{\omega^2}{c^2} p = 0$$

Using the wave equation in cylindrical coordinates allows for the elimination of $\dfrac{\partial^2 p}{\partial r^2}$, so that the former condition also reads as follows:

$$\frac{\partial p}{\partial r} = -\left(\frac{1}{r} + \frac{i\omega}{c} \right) p + \frac{1}{2r^2(1/r + i\omega/c)} \left(\frac{1}{\sin\theta} \frac{\partial}{\partial\theta} \left(\frac{\partial p}{\partial\theta} \sin\theta \right) + \frac{1}{\sin^2\theta} \frac{\partial^2 p}{\partial\varphi^2} \right)$$

Table 3.4 Accuracy of the pressure field computations with BGT-type boundary condition. The numerical data refer to a solid sphere ($R = 1$ m) immersed within a concentric sphere ($R' = 10$ m) of water ($\rho = 1,000$ kg/m^3, $c = 1,500$ m/s), as represented in Figure 3.11. The finite element computations are performed for $\omega r/c$ varying between 0 and 5. Over the frequency range of interest, ε quantifies the relative error between analytical and numerical results, either on the real part or the imaginary part of the pressure field. For the problem at hand, the BGT-0 condition does not yield a satisfying accuracy: designed for plane wave propagation, it is not suited to describe the propagation of the acoustic wave triggered by the pulsation of the sphere. The approximation induced by the BGT-1 condition proves however more accurate for such a problem.

	BGT-0					BGT-1				
r	R	$2R$	$3R$	$5R$	$10R$	R	$2R$	$3R$	$5R$	$10R$
$\varepsilon_{\Re(p)}(r)$	5.55	7.36	6.71	7.28	7.44	0.68	0.19	0.15	0.23	0.22
$\varepsilon_{Im(p)}(r)$	6.30	7.01	8.31	7.17	6.77	0.07	0.09	0.19	0.20	0.59

The BGT conditions stand for the Sommerfeld condition: hence, questions naturally arise concerning the accuracy and the convergence of the numerical solutions produced with such boundary conditions. When a truncated fluid domain is considered, what is the error committed on the evaluation of the pressure field – depending on the order of approximation, the frequency of interest, the shape and size of the truncated domain?

Some convergence results are proposed, for instance, by Bayliss *et al.* (1982); as no general results to the question are however reported, it leaves an open field to numerical experimentations for each specific case, as illustrated in Figure 3.11 and Table 3.4.

An example of the performance of various local absorbing conditions, among which BGT conditions, may be found in Harari and Djellouli (2004), while an example of acoustic radiation using the BGT condition is proposed in Section 3.8.

Remark 3.4 Wave equations in terms of pressure or in terms of displacement vector or potential   *Using the pressure field in the statement of the wave equations is of engineering relevance, in particular because its physical interpretation is straightforward. A formulation of the propagation equation in terms of displacement is also possible. Although both formulations are equivalent from the physical standpoint, they do not exhibit the same numerical robustness, as emphasised in Chapter 5.*

Here, the fluid is assumed to be governed by the following law of elasticity:

$$\sigma_{ij}(\xi) = -p\delta_{ij} \ \text{ with } p = -\rho c^2 \frac{\partial \xi_k}{\partial x_k}\delta_{ij}$$

where ξ is the displacement *field. $\nabla \cdot \xi = \dfrac{\partial \xi_k}{\partial x_k}$, with $\nabla \cdot (\bullet)$ the divergence operator, is a measure of the rate of change of the local volume: the previous relation states that when the fluid locally expands, the pressure decreases, so that the acoustic pressure is locally negative.*

The equation of motion in the fluid formulated in terms of displacement is expressed as follows:

$$-\rho \omega^2 \xi_i - \rho c^2 \frac{\partial \xi_j}{\partial x_i \partial x_j} = 0$$

The coupling with a fixed wall is expressed as follows:

$$\xi_j n_j = 0$$

while an isobar condition reads as follows:

$$\frac{\partial \xi_j}{\partial x_j} = 0$$

which is equivalent to $p = 0$.

The free surface condition with gravity reads as follows:

$$\sigma_{ij}(\xi)n_j = \rho g \xi_k n_k n_i$$

which is equivalent to $p = \rho g \eta$, with $\eta = \xi_k n_k$.

As for small transformations, the time derivation $\dfrac{d(\bullet)}{dt}$ reduces to $\dfrac{\partial(\bullet)}{\partial t}$ because the advection $v_i \dfrac{\partial(\bullet)}{\partial x_i}$ is negligible. Thus, the velocity in the fluid is $v_i = \dfrac{\partial \xi_i}{\partial t}$ and the pressure gradient is found to be

$$\frac{\partial p}{\partial x_i} = -\rho \frac{\partial v_i}{\partial t} = -\rho \frac{\partial^2 \xi_i}{\partial t^2}$$

For harmonic vibrations, the latter relation is stated as follows:

$$\frac{\partial p}{\partial x_i} = \rho \omega^2 \xi_i$$

so that when $\omega \neq 0$, ξ_i may be calculated as

$$\xi_i = \frac{\partial \varphi}{\partial x_i} \qquad \varphi = \frac{p}{\rho \omega^2}$$

where φ is the displacement potential. *It is related to the pressure according to*

$$p - \rho \omega^2 \varphi = 0$$

so that p is also interpreted as the acceleration potential in the fluid. Since ξ derives from the potential φ, the fluid flow is irrotational:

$$\nabla \times \xi = 0$$

with $\nabla \times (\bullet)$ the rotational operator.

The constitutive law $p = -\rho c^2 \dfrac{\partial \xi_k}{\partial x_k}$ is also expressed as:

$$p + \rho c^2 \frac{\partial^2 \varphi}{\partial x_i \partial x_i} = 0$$

The acoustic wave propagation may also be formulated in terms of the displacement potential:

$$-\frac{\omega^2}{c^2} \varphi - \frac{\partial^2 \varphi}{\partial x_i \partial x_i} = 0$$

The isobar boundary condition is stated as follows:

$$\varphi = 0$$

which is equivalent to $p = 0$ and the rigid wall coupling condition is stated as

$$\frac{\partial \varphi}{\partial x_j} n_j = 0$$

which is equivalent to $\xi_i n_i = 0$. The free surface condition with gravity is therefore expressed as follows:

$$-\rho \omega^2 \varphi + \rho g \eta = 0$$

which is equivalent to $p = \rho g \eta$.

Displacement-based or displacement potential-based formulations can be used in the context of fluid sloshing or acoustic modes (as illustrated in Section 3.6) or fluid–structure coupled modes (as illustrated in Chapter 5).   ∎

3.3   Finite Element Method

3.3.1   Pressure-Based Formulation

Compressibility and gravity effects are distinct in their physical nature and involve different wave numbers, as discussed further in Remark 3.5. From the engineering standpoint, a unified pressure-based formulation which accounts both for sloshing and acoustic modes may however be proposed. In the harmonic regime, the acoustic waves are governed by the Helmholtz equation:

$$-\frac{\omega^2}{c^2}p + \frac{\partial^2 p}{\partial x_i \partial x_i} = 0 \text{ in } \Omega \tag{3.22}$$

while the sloshing wave equation is expressed as follows:

$$\frac{\partial p}{\partial n} = \frac{\omega^2}{g}p \text{ on } \Gamma \tag{3.23}$$

The reflection conditions associated with zero or infinite wall impedance are stated as follows:

$$p = 0 \text{ on } \Gamma_o \tag{3.24}$$

$$\frac{\partial p}{\partial n} = 0 \text{ on } \Gamma_\pi \tag{3.25}$$

where $\partial\Omega = \Gamma \cup \Gamma_o \cup \Gamma_\pi$ is the domain boundary.

3.3.1.1   Weighted Integral Formulation

Let δp be a pressure field which complies with $p|_{\Gamma_o} = 0$, so that multiplying Equation (3.22) by δp and integrating over Ω gives:

$$-\omega^2 \int_\Omega \frac{1}{c^2}p \, \delta p \, d\Omega - \int_\Omega \frac{\partial^2 p}{\partial x_i \partial x_i} \, \delta p \, d\Omega = 0$$

Integrating by parts and taking into account boundary conditions (3.23 –3.25) yields:

$$\int_\Omega \frac{\partial^2 p}{\partial x_i \partial x_i} \, \delta p \, d\Omega = -\int_\Omega \frac{\partial p}{\partial x_i} \frac{\partial \delta p}{\partial x_i} \, d\Omega + \int_\Gamma \frac{\partial p}{\partial x_i} \, n_i \, \delta p \, d\Gamma$$

Hence, the integral formulation:

$$-\omega^2 \int_\Omega \frac{1}{c^2}p \, \delta p \, d\Omega - \omega^2 \int_\Gamma \frac{1}{g}p \, \delta p \, d\Gamma + \int_\Omega \frac{\partial p}{\partial x_i} \frac{\partial \delta p}{\partial x_i} \, d\Omega = 0 \tag{3.26}$$

An appropriate functional space for this problem is

$$\mathcal{P} = \{p \in H^1(\Omega), p|_{\Gamma_o} = 0\} \tag{3.27}$$

The weighted integral formulation is termed as follows:

Find $p \in \mathcal{P}$ and ω such that:

$$-\omega^2 \int_\Omega \frac{1}{c^2}p \, \delta p \, d\Omega - \omega^2 \int_\Gamma \frac{1}{g}p \, \delta p \, d\Gamma + \int_\Omega \frac{\partial p}{\partial x_i} \frac{\partial \delta p}{\partial x_i} \, d\Omega = 0$$

for all $\delta p \in \mathcal{P}$.

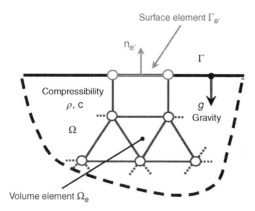

Surface element $\Gamma_{e'}$

$n_{e'}$

Γ

Compressibility
ρ, c

g
Gravity

Ω

Volume element Ω_e

Figure 3.12   Finite element mesh with volume elements Ω_e and surface elements $\Gamma_{e'}$ for compressibility and gravity waves

It is worthwhile to mention that p and δp belonging in $\mathcal{P}$ is a condition which ensures the existence of all the integrals in Equation (3.26), whether of volume or of surface nature[11].

3.3.1.2   Mass and Stiffness Matrices

As sketched in Figure 3.12, the geometrical approximation of Ω is performed with a finite element mesh $\Omega \simeq \bigcup\limits_{e=1}^{e=E} \Omega_e$, which also produces an approximation of Γ according to $\Gamma \simeq \bigcup\limits_{e'=1}^{e=E'} \Gamma_{e'}$.

The volume and surface integrals in the weighted integral formulation are evaluated according to

$$\int_{\Omega}(\bullet)\, d\Omega \simeq \sum_{e=1}^{e=E}\int_{\Omega_e}(\bullet)\, d\Omega_e \qquad \int_{\Gamma}(\bullet)\, d\Gamma \simeq \sum_{e'=1}^{e=E'}\int_{\Gamma_{e'}}(\bullet)\, d\Gamma_{e'}$$

On each element Ω_e or $\Gamma_{e'}$, the approximation of the pressure field is written as follows:

$$p|_{\Omega_e}(\mathbf{x}) = \mathbf{N}_e(\mathbf{x})\mathbf{P}_e \qquad p|_{\Gamma_{e'}}(\mathbf{x}) = \mathbf{N}_{e'}(\mathbf{x})\mathbf{P}_{e'}$$

where

- $\mathbf{P}_{e/e'}$ is the vector of the pressure nodal values on the element;
- $\mathbf{N}_{e/e'}$ is the vector of shape functions associated with the nodes of the element.

The approximation of the virtual pressure field is performed in a similar manner.

Elementary 'mass' matrices are generated by the product $p\delta p$, which is calculated as $\delta \mathbf{P}_{e,e'}^{\mathsf{T}} \mathbf{N}_{e,e'}^{\mathsf{T}} \mathbf{N}_{e,e'} \mathbf{P}_{e,e'}$ so that

$$\int_{\Omega_e}\frac{1}{c^2}p\delta p\, d\Omega_e = \delta \mathbf{P}_e^{\mathsf{T}} \mathbf{m}_F^e \mathbf{P}_e$$

---

[11] Any function $\psi \in H^1(\Omega)$ is in nature square-integrable over Ω. For any $\psi \in H^1(\Omega)$, it is possible to define the value of ψ on Γ, denoted $\psi|_{\Gamma}$, which is also proved to be square-integrable on Γ.

where $\mathbf{m}_F^e$ is the 'mass' matrix associated with compressibility effects:

$$\mathbf{m}_F^e = \int_{\Omega_e} \frac{1}{c^2} \mathbf{N}_e^{\mathsf{T}} \mathbf{N}_e \, d\Omega_e \qquad (3.28)$$

and

$$\int_{\Gamma_{e'}} \frac{1}{g} p \delta p \, d\Gamma_{e'} = \delta \mathbf{P}_{e'}^{\mathsf{T}} \mathbf{m}_F^{e'} \mathbf{P}_{e'}$$

where $\mathbf{m'}_F^{e'}$ is the 'mass' matrix associated with gravity effects:

$$\mathbf{m'}_F^{e'} = \int_{\Gamma_{e'}} \frac{1}{g} \mathbf{N}_{e'}^{\mathsf{T}} \mathbf{N}_{e'} \, d\Gamma_{e'} \qquad (3.29)$$

Elementary 'stiffness' matrix originates from the scalar product $\nabla p \cdot \nabla \delta p$, where the components of ∇p are $\left\{ \dfrac{\partial p}{\partial x_i} \right\}$; the gradient of the pressure field is calculated on the element according to

$$\nabla \mathbf{p} = \mathbf{G}_e \mathbf{P}_e$$

where $\mathbf{G}_e$ assembles the partial derivatives of the shape functions:

$$\mathbf{G}_e = \left\{ \frac{\partial \mathbf{N}_e}{\partial x_i} \right\}$$

Hence,

$$\int_{\Omega_e} \frac{\partial p}{\partial x_i} \frac{\partial \delta p}{\partial x_i} \, d\Omega_e = \delta \mathbf{P}_e^{\mathsf{T}} \mathbf{k}_F^e \mathbf{P}_e$$

where $\mathbf{k}_F^e$ is the 'stiffness' matrix of the fluid:

$$\mathbf{k}_F^e = \int_{\Omega_e} \mathbf{G}_e^{\mathsf{T}} \mathbf{G}_e \, d\Omega_e \qquad (3.30)$$

The discrete form of Equation (3.26) is then

$$-\omega^2 \sum_{e=1}^{e=E} \delta \mathbf{P}_e^{\mathsf{T}} \mathbf{m}_F^e \mathbf{P}_e - \omega^2 \sum_{e'=1}^{e'=E'} \delta \mathbf{P}_{e'}^{\mathsf{T}} \mathbf{m'}_F^{e'} \mathbf{P}_{e'} + \sum_{e=1}^{e=E} \delta \mathbf{P}_e^{\mathsf{T}} \mathbf{k}_F^e \mathbf{P}_e = 0 \qquad \forall \delta \mathbf{P}$$

It is also formulated as a generalised eigenvalue problem:

$$(-\omega^2 (\mathbf{M}_F + \mathbf{M'}_F) + \mathbf{K}_F) \mathbf{P} = \mathbf{0} \qquad (3.31)$$

where the matrices are assembled from elementary matrices using the localisation matrix, according to $\mathbf{P}_{e/e'} = \mathbf{\Lambda}_{e/e'} \mathbf{P}$.

- $\mathbf{M}_F = \sum\limits_{e=1}^{e=E} \mathbf{\Lambda}_e^{\mathsf{T}} \mathbf{m}_F^e \mathbf{\Lambda}_e$ and $\mathbf{M'}_F = \sum\limits_{e'=1}^{e'=E'} \mathbf{\Lambda}_{e'}^{\mathsf{T}} \mathbf{m'}_F^{e'} \mathbf{\Lambda}_{e'}$ are the fluid 'mass' matrices of the fluid. They assemble the contribution of the volume and surface elements.

- $\mathbf{K}_F = \sum\limits_{e=1}^{e=E} \mathbf{\Lambda}_e{}^\mathsf{T}\mathbf{k}_F^e\mathbf{\Lambda}_e$ is the fluid 'stiffness' matrix of the fluid, which assembles the contribution of all elements.

Remark 3.5 Acoustic and sloshing modes   *The formulation expressed in Equation (3.31) gathers compressibility and gravity effects in order to cover an extended range of wave numbers. However, in many practical applications, gravity and compressibility have a marked influence on distinct wave numbers. Let L and P be a scale for length and for pressure, and let $x_i^* = x_i/L$ and $p^* = p/P$, so that the non-dimensional wave equations may be recast as follows:*

$$-\frac{\omega^2 L^2}{c^2}p^* - \frac{\partial^2 p^*}{\partial x_i^* \partial x_i^*} = 0 \qquad \frac{\partial p^*}{\partial x_i^*} = \frac{\omega^2 L}{g}p^*$$

As highlighted in Figure 3.13, the compressibility number $C_o = \dfrac{\omega L}{c}$ and the Froude number $\mathcal{F}_r = \omega\sqrt{\dfrac{L}{g}}$ define bounds for compressibility and gravity influence.

This also indicates that for $\omega \to 0$, Equation (3.22) with boundary conditions (3.23) and (3.25) becomes $\Delta p = 0$ on Ω with $\dfrac{\partial p}{\partial n} = 0$ on $\partial\Omega$ (only coupling with a fixed wall is considered). The problem is singular since any constant pressure field is a solution: the rank of $\mathbf{K}_F$ is $\mathcal{N} - 1$, while its dimension is $\mathcal{N}$. A zero-frequency mode is solution of Equation (3.31): it is equivalent to structural rigid modes obtained with a displacement-based formulation.

When an acoustic-free surface is present, the boundary condition is $p = 0$ on Γ_o. In such cases, the eigenvalue problem is not singular for $\omega \to 0$ since the unique solution is $p = 0$ on Ω. Such a boundary condition may be accounted for by the elimination of the lines and rows of $\mathbf{K}_F$ which are associated with the constrained pressure nodes. Since there is at least one constrained node, the dimension of the stiffness matrix extracted from $\mathbf{K}_F$ is less than or equal to $\mathcal{N} - 1$ and it is therefore always regular.

The constraining relation $p = 0$ may also be enforced with Lagrange multipliers, according to Equation (2.20). As illustrated in Section 3.6, Lagrange multipliers may be interpreted in this case as the fluid displacement on Γ_o.

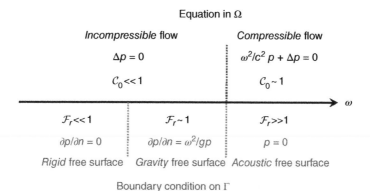

Figure 3.13   *Influence of compressibility and gravity*

When sloshing modes are studied, the eigenvalue problem can be conveniently formulated using the pressure degrees-of-freedom for the fluid-free level solely. To that end, let $\mathbf{P}$ *be written as follows:*

$$\mathbf{P} = \left\{ \begin{matrix} \mathbf{P}' \\ \bar{\mathbf{P}} \end{matrix} \right\}$$

with $\mathbf{P}'$ *and* $\bar{\mathbf{P}}$ *the pressure values for nodes belonging, respectively, to* Γ *(on the free level) and* $\Omega\backslash\Gamma$ *(in the remainder of the fluid domain). Since compressibility effects are not accounted for, the mass matrix contains only the contribution of* $\mathbf{P}'$. *Equation (3.31) is readily:*

$$\left(-\omega^2 \begin{bmatrix} \mathbf{M}'_F & \mathbf{0} \\ \mathbf{0} & \mathbf{0} \end{bmatrix} + \begin{bmatrix} \mathbf{K}'_F & \tilde{\mathbf{K}}_F \\ \tilde{\mathbf{K}}_F{}^\mathsf{T} & \bar{\mathbf{K}}_F \end{bmatrix} \right) \left\{ \begin{matrix} \mathbf{P}' \\ \bar{\mathbf{P}} \end{matrix} \right\} = \left\{ \begin{matrix} \mathbf{0} \\ \mathbf{0} \end{matrix} \right\}$$

Eliminating $\bar{\mathbf{P}}$ *according to* $\bar{\mathbf{P}} = -\bar{\mathbf{K}}_F^{-1}\tilde{\mathbf{K}}_F^\mathsf{T}\mathbf{P}'$ *produces an eigenvalue problem in terms of pressure at the free level solely:*

$$(-\omega^2 \mathbf{M}_F + (\mathbf{K}'_F - \tilde{\mathbf{K}}_F\bar{\mathbf{K}}_F^{-1}\tilde{\mathbf{K}}_F^\mathsf{T}))\mathbf{P}' = \mathbf{0} \tag{3.32}$$

$\mathbf{K}'_F - \tilde{\mathbf{K}}_F\bar{\mathbf{K}}_F^{-1}\tilde{\mathbf{K}}_F^\mathsf{T}$ *is named a* static condensation *of*

$$\begin{bmatrix} \mathbf{K}'_F & \tilde{\mathbf{K}}_F \\ \tilde{\mathbf{K}}_F{}^\mathsf{T} & \bar{\mathbf{K}}_F \end{bmatrix}$$

onto the degrees-of-freedom $\mathbf{P}'$. *The condensation is in any case always possible since* $\bar{\mathbf{K}}_F$, *whose size is less than or equal to* $\mathcal{N}-1$, *is non-singular.* ∎

3.3.2 Displacement-Based Formulations

A unified displacement-based formulation which takes into account both sloshing and acoustic waves may be stated as follows. In the harmonic regime, the wave equation in the fluid is given as follows:

$$-\omega^2 \rho\xi_i - \frac{\partial\sigma_{ij}(\xi)}{\partial x_j} = 0 \text{ in } \Omega \tag{3.33}$$

As detailed in Remark 3.4, the stress tensor in the fluid is spherical, so that $\sigma_{ij}(\xi) = -p(\xi)\delta_{ij}$, where the pressure fluctuation describes either compressibility effects ($p = -\rho c^2 \frac{\partial\xi_k}{\partial x_k}$ throughout the fluid domain Ω) or gravity effects ($p = \rho g\xi_k n_k$ on the fluid free level Γ, with outer unit normal $\mathbf{n}$).

Hence, Equation (3.33) is satisfied with

$$\sigma_{ij}(\xi) = \rho c^2 \frac{\partial\xi_k}{\partial x_k}$$

It is associated with boundary conditions representing the weighting surface Γ, an isobar Γ_o and/or a fixed wall Γ_π:

$$\sigma_{ij}(\xi)n_j = -\rho g\xi_k n_k n_i \text{ on } \Gamma \tag{3.34}$$

$$\sigma_{ij}(\xi)n_j = 0 \text{ on } \Gamma_o \tag{3.35}$$

$$\xi_j n_j = 0 \text{ on } \Gamma_\pi \tag{3.36}$$

The displacement field also complies with the irrotationality constraint:

$$\nabla \times \boldsymbol{\xi} = \mathbf{0} \qquad (3.37)$$

The weighted integral formulation of the problem is formulated as follows:

Find $\boldsymbol{\xi} \in \mathcal{X}$ and ω such that

$$-\omega^2 \int_\Omega \rho \xi_i \delta\xi_i \, d\Omega + \int_\Omega \rho c^2 \frac{\partial \xi_i}{\partial x_i} \frac{\partial \delta\xi_j}{\partial x_j} \, d\Omega + \int_\Gamma \rho g \xi_i n_i \delta\xi_j n_j \, d\Gamma = 0$$

for all $\delta\boldsymbol{\xi} \in \mathcal{X}$, with $\mathcal{X} = \{\boldsymbol{\xi} \in H^1(\Omega), \boldsymbol{\xi} \cdot \mathbf{n}|_{\Gamma_\pi} = 0, \nabla \times \boldsymbol{\xi}|_\Omega = \mathbf{0}\}$.

It is obtained for any virtual displacement field which complies with Equation (3.36), multiplying and integrating by parts Equation (3.33) and taking into account Equations (3.34) and (3.35).

As for the pressure-based formulation, discretisation of the weighted integral formulation lies on a volume and surface mesh of Ω and Γ, as depicted in Figure 3.12. Similarly to structure finite elements, an approximation of the fluid displacement is written $\boldsymbol{\xi}|_{\Omega_e} = \mathbf{N}_e \mathbf{X}_e$, where the matrix of the shape functions reads as follows:

$$\mathbf{N}^e = \begin{bmatrix} \mathbf{N}_i^e & \mathbf{0} \\ \mathbf{0} & \mathbf{N}_j^e \end{bmatrix}$$

when two directions i and j are considered.

The elementary mass matrix $\mathbf{m}_F^e$ is calculated as follows:

$$\mathbf{m}_F^e = \int_{\Omega_e} \rho \mathbf{N}_e^\mathsf{T} \mathbf{N}_e \, d\Omega_e$$

The elementary stiffness matrices arise from the volume and surface integrals.

- The elementary *stiffness matrix* accounting for compressibility effects is calculated on Ω_e:

$$\int_{\Omega_e} \rho c^2 \frac{\partial \xi_i}{\partial x_i} \frac{\partial \delta\xi_j}{\partial x_j} \, d\Omega_e = \delta\mathbf{X}_e^\mathsf{T} \mathbf{k}_F^e \mathbf{X}_e$$

with:

$$\mathbf{k}_F^e = \int_{\Omega_e} \rho c^2 \mathbf{D}_F^{e\,\mathsf{T}} \mathbf{D}_F^e \, d\Omega_e$$

$\mathbf{D}_F^e$ stands for the *divergence* operator and is calculated as

$$\mathbf{D}_F^e = \left\langle \frac{\partial \mathbf{N}_i^e}{\partial x_i}, \frac{\partial \mathbf{N}_j^e}{\partial x_j} \right\rangle$$

for two directions i and j.

- The elementary *stiffness matrix* accounting for gravity effects is calculated on $\Gamma_{e'}$:

$$\int_{\Gamma_{e'}} \rho g \xi_i n_i \delta\xi_j n_j \, d\Gamma e' = \delta\mathbf{X}_e^\mathsf{T} \mathbf{k}_F'^e \mathbf{X}_e$$

with:

$$\mathbf{k}'^{e}_{F} = \int_{\Gamma_{e'}} \rho g \mathbf{N}^{\mathsf{T}}_{e'} \mathbf{n}^{\mathsf{T}}_{e'} \mathbf{n}_{e'} \mathbf{N}_{e'}\, d\Gamma e'$$

where $\mathbf{n}_{e'}$ stands for the unit normal on $\Gamma_{e'}$.

Assembling these matrices on Ω and Γ generates the mass and stiffness matrices of the fluid $\mathbf{M}_F$, $\mathbf{K}_F$ and $\mathbf{K}'_F$, so that the discrete weighted integral formulation produces the following eigenvalue problem:

$$(-\omega^2 \mathbf{M}_F + \mathbf{K}_F + \mathbf{K}'_F)\mathbf{X} = \mathbf{0} \tag{3.38}$$

In most finite element codes, the displacement-formulated matrices do not take into account the constraint (3.37), which is a complex numerical issue. As a result, the modal analysis yields spurious modes (i.e. non-physical modes with non-null rotational) among the physical eigenmodes. In the context of fluid–structure coupling, this is a major drawback of the displacement-based method. It requires a specific numerical approach whose implementation is not straightforward (Bermudez *et al.*, 1995).

Remark 3.6 Alternative formulation for sloshing modes *The gravity wave equations may also be established using the φ and η variables – namely the displacement potential in the fluid volume and the elevation of the free level. These equations are stated as follows:*

$$\frac{\partial^2 \varphi}{\partial x_i \partial x_i} = 0 \;\; on\; \Omega$$

$$\frac{\partial \varphi}{\partial x_j} n_j = \eta \;\; on\; \Gamma$$

$$-\omega^2 \rho \varphi + \rho g\, \eta = \;\; on\; \Gamma$$

and are associated with the following boundary conditions:

$$\frac{\partial \varphi}{\partial x_j} n_j = 0 \;\; on\; \Gamma_\pi$$

$$\varphi = 0 \;\; on\; \Gamma_o$$

The weighted integral formulation of the problem is termed as follows:

Find $\varphi \in \mathcal{F}$, $\eta \in \mathcal{H}$ and ω such that:

$$\int_\Omega \rho \frac{\partial \varphi}{\partial x_i} \frac{\partial \delta \varphi}{\partial x_i}\, d\Omega - \int_\Gamma \rho \eta\, \delta\varphi\, d\Gamma = 0$$

and:

$$-\omega^2 \int_\Gamma \rho \varphi\, \delta\eta\, d\Gamma_o + \int_\Gamma \rho g\, \eta \delta\eta\, d\Gamma = 0$$

for all $\delta\varphi \in \mathcal{F}$ and $\delta\eta \in \mathcal{H}$, with $\mathcal{F} = \{\varphi \in H^1(\Omega), \varphi|_{\Gamma_o} = 0\}$ and $\mathcal{H} = \{\eta \in L^2(\Gamma)\}$.

The finite element approximation of φ on Ω_e and η on $\Gamma_{e'}$ are

$$\varphi = \mathbf{N}_e \mathbf{\Phi}_e \qquad \eta = \mathbf{N}_{e'} \mathbf{H}_{e'}$$

The shape functions $\mathbf{N}_e$ and $\mathbf{N}_{e'}$ and the nodal values $\mathbf{\Phi}_e$ and $\mathbf{H}_{e'}$ are associated with volume or surface elements. Hence,

$$\int_{\Omega_e} \frac{\partial \varphi}{\partial x_i} \frac{\partial \delta \varphi}{\partial x_i} \, d\Omega_e = \delta \mathbf{\Phi}_e^T \mathbf{k}_F^e \mathbf{\Phi}_e$$

where $\mathbf{k}_F^e$ is the 'stiffness' matrix introduced for the pressure-based formulation and

$$\int_{\Gamma_{e'}} \rho \eta \, \delta \varphi \, d\Gamma_{e'} = \delta \mathbf{\Phi}_{e'}^T \mathbf{m'}_F^{e'} \mathbf{H}_{e'} \qquad \int_{\Gamma_{e'}} \rho g \eta \, \delta \eta \, d\Gamma_{e'} = \delta \mathbf{H}_{e'}^T \mathbf{k'}_F^{e'} \mathbf{H}_{e'}$$

The 'mass' and 'stiffness' matrices $\mathbf{m'}_F^{e'}$ and $\mathbf{k'}_F^{e'}$ are expressed as[12]:

$$\mathbf{m'}_F^{e'} = \int_{\Gamma_{e'}} \rho \mathbf{N}_F^{e'}{}^T \mathbf{N}_F^{e'} \, d\Gamma_{e'} \qquad \mathbf{k'}_F^{e'} = \int_{\Gamma_{e'}} \rho g \mathbf{N}_F^{e'}{}^T \mathbf{N}_F^{e'} \, d\Gamma_{e'}$$

The finite element discretisation of the weighted integral formulation is expressed as follows:

$$\rho \mathbf{K'}_F \mathbf{\Phi} - \mathbf{M'}_F^T \mathbf{H} = 0 \qquad -\omega^2 \mathbf{M'}_F \mathbf{\Phi} + \mathbf{K'}_F \mathbf{H} = 0$$

A formulation in terms of the free level elevation solely may be derived by eliminating $\mathbf{\Phi}$ and it reads as follows:

$$(-\omega^2 \mathbf{M'}_F \mathbf{K}_F^{-1} \mathbf{M'}_F^T / \rho + \mathbf{K'}_F) \mathbf{H} = 0 \tag{3.39}$$

■

3.3.3   Finite Element Matrices

Whether formulated in terms of pressure or displacement, the eigenvalue problem associated with acoustic and sloshing modes is written $(-\omega^2 \mathbf{M} + \mathbf{K})\boldsymbol{\psi} = 0$, in which matrices $\mathbf{M}$ and $\mathbf{K}$ are symmetric[13].

Non-null solutions of $\mathbf{K}\boldsymbol{\psi} = 0$ are fluid *rigid modes*, which have been discussed above for the pressure-based formulation, and the eigenmodes $(\boldsymbol{\psi}_n)_{n \geq 1}$ constitute an orthogonal basis for the solutions of the time (or frequency) problem $\mathbf{M}\ddot{\boldsymbol{\psi}}(t) + \mathbf{K}\boldsymbol{\psi}(t) = \mathbf{F}(t)$ (or $(-\omega^2 \mathbf{M} + \mathbf{K})\boldsymbol{\psi}(\omega) = \mathbf{F}(\omega)$).

The calculation of kinetic and potential energies is associated with $\mathbf{M}$ and $\mathbf{K}$, respectively. However, the physical signification of the matrices depends on the formulation which is adopted. The kinetic and the potential energies of the fluid are given as follows:

$$\mathcal{E}_c^F = \frac{1}{2} \int_\Omega \rho \left(\frac{\partial \xi}{\partial t} \right)^2 \, d\Omega \qquad \mathcal{E}_p^F = \frac{1}{2} \int_\Omega \varepsilon(\xi) : \sigma(\xi) \, d\Omega$$

For an harmonic vibration $\xi(t) = \mathbf{X} \exp(i\omega t)$, the kinetic energy is proportional to

$$-\frac{\omega^2}{2} \int_\Omega \rho \xi_i \xi_i d\Omega$$

---

[12] These matrices are identical, less than the factor g in the integral; in order to unify the notations, they are however supposed distinct – this choice will be clarified in Chapter 5.

[13] It is recalled that in addition, $\mathbf{M}$ is positive definite and $\mathbf{K}$ is non-negative definite.

According to the pressure/displacement relation $\xi_i = \dfrac{1}{\rho\omega^2}\dfrac{\partial p}{\partial x_i}$, it may also be calculated as

$$-\frac{1}{2\rho\omega^2}\int_\Omega \frac{\partial p}{\partial x_i}\frac{\partial p}{\partial x_i}\,d\Omega$$

Hence, using the pressure-based formulation (according to the definition of the 'stiffness' matrix in Section 3.3), the kinetic energy in the fluid is found to be

$$\mathcal{E}_c^F \propto -\frac{1}{2\rho\omega^2}\mathbf{P}^\mathsf{T}\mathbf{K}_F\mathbf{P} \tag{3.40}$$

Using the displacement-based formulation (following the definition of the mass matrix in Section 3.3.2), it is also

$$\mathcal{E}_c^F \propto -\frac{\omega^2}{2}\mathbf{X}^\mathsf{T}\mathbf{M}_F\mathbf{X} \tag{3.41}$$

In the same manner, the potential energy associated with compressibility and gravity effects is proportional to

$$\frac{1}{2}\int_\Omega \varepsilon_{ij}(\xi)\sigma_{ij}(\xi)\,d\Omega + \frac{1}{2}\int_\Gamma \rho g(\xi_i n_i)^2\,d\Gamma$$

with $\xi_i n_i = \eta$.

Using the expression of the stress tensor $\sigma_{ij}(\xi) = -p\delta_{ij}$, of the pressure $p = -\rho c^2 \dfrac{\partial \xi_k}{\partial x_k}$ in Ω

and $p = \rho g\xi_k n_k$ on Γ, together with the expression of the strain tensor $\varepsilon_{ij}(\xi) = \dfrac{1}{2}\left(\dfrac{\partial \xi_i}{\partial x_j} + \dfrac{\partial \xi_j}{\partial x_i}\right)$,

the former relation may be recast as follows:

$$\frac{1}{2\rho}\int_\Omega \frac{p^2}{c^2}\,d\Omega + \frac{1}{2\rho}\int_\Gamma \frac{p^2}{g}\,d\Gamma$$

Hence,

$$\mathcal{E}_p^F \propto \frac{1}{2\rho}\mathbf{P}^\mathsf{T}\mathbf{M}_F\mathbf{P} + \frac{1}{2\rho}\mathbf{P}^\mathsf{T}\mathbf{M}'_F\mathbf{P} \tag{3.42}$$

and

$$\mathcal{E}_p^F \propto \frac{1}{2}\mathbf{X}^\mathsf{T}\mathbf{K}_F\mathbf{X} + \frac{1}{2}\mathbf{X}^\mathsf{T}\mathbf{K}'_F\mathbf{X} \tag{3.43}$$

As suggested by Equations (3.40 and 3.41) and (3.42 and 3.43), and as summarised in Table 3.5, matrices $\mathbf{M}$ and $\mathbf{K}$ play a dual role as far as energy calculation is concerned.

Table 3.5 Kinetic and potential energies with displacement-based or pressure-based formulations

	Displacement formulation ξ	Pressure formulation p
Degree-of-freedom	$\mathbf{X}$	$\mathbf{P}$
Kinetic energy	Mass matrix $\mathbf{M}_F$	'Stiffness' matrix $\mathbf{K}_F$
Potential energy	Stiffness matrices $\mathbf{K}_F$, $\mathbf{K}'_F$	'Mass' matrices $\mathbf{M}_F$, $\mathbf{M}'_F$

3.4   Boundary Element Method

3.4.1   Green Function and Green's Integral Theorem

Numerical methods based on an integral representation of the wave equation – such as derived with the boundary element method (BEM) – are particularly suited for problems stated in large fluid domains, whether bounded or unbounded. The resulting numerical model involves a smaller number of degrees-of-freedom, which makes BEM particularly appealing for tri-dimensional problems.

The BEM allows for a computation of the pressure and velocity fields at arbitrary interior/exterior points with equal accuracy to FEM computations. Mesh generation is in addition facilitated by the surface description. Radiation modelling is besides implemented without difficulty because the integral representation constitutionally complies with the Sommerfeld condition[14].

The essential features of the BEM are introduced hereafter; further reading on BEM may be found, for instance, in Aliabadi (2001), Beer (2001) and Manolis and Polyzos (2009).

To begin with, an integral representation of the wave equation is set down. It originates from the second Green identity:

$$\int_\Gamma \psi \frac{\partial \delta\psi}{\partial n} - \delta\psi \frac{\partial \psi}{\partial n} \, d\Gamma = \int_\Omega \psi \Delta\delta\psi - \delta\psi \Delta\psi \, d\Omega \tag{3.44}$$

This relation holds for sufficiently smoothed functions ψ and $\delta\psi$, these functions being defined on a domain Ω of boundary Γ with outer normal $\mathbf{n}$[15].

Equation (3.44) is formulated with the pressure field $p(M)$, which is solution of the wave equation, and with the *Green function* of the problem, denoted hereafter $G(M_o, M)$. For the Helmholtz equation, it is defined as the solution of

$$\Delta G(M_o, M) + \frac{\omega^2}{c^2} G(M_o, M) = -\delta(M_o) \; \forall \, M \in \Omega \tag{3.45}$$

endowed with appropriate boundary conditions. $\delta(M_o)$ stands for the Dirac distribution: the Green function represents the pressure field $p(M_o, M)$, which is triggered at point M by an acoustic source placed at point M_o.

---

[14] The integral representation of wave equations is discussed here in the context of acoustic waves. Issues regarding the damping induced by the radiation of gravity waves are, for instance, of importance when describing the dynamics of floating structures: a BEM representation of the wave propagation also applies in this case – see for instance Section 7.3.2 of Axisa and Antunes (2007).

[15] Green's identity is established as follows. Let ψ and $\delta\psi$ be smoothed scalar functions defined on Ω. Applying the divergence theorem to function $\psi \Delta\delta\psi - \delta\psi \Delta\psi$ yields the following:

$$\int_\Omega \nabla \cdot (\psi \Delta\delta\psi - \delta\psi \Delta\psi) \, d\Omega = \int_\Gamma \psi \frac{\partial \delta\psi}{\partial n} - \delta\psi \frac{\partial \psi}{\partial n} \, d\Gamma$$

where $\frac{\partial \psi}{\partial n} = \nabla\psi \cdot \mathbf{n}$. Using vectorial identities of the divergence $\nabla \cdot (\bullet)$ and gradient $\nabla\bullet$ operators gives

$$\nabla \cdot (\psi \nabla \delta\psi) = \nabla\psi \cdot \nabla\delta\psi + \psi \Delta\delta\psi \qquad \nabla \cdot (\delta\psi \nabla\psi) = \nabla\delta\psi \cdot \nabla\psi + \delta\psi \Delta\psi$$

Thus,

$$\int_\Omega \nabla \cdot (\psi \Delta\delta\psi - \delta\psi \Delta\psi) \, d\Omega = \int_\Omega \psi \Delta\delta\psi - \delta\psi \Delta\psi \, d\Omega$$

Combining the former integral expressions finally produces Equation (3.44).

Table 3.6   Green functions of the Helmholtz and Laplace equations. r denotes the distance between points M and M_o. $H_m^\pm$ is the Hankel function of order m, and it is defined as $H_m^\pm(\omega r/c) = J_m(\omega r/c) \pm iY_m(\omega r/c)$, for functions of the first or the second kind (Abramowitz and Stegun, 1970).

Equation / dimension	3	2
$\Delta + \dfrac{\omega^2}{c^2}(\bullet)$	$\dfrac{1}{4\pi r}\exp(i\omega r/c)$	$-\dfrac{i}{4}H_0^-(\omega r/c)$
$\Delta(\bullet)$	$\dfrac{1}{4\pi r}$	$-\dfrac{\ln r}{2\pi}$

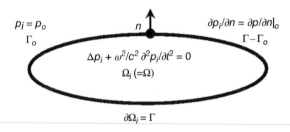

$p_i = p_o$
Γ_o

n

$\partial p_i/\partial n = \partial p/\partial n|_o$
$\Gamma - \Gamma_o$

$\Delta p_i + \omega^2/c^2\, \partial^2 p_i/\partial t^2 = 0$
$\Omega_i\, (=\Omega)$

$\partial \Omega_i = \Gamma$

Figure 3.14   Acoustics interior problem

Table 3.6 gives the Green function for the Helmholtz and Laplace equations formulated on $\mathbb{R}^d$.

3.4.2   Interior and Exterior Problems

The integral formulation of the wave equation for interior or exterior problems is as follows:

Interior problem With the notations of Figure 3.14, the wave equation for interior problems is stated as follows:

$$\Delta p_i + \frac{\omega^2}{c^2}p_i = 0 \text{ in } \Omega \qquad p_i = p_o \text{ on } \Gamma_o \qquad \frac{\partial p_i}{\partial n} = \frac{\partial p}{\partial n}\bigg|_o \text{ on } \Gamma - \Gamma_o \qquad (3.46)$$

An equivalent formulation of the problem is derived from Equation (3.44) for the function p_i and the Green function of the problem. The integral representation of Equation (3.46) is shown to be

$$\int_\Gamma \left[-p_i(M)\frac{\partial G}{\partial n}(M_o, M) + G(M_o, M)\frac{\partial p_i}{\partial n}(M) \right] d\Gamma(M)$$

$$= \begin{cases} p_i(M_o) & M_o \in \Omega \\ p_i(M_o)/2 & M_0 \in \Gamma \\ 0 & M_o \notin \Omega \cup \Gamma \end{cases} \qquad (3.47)$$

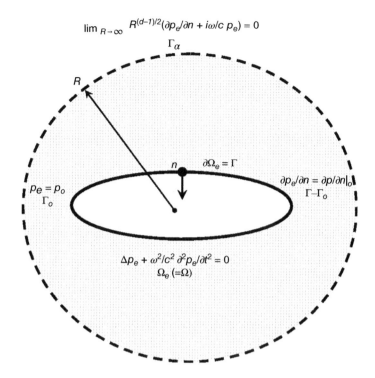

$$\lim_{R \to \infty} R^{(d-1)/2}(\partial p_e / \partial n + i\omega/c \ p_e) = 0$$

$$\Gamma_\alpha$$

$$R$$

$$n \qquad \partial\Omega_e = \Gamma$$

$$p_e = p_0 \qquad\qquad\qquad\qquad \partial p_e / \partial n = \partial p / \partial n|_o$$
$$\Gamma_o \qquad\qquad\qquad\qquad\qquad \Gamma - \Gamma_o$$

$$\Delta p_e + \omega^2/c^2 \ \partial^2 p_e / \partial t^2 = 0$$
$$\Omega_e \ (=\Omega)$$

Figure 3.15   Acoustics exterior problem

Exterior problem The wave equation for exterior problems is stated as follows:

$$\Delta p_e + \frac{\omega^2}{c^2} p_e = 0 \text{ in } \Omega \qquad p_e = p_0 \text{ on } \Gamma_o \qquad \frac{\partial p_e}{\partial n} = \frac{\partial p}{\partial n}\bigg|_o \text{ on } \Gamma - \Gamma_o \qquad (3.48)$$

It is complemented with the Sommerfeld condition, formulated as follows:

$$\lim_{\|\mathbf{r}\| \mapsto \infty} \|\mathbf{r}\|^{(d-1)/2} \left(\frac{\partial p_e}{\partial r} + \frac{i\omega}{c} p_e \right) = 0 \qquad (3.49)$$

The integral formulation of interior problems is stated for an artificially bounded domain at $r = R$, as depicted in Figure 3.15. It is extended for $R \to \infty$ by taking into account the following relations, which are brought about by Equation (3.49) in a spherical coordinate system:

$$\lim_{R \to \infty} p_e(R, \theta, \phi) = 0 \qquad \lim_{R \to \infty} R \left(\frac{\partial p_e}{\partial R}(R, \theta, \phi) + \frac{i\omega}{c} p_e(R, \theta, \phi) \right) = 0$$

An integral representation for the exterior problem (3.48 and 3.49) is then arrived at:

$$\int_\Gamma \left[p_e(M) \frac{\partial G}{\partial n}(M_o, M) - G(M_o, M) \frac{\partial p_e}{\partial n}(M) \right] d\Gamma(M)$$

$$= \begin{cases} p_e(M_o) & M_o \in \Omega \\ p_e(M_o)/2 & M_0 \in \Gamma \\ 0 & M_o \notin \Omega \cup \Gamma \end{cases} \qquad (3.50)$$

3.4.3 Direct and Indirect Boundary Element Method

A discrete representation of the integral formulation is derived, using either the *direct* or the *indirect* BEM.

Direct BEM is adapted to exterior and/or interior problems. The pressure at any point $M_o \in \Gamma$ may in such cases be derived from Equations (3.47) and (3.50), and be expressed as follows:

$$\pm \frac{p(M_o)}{2} = \int_\Gamma \left[p(M)\frac{\partial G}{\partial n}(M_o, M) - G(M_o, M)\frac{\partial p}{\partial n}(M) \right] d\Gamma(M) \qquad (3.51)$$

with the $\pm$ sign standing for exterior/interior problem.

The direct BEM generates a discretisation of Equation (3.51) as stated above. It is written for a set of the so-called *collocation points*, which define the approximation schemes to derive the discretisation of the integrals – an example with a zero-order approximation scheme is proposed in Figure 3.16.

Let $\Gamma = \bigcup_{e=1}^{e=E} \Gamma_e$ be a discretisation of the boundary Γ. On each boundary element Γ_e, p and $\dfrac{\partial p}{\partial n}$ are approximated according to

$$p|_{\Gamma_e} \simeq \mathbf{N}^e_\mu \mathbf{P}_e \qquad \frac{\partial p}{\partial n}\bigg|_{\Gamma_e} \simeq \mathbf{N}^e_\sigma \frac{\partial \mathbf{P}}{\partial n}_e$$

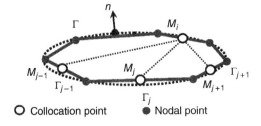

O Collocation point   ● Nodal point

Figure 3.16 Direct BEM discretisation with a zero-order scheme. $\bigcup_{j=1}^{j=J} \Gamma_j$ is a geometrical approximation of Γ. A collocation of Equation (3.51) at the centre of each surface Γ_j is considered here, so that the pressure and its gradient are constant on Γ_j: for all j, $p|_{\Gamma_j} \simeq p(M_j)$ and $\dfrac{\partial p}{\partial n}\bigg|_{\Gamma_j} \simeq \dfrac{\partial p}{\partial n}(M_j)$. The discretisa-

tion of Equation (3.51) is $\pm\dfrac{p(M_i)}{2} = \displaystyle\sum_{j=1}^{j=J} p(M_j)\int_{\Gamma_j}\dfrac{\partial G}{\partial n}(M_i, M)\,d\Gamma_j(M) - \sum_{j=1}^{j=J}\dfrac{\partial p}{\partial n}(M_j)\int_{\Gamma_j} G(M_i, M)\,d\Gamma_j(M)$,

for all point M_i, $i \in [1, J]$. The former expression is Equation (3.53) with $S_{i,j} = \int_{\Gamma_j} G(M_i, M)\,d\Gamma_j(M)$ and $D_{i,j} = \mp\frac{1}{2}\delta_{i,j} + \int_{\Gamma_j}\frac{\partial G}{\partial n}(M_i, M)\,d\Gamma_j(M)$, where the $\mp$ sign stands for exterior/interior problems

$\mathbf{P}_e$ and $\dfrac{\partial \mathbf{P}}{\partial n}_e$ are the vectors composed of the values of the pressure and the pressure gradient at the $\mathcal{N}_e$ collocation points and $\mathbf{N}^e_\mu$ and $\mathbf{N}^e_\sigma$ are the associated shape functions. The nodal values are assembled in the vectors $\mathbf{P}$ and $\dfrac{\partial \mathbf{P}}{\partial n}$ according to

$$\mathbf{P}_e = \mathbf{\Lambda}_e \mathbf{P} \qquad \frac{\partial \mathbf{P}}{\partial n}_e = \mathbf{\Lambda}_e \frac{\partial \mathbf{P}}{\partial n}$$

where $\mathbf{\Lambda}_e$ is the localisation matrix associated with the boundary element Γ_e. $\mathcal{N}$ is the total number of collocation points. The discretisation of Equation (3.51) is written as follows:

$$\pm \frac{p(M_n)}{2} = \sum_{e=1}^{e=E} \int_{\Gamma_e} \mathbf{N}^e_\mu(M) \frac{\partial G}{\partial n}(M_n, M) \, d\Gamma_e(M) \mathbf{\Lambda}_e \mathbf{P}$$

$$- \sum_{e=1}^{e=E} \int_{\Gamma_e} \mathbf{N}^e_\sigma(M) G(M_n, M) \, d\Gamma_e \mathbf{\Lambda}_e(M) \frac{\partial \mathbf{P}}{\partial n} \qquad \forall M_n, n \in [1, \mathcal{N}] \qquad (3.52)$$

It is also formulated as follows:

$$\mathbf{D}\mathbf{P} = \mathbf{S} \frac{\partial \mathbf{P}}{\partial n} \qquad (3.53)$$

Matrices $\mathbf{D}$ and $\mathbf{S}$ are assembled from row vectors:

$$\mathbf{D} = \{\mathbf{d}_n\}_{n \in [1, \mathcal{N}]} \qquad \mathbf{S} = \{\mathbf{s}_n\}_{n \in [1, \mathcal{N}]}$$

with:

$$\mathbf{d}_n = \mp \frac{1}{2} \delta_n + \sum_{e=1}^{e=E} \int_{\Gamma_e} \mathbf{N}^e_\mu(M) \frac{\partial G}{\partial n}(M_n, M) \, d\Gamma_e(M) \mathbf{\Lambda}_e$$

where the $\mp$ sign stands for exterior/interior problems, and

$$\mathbf{s}_n = \sum_{e=1}^{e=E} \int_{\Gamma_e} \mathbf{N}^e_\sigma(M) G(M_n, M) \, d\Gamma_e(M) \mathbf{\Lambda}_e$$

δ_n is a row vector of size $\mathcal{N}$, with component n equals to 1 and all other components equal to 0.

The evaluation of the pressure field at any point $M_o \in \Omega$ may be achieved using the integral representation:

$$p(M_o) = \pm \int_\Gamma \left[p(M) \frac{\partial G}{\partial n}(M_o, M) - G(M_o, M) \frac{\partial p}{\partial n}(M) \right] d\Gamma(M)$$

With a single point collocation, the discrete equation reads as follows:

$$p(M_o) = \mathbf{\Pi}(M_o)\mathbf{P} + \mathbf{\Gamma}(M_o) \frac{\partial \mathbf{P}}{\partial n} \qquad (3.54)$$

where $\mathbf{\Pi}$ and $\mathbf{\Gamma}$ are calculated according to

$$\mathbf{\Pi}(M_o) = \pm \sum_{e=1}^{e=E} \int_{\Gamma_e} \frac{\partial G}{\partial n}(M_o, M) \mathbf{N}^e_\mu(M) \, d\Gamma_e(M) \mathbf{\Lambda}_e$$

and

$$\Gamma(M_o) = \mp \sum_{e=1}^{e=E} \int_{\Gamma_e} G(M_o, M) N_\sigma^e(M) \, d\Gamma_e(M) \Lambda_e$$

Indirect BEM is suited to exterior and interior problems. The pressure at any point $M_o \in \Omega$ may in such cases be derived from Equations (3.47) and (3.50) and be expressed as follows:

$$p(M_o) = \int_\Gamma \left[\mu(M) \frac{\partial G}{\partial n}(M_o, M) - G(M_o, M) \sigma(M) \right] \, d\Gamma(M) \tag{3.55}$$

where μ and σ denote the so-called *double* and *single potential layers* (Sayhi *et al.*, 1981). They are the difference between the interior and exterior values of the pressure and its gradient:

$$\mu(M) = p_e(M) - p_i(M) \qquad \sigma(M) = \frac{\partial p_e}{\partial n}(M) - \frac{\partial p_i}{\partial n}(M)$$

The integral problem formulated on Ω in terms of μ and σ is endowed with the following boundary conditions on Γ:

$$\mu = 0 \ \text{ on } \Gamma_\sigma \qquad \sigma = 0 \ \text{ on } \Gamma_\mu$$

A weighted integral formulation of Equation (3.55) is then stated on Γ using μ and σ as unknowns of the problems. It is expressed as follows:

$$\int_{\Gamma_\sigma} \int_{\Gamma_\sigma} \sigma(M) G(M_o, M) \delta\sigma(M_o) \, d\Gamma_\sigma(M) d\Gamma_\sigma(M_o)$$

$$- \int_{\Gamma_\sigma} \int_{\Gamma_\mu} \mu(M) \frac{\partial G}{\partial n_M}(M_o, M) \delta\sigma(M_o) d\Gamma_\mu(M) d\Gamma_\sigma(M_o)$$

$$= - \int_{\Gamma_\sigma} p_o(M_o) \delta\sigma(M_o) \, d\Gamma_\sigma(M_o) \tag{3.56}$$

for all $\delta\sigma$, and

$$\int_{\Gamma_\mu} \int_{\Gamma_\mu} \mu(M) \frac{\partial^2 G}{\partial n_{(M_o)} \partial n_{(M)}}(M_o, M) \delta\mu(M_o) \, d\Gamma_\mu(M) d\Gamma_\mu(M_o)$$

$$- \int_{\Gamma_\mu} \int_{\Gamma_\sigma} \sigma(M) \frac{\partial G}{\partial n_{(M_o)}}(M_o, M) \delta\mu(M_o) \, d\Gamma_\sigma(M) d\Gamma_\mu(M_o)$$

$$= \int_{\Gamma_\mu} \frac{\partial p}{\partial n}\bigg|_o (M_o) \delta\mu(M_o) \, d\Gamma_\mu(M_o) \tag{3.57}$$

for all $\delta\mu$.

The finite element approximation of μ and σ is finally arrived at; it reads as follows:

$$\sigma(M) = \mathbf{N}_\sigma(M) \sigma \qquad \mu(M) = \mathbf{N}_\mu(M) \mu \qquad \forall M \in \Gamma$$

where $\mathbf{N}_\sigma$ and $\mathbf{N}_\mu$ are the shape functions associated with the boundary element.

The discrete form of Equations (3.56) and (3.57) is then given as follows:

$$\begin{bmatrix} \mathbf{\Lambda}_\sigma & \mathbf{\Sigma} \\ \mathbf{\Sigma}^\mathsf{T} & \mathbf{\Lambda}_\mu \end{bmatrix} \begin{Bmatrix} \sigma \\ \mu \end{Bmatrix} = \begin{Bmatrix} \boldsymbol{\varphi}_\sigma \\ \boldsymbol{\varphi}_\mu \end{Bmatrix}$$

where matrices $\mathbf{\Lambda}_\sigma$, $\mathbf{\Lambda}_\mu$ and $\mathbf{\Sigma}$ are calculated as:

$$\mathbf{\Lambda}_\sigma = \int_{\Gamma_\sigma}\int_{\Gamma_\sigma} \mathbf{N}_\sigma^\mathsf{T}(M')G(M,M')\mathbf{N}_\sigma(M)\,d\Gamma_\sigma(M)d\Gamma_\sigma(M')$$

$$\mathbf{\Lambda}_\mu = \int_{\Gamma_\mu}\int_{\Gamma_\mu} \mathbf{N}_\mu^\mathsf{T}(M')\frac{\partial^2 G}{\partial n \partial n'}(M,M')\mathbf{N}_\mu\,d\Gamma_\mu(M)d\Gamma_\mu(M')$$

$$\mathbf{\Sigma} = -\int_{\Gamma_\sigma}\int_{\Gamma_\mu} \mathbf{N}_\sigma^\mathsf{T}(M')\frac{\partial G}{\partial n}(M,M')\mathbf{N}_\mu d\Gamma_\mu(M)d\Gamma_\sigma(M')$$

and where vectors $\boldsymbol{\varphi}_\sigma$, $\boldsymbol{\varphi}_\mu$ are defined by:

$$\boldsymbol{\varphi}_\sigma = -\int_{\Gamma_\sigma} p_o\mathbf{N}_\sigma(M_o)\,d\Gamma_\sigma(M_o) \qquad \boldsymbol{\varphi}_\mu = \int_{\Gamma_\mu} \frac{\partial p}{\partial n}\bigg|_o(M_o)\mathbf{N}_\mu(M_o)\,d\Gamma_\mu(M_o)$$

The calculation of the pressure in the domain ensues from Equation (3.55), once the single and double layer potentials are known:

$$p(M_o) = \mathbf{\Pi}_\sigma(M_o)\sigma + \mathbf{\Pi}_\mu(M_o)\mu \tag{3.58}$$

with:

$$\Pi_\sigma(M_o)_i = -\int_{\Gamma_\sigma} N_\sigma^i(M_o)G(M_o,M)\,d\Gamma_\sigma(M)$$

and:

$$\Pi_\mu(M_o)_i = \int_{\Gamma_\mu} N_\mu^i(M_o)\frac{\partial G}{\partial n}(M_o,M)\,d\Gamma_\mu(M)$$

3.4.4 Boundary Element Matrices

As indicated in Table 3.7, matrices arising from the BEM are fully populated and they are non-symmetric for the direct method, while symmetric for the indirect method. The matrices are also frequency dependent: as illustrated in Chapters 5 and 6, the computation of response functions may therefore be computationally expensive since an evaluation of the matrices is required at each frequency of interest. In the same manner, calculating the fields at interior/exterior points depends on their location: as indicated by Equation (3.54), the integrals have to be computed for each point of interest, which can become quite expensive when a large number of points is considered.

The integral involved in the BEM are in addition singular or nearly singular, because $G(M,M_o)$ and $\frac{\partial G}{\partial n}(M_o,M)$ are not defined for $M = M_o$. Appropriate numerical schemes have to be used in order for a BEM code to be robust: a discussion on the matter can be found, for instance, in the study by Rosen and Cormack (1993) and Scuderi (2008).

Table 3.7 Comparison of direct and indirect BEM

	Direct BEM	Indirect BEM
Problem type	Interior and/or exterior	Interior *and* exterior
Problem Unknowns	p_e or p_i $\dfrac{\partial p_e}{\partial n}$ or $\dfrac{\partial p_i}{\partial n}$	$p_e - p_i$ $\dfrac{\partial p_e}{\partial n} - \dfrac{\partial p_i}{\partial n}$
Matrices	Non-symmetric	Symmetric
Elements	Discontinuous–continuous	Continuous
Computational cost	High	Low

The integral formulation of the wave equation is equivalent to its original form only for interior problems: in such a case, Equation (3.47) has a unique solution for each frequency. For exterior problems however, the integral formulation is not equivalent to the original problem. Equation (3.50) is indeed singular for an infinite set of frequencies, which correspond to the eigenfrequencies of the interior problem with a prescribed null value of the pressure at the boundary (Colton and Kress, 1983; Qing and Bo-Hou, 1990).

A simple and efficient method to overcome the non-uniqueness problem of the exterior integral formulation is the combined Helmholtz integral equation formulation (CHIEF), which has been proposed by Schenck (1968) and is implemented in most boundary element codes[16]. The integral formulation is written for points in the interior domain (the so-called *CHIEF points*): doing so provides additional linearly independent relations in Equation (3.53). As the resulting linear system involves more equations than unknowns, a solution may be obtained by using the least square method[17].

The location and number of CHIEF point depends on the problem at stake: as a general rule, CHIEF points should not be placed at the vicinity of a nodal surface of interior acoustic modes, that is, should be away from points where the pressure is null. CHIEF points are usually chosen at random within the interior domain in order to yield a conditioning of matrices **D** and **S** which ensures the accuracy of the computation (Seybert, 1987). As the size of the nodal surface increases with frequency and as more CHIEF points are required when high wave numbers are investigated, the efficiency and accuracy of the BEM are usually better in the 'low-frequency range'.

As a concluding remark on the subject, it is emphasised that the frequency-domain formulation of (vibro-)acoustic problems using an integral representation is particularly well suited for massively parallel computing techniques and that the computational cost may be dramatically reduced by resorting to the so-called *Fast Multipole Method*, as presented, for instance, by Brunner *et al.* (2009). An illustration of the BEM direct method for exterior problems is

---

[16] Other techniques are of course available to tackle the issue of spurious frequencies in BEM; see, for instance, Chapter 10 of Soize and Ohayon (2014).

[17] Let A be a $M \times N$ matrix and let **b** be a vector of length M, with $M > N$. The linear system $\mathbf{Au} = \mathbf{b}$, where **u** is a vector of length N is *over-determined*: it involves more equations than unknowns. With the least square method, an approximation of the solution is found to be minimising the error made in the results of each single equation: in other words, **u** is the solution of the minimisation problem $\min_{\delta\mathbf{u}} \|\mathbf{A}\delta\mathbf{u} - \mathbf{b}\|$.

proposed in this chapter for the modelling of acoustic radiation, as well as in Chapter 4 for the computation of added mass and in Chapter 5 for the description of vibro-acoustic coupling.

3.5 Example: Sloshing Modes

3.5.1 Circular Reservoir with Fluid-Free Surface

As described in Figure 3.17, the sloshing modes in a closed cylindrical reservoir are investigated. The inner and outer radii are R and $R' = \alpha R$, while the height is L. The reservoir is filled up to level $H \leq L$ with a fluid of density ρ under constant gravity g. The pressure in the fluid is denoted $p(r, \theta, z)$ using the cylindrical coordinate system and it complies with the Laplace equation:

$$\frac{\partial^2 p}{\partial r^2} + \frac{1}{r}\frac{\partial p}{\partial r} + \frac{1}{r^2}\frac{\partial^2 p}{\partial \theta^2} + \frac{\partial^2 p}{\partial z^2} = 0 \text{ for } (r, \theta, z) \in [R, R'] \times [0, 2\pi] \times [0, H]$$

The coupling condition on fixed walls are given as follows:

$$\left.\frac{\partial p}{\partial z}\right|_{z=0} = 0 \qquad \left.\frac{\partial p}{\partial r}\right|_{r=R} = 0 \qquad \left.\frac{\partial p}{\partial r}\right|_{r=R'} = 0$$

while the free surface condition with gravity effects is given as follows:

$$\left.\frac{\partial p}{\partial z}\right|_{z=H} = \frac{\omega^2}{g}p\Big|_{z=H}$$

The solution of the problem is written in a separable form by using a Fourier series, that is, with the pressure stated as $p(r, \theta, z) = p_R(r)\cos(m\theta)p_Z(z)$ (for symmetric Fourier components) or $p(r, \theta, z) = p_R(r)\sin(m\theta)p_Z(z)$ (for antisymmetric Fourier components). $p_Z(z)$ is

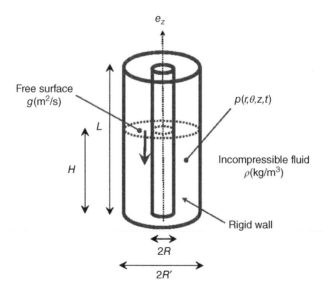

Figure 3.17   Incompressible fluid with free surface within a cylindrical reservoir

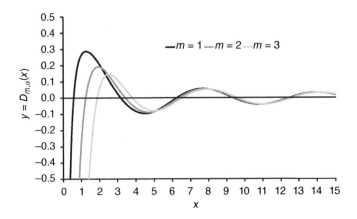

Figure 3.18   Function $x \mapsto D_{m,\alpha}(x)$ for $\alpha = 2$ and $m = 1, 2, 3$

Table 3.8   First roots of function $D_{m,\alpha}(x)$ for $m = 1, 2, 3$ and $\alpha = 2$

n	1	2	3	4
$q_{0,n}$	3.19658	6.31235	9.44446	12.58121
$q_{1,n}$	0.67734	3.28247	6.35321	9.47133
$q_{2,n}$	1.34060	3.53129	6.47469	9.55158
$q_{3,n}$	1.97888	3.92005	6.67380	9.68421

expressed as a linear superposition of $\cosh(qz)$ and $\sinh(qz)$ and $p_R(r)$ as a linear superposition of $J_m(qr)$ and $Y_m(qr)$[18]. Substituting the latter expression in the former equations and boundary conditions, it is established that $q = q_{m,n}/R$, where $q_{m,n}$ is the nth root of function $D_{m,\alpha}(x) = J'_m(x)Y'_m(\alpha x) - J'_m(\alpha x)Y'_m(x)$. Figure 3.18 plots $D_{m,\alpha}$ for a few values of m, while some of the first roots of $D_{m,\alpha}$ are gathered in Table 3.8.

The sloshing mode shapes expressed in terms of pressure are found to be

$$p_{m,n}(r, \theta, z) =$$
$$(J_m(q_{m,n}r/R)Y'_m(q_{m,n}) - Y_m(q_{m,n}r/R)J'_m(q_{m,n})) \cos(m\theta) \cosh(q_{m,n}z/R) \quad (3.59)$$

and the corresponding pulsations are

$$\omega_{m,n} = \sqrt{g/R \, q_{m,n} \tanh(q_{m,n}H/R)} \quad (3.60)$$

It is noted that the axisymmetric Fourier component ($m = 0$) yields a zero-frequency mode, while the symmetric and antisymmetric Fourier components ($m, m' \neq 0$) intrinsically have a zero pressure mean value on $[R, R'] \times [0, 2\pi] \times [0, H]$.

---

[18] J_m and Y_m are the so-called *Bessel functions* of the first and second kind of order m. A definition and some properties of these functions may be found, for instance, in the study by Abramowitz and Stegun (1970).

3.5.2   2D Axisymmetric Elements with Gravity

The weighted integral formulation of the problem is expressed as follows:

$$-\omega^2 \int_R^{R'} \int_0^{2\pi} \frac{1}{g} p|_{z=H} \delta p|_{z=H} \, r dr d\theta + \int_R^{R'} \int_0^{2\pi} \int_0^H \nabla p \, \nabla \delta p \, r dr d\theta dz = 0$$

Given the 2π–periodicity, a Fourier expansion of $p(r, \theta, z)$ in harmonic angular functions is convenient for the problem at hand and allows for a bi-dimensional finite element discretisation, which is numerically convenient in the present case. The pressure field is expressed as follows:

$$p\,(r, \theta, z, \omega) = \underbrace{\sum_{m=0}^{m=+\infty} p_m(r, z, \omega) \cos(m\theta)}_{\text{Symmetric components}} + \underbrace{\sum_{m'=1}^{m'=+\infty} p_{m'}(r, z, \omega) \sin(m'\theta)}_{\text{Anti-symmetric components}} \qquad (3.61)$$

The weighted integral formulation may be stated for any component of order m as follows:

$$-\omega^2 \int_R^{R'} \frac{1}{g} p_m|_{z=H} \delta p_m|_{z=H} \, r dr + \int_R^{R'} \int_0^H \nabla p_m \, \nabla \delta p_m \, r dr dz = 0$$

where $\nabla p_m \, \nabla \delta p_m$ is calculated according to

$$\nabla p_m \, \nabla \delta p_m = \frac{\partial p_m}{\partial r} \frac{\partial \delta p_m}{\partial r} + \frac{m^2}{r^2} p_m \delta p_m + \frac{\partial p_m}{\partial z} \frac{\partial \delta p_m}{\partial z}$$

The finite element discretisation is achieved with $J \times I$ bi-dimensional elements $[r_j, r_{j+1}] \times [z_i, z_{i+1}]$, arranged in such a manner that:

$$\int_R^{R'} \int_0^H (\bullet) \, r dr dz = \sum_{j=1}^{j=J} \sum_{i=1}^{i=I} \int_{r_j}^{r_{j+1}} \int_{z_i}^{z_{i+1}} (\bullet) \, r dr dz$$

Using finite elements with four nodes and one degree-of-freedom per node, as represented in Figure 3.19, the approximation of the pressure field is stated as follows:

$$p_m|_{[r_j, r_{j+1}] \times [z_i, z_{i+1}]} = \mathbf{N}_{i,j}^P \mathbf{P}_{i,j}^m \qquad \forall m \geq 1$$

- $\mathbf{N}_{i,j}^P$ is the vector of shape functions. Linear functions are used here, which allow for a representation of the pressure gradient with constant values on each element. With $r = r_o + \xi a_j$ for $r \in [r_j, r_{j+1}]$ and $z = z_o + \eta b_i$ for $z \in [z_i, z_{i+1}]$, $2a_j = r_{j+1} - r_j$ and $2b_i = z_{i+1} - z_i$, the shape function associated with node $k \in [1, 4]$ is given as follows:

$$N_k^P(\xi, \eta) = \frac{1}{4}(1 + \xi_k \xi)(1 + \eta_k \eta)$$

where $\xi_k = \pm 1$, $\eta_k = \pm 1$ and $(\xi, \eta) \in [-1, +1] \times [-1, +1]$.

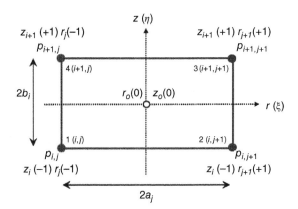

Figure 3.19   2D axisymmetric fluid element with four nodes and one degree-of-freedom per node

- $\mathbf{P}^m_{i,j}$ is the vector of the pressure unknowns of the element:

$$\mathbf{P}^m_{i,j} = \langle p^m_{i,j}, \ p^m_{i,j+1}, \ p^m_{i+1,j+1}, \ p^m_{i+1,j}\rangle^{\mathsf{T}}$$

An analytical expression of the elementary mass and stiffness matrices may be obtained here. After a few mathematical operations, the following expressions of the matrices elements are arrived at:

$$\mathbf{m}'_{p(k,l)} = \frac{1}{g}\frac{a}{8}\left(r_0 + \frac{\xi_k\xi_l r_0}{3} + \frac{a}{3}(\xi_k + \xi_l)\right)(1 + \eta_k)(1 + \eta_l) \tag{3.62}$$

and $\mathbf{k}_p = \mathbf{k}_{p,o} + m^2\mathbf{k}_{p,m}$ with:

$$\mathbf{k}_{p,o\ (k,l)} = abr_0\left(\frac{\xi_k\xi_l}{4a^2}\left(1 + \frac{\eta_k\eta_l}{3}\right) + \frac{\eta_k\eta_l}{4b^2}\left(1 + \frac{\xi_k\xi_l}{3}\right)\right) + \frac{a^2}{12b}\eta_k\eta_l(\xi_k + \xi_l) \tag{3.63}$$

$$\mathbf{k}_{p,m\ (k,l)} =$$

$$\frac{b}{8}\left(2(\xi_k + \xi_l) - \frac{2r_0}{a}\xi_k\xi_l + \left(1 - \frac{r_0}{a}(\xi_k + \xi_l) + \frac{r_0^2}{a^2}\xi_k\xi_l\right)\ln\left(\frac{r_0 + a}{r_0 - a}\right)\right) \tag{3.64}$$

The global 'mass' and 'stiffness' matrices are assembled according to

$$[\bullet] = \sum_{j=1}^{j=J}\sum_{i=1}^{i=I}\mathbf{\Lambda}_{i,j}{}^{\mathsf{T}}[\bullet]^i_u\mathbf{\Lambda}_{i,j}$$

where $\mathbf{\Lambda}_{i,j}$ is the connectivity matrix for the fluid element i,j. As detailed in Table 3.9, $\mathbf{\Lambda}_{i,j}$ is a $4 \times (I + 1)(J + 1)$ matrix, which relates the element nodes to the mesh nodes.

The discrete weighted integral formulation yields the eigenvalue problem (3.31), with the 'mass' matrix accounting only for gravity effects. Equivalent formulations, either in terms of pressure or elevation, may also be obtained.

Table 3.9 Connectivity matrix of 2D fluid element i, j

Node	Element Degree-of-freedom	Number	Node	Mesh Degree-of-freedom	Number
1	p_1	1	(i, j)	$p_{i,j}$	$j + (i - 1)J$
2	p_2	2	$(i, j + 1)$	$p_{i,j+1}$	$j + 1 + (i - 1)J$
3	p_3	3	$(i + 1, j + 1)$	$p_{i+1,j+1}$	$j + 1 + iJ$
4	p_4	4	$(i + 1, j)$	$p_{i+1,j}$	$j + iJ$

Table 3.10 Eigenfrequencies of the fluid reservoir: analytical solution and numerical computation

Frequency	Analytical (Hz) $m = 0$	Numerical (Hz)	Analytical (Hz) $m = 1$	Numerical (Hz)
f_1	1.78	1.80	0.81	0.81
f_2	2.50	2.59	1.81	1.82
f_3	3.06	3.31	2.51	2.60
f_4	3.54	4.05	3.07	3.31
	$m = 2$		$m = 3$	
f_1	1.15	1.16	1.40	1.41
f_2	1.87	1.89	1.97	1.99
f_3	2.54	2.62	2.58	2.67
f_4	3.08	3.33	3.10	3.35

3.5.3 Sloshing Modes

Dimensions of the problem are as follows: $R = 0.25$ m, $R' = 0.5$ m, $L = 1$ m, $H/L = 75\%$ and $g = 9.81$ m/s^2. Numerical computations are performed for Fourier components of order $m = 0, 1, 2$ and $m = 3$, with $I \times J = 7 \times 20$ elements. As displayed in Table 3.10, the agreement with the analytical solution is notable for the first three frequencies, with a relative error being less than 3%. The accuracy is however lower for the subsequent frequencies, as evidenced for Fourier component $m = 0$: f_3 and f_4 denoting the fourth and fifth modes (the first mode being $f_0 = 0$ Hz); the discrepancies between numerical and analytical values for these modes are about 7% and 12%.

The convergence of the finite element computation is underlined in Table 3.11. It is noted that the convergence rate on the evaluation of the sloshing frequencies is not as steep as for acoustic modes for a similar geometry – see the following two examples. A grid refinement is possible for a simple enough geometry, but for many industrial applications, a convergence rate such as reported in Table 3.11 may not be affordable.

As discussed in Section 3.3, the finite element computation of the sloshing modes can be performed with pressure-based or displacement-based formulations, using full or condensed matrices. Figure 3.20 sketches the first four mode shapes for Fourier component $m = 1$, in terms of the p or η variables. The mode shape is in accordance with the analytical expression, and, as expected, identical results are obtained with the two formulations.

Table 3.11 Convergence of finite element computation of sloshing frequencies

Frequency	5 × 15 elements	10 × 30 elements	15 × 45 elements
f_1	0.810 Hz (0.04 %)	0.807 Hz (0.01 %)	0.807 Hz (0.01 %)
f_2	1.837 Hz (1.69 %)	1.814 Hz (0.42 %)	1.810 Hz (0.18 %)
f_3	2.687 Hz (6.93 %)	2.565 Hz (1.67 %)	2.532 Hz (0.74 %)
f_4	3.560 Hz (16.0 %)	3.196 Hz (3.82 %)	3.120 Hz (1.67 %)

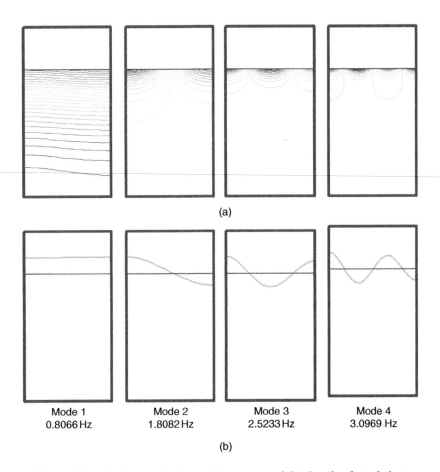

(a)

Mode 1	Mode 2	Mode 3	Mode 4
0.8066 Hz	1.8082 Hz	2.5233 Hz	3.0969 Hz

(b)

Figure 3.20 Sloshing mode shapes: (a) pressure and (b) elevation formulations

3.6   Example: Acoustic Modes in an Open Reservoir

3.6.1   *Cylindrical Acoustic Opened Cavity*

As represented in Figure 3.21, a cylindrical cavity with inner radius R, outer radius R' and height L is considered. It is filled with a compressible fluid of density ρ and speed of sound c, and it opens on the top to the ambient pressure.

The wave equation is formulated in the cylindrical coordinate system:

$$\frac{\omega^2}{c^2}p + \frac{\partial^2 p}{\partial r^2} + \frac{1}{r}\frac{\partial p}{\partial r} + \frac{1}{r^2}\frac{\partial^2 p}{\partial \theta^2} + \frac{\partial^2 p}{\partial z^2} = 0 \ \text{ for } (r,\theta,z) \in [R,R'] \times [0,2\pi] \times [0,L]$$

It is complemented with the following boundary conditions:

$$\left.\frac{\partial p}{\partial r}\right|_{r=R} = 0 \qquad \left.\frac{\partial p}{\partial r}\right|_{r=R'} = 0 \qquad \left.\frac{\partial p}{\partial z}\right|_{z=0} = 0 \qquad p|_{z=L} = 0$$

The acoustic mode shape in terms of pressure is shown to be

$$p_{l,m,n}(r,\theta,z) =$$

$$(J_m(q_{m,n}r/R)Y'_m(q_{m,n}) - Y_m(q_{m,n}r/R)J'_m(q_{m,n}))\cos(m\theta)\cos((l+1/2)\pi z/L)$$

and the associated pulsation is found to be

$$\omega_{l,m,n} = c\sqrt{\left(\frac{(l+1/2)\pi}{L}\right)^2 + \left(\frac{q_{m,n}}{R}\right)^2}$$

where $q_{m,n}$ has been introduced in the previous section.

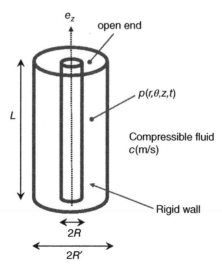

Figure 3.21   Compressible fluid filling a cylindrical opened cavity

3.6.2   2D Axisymmetric Elements with Compressibility

The weighted integral formulation of the problem is proposed for a Fourier expansion of the pressure field, as in the previous example. For any component p_m, it is stated as follows:

$$-\omega^2 \int_R^{R'} \int_0^L \frac{p_m \delta p_m}{c^2} \, r \, drdz + \int_R^{R'} \int_0^L \frac{\partial p_m}{\partial r} \frac{\partial \delta p_m}{\partial r} \, r \, drdz$$

$$+ \int_R^{R'} \int_0^L \frac{\partial p_m}{\partial z} \frac{\partial \delta p_m}{\partial z} \, r \, drdz + m^2 \int_R^{R'} \int_0^L p_m \delta p_m \frac{dr}{r} dz = 0$$

for all δp_m complying with the constraining relation $p(r, z = L) = 0$.

The finite element discretisation is performed with 2D axisymmetric elements, with four nodes and linear shape functions, as presented in Figure 3.19. The 'stiffness' matrices are expressed by Equations (3.63) and (3.64), and the elements of the 'mass' matrix are found to be:

$$\mathbf{m}_{p(k,l)} = \frac{1}{c^2} \frac{ab}{4} \left(r_0 + \frac{\xi_k \xi_l r_0}{3} + \frac{a}{3} (\xi_k + \xi_l) \right) \left(1 + \frac{\eta_k \eta_l}{3} \right)$$

The elementary matrices are assembled using the connectivity matrix presented in Table 3.9. The boundary condition $p_m = 0$ is enforced either by elimination of the degrees-of-freedom associated with the nodes with constrained pressure or by means of Lagrange multipliers. In this case, the weighted integral formulation of the problem is stated as follows:

$$-\omega^2 \int_R^{R'} \int_0^L \frac{1}{c^2} p_m \delta p_m \, rdrdz + \int_R^{R'} \int_0^L \nabla p_m \nabla \delta p_m \, rdrdz$$

$$-\rho\omega^2 \int_R^{R'} \lambda_m |\delta p_m|_{z=L} \, rdr = 0$$

λ_m is the Lagrange multiplier associated with the constraining relation $p_m|_{z=L} = 0$, and it represents the elevation of the free surface. On the other hand, the isobar condition is termed as a weighted integral:

$$\int_R^{R'} \delta \lambda_m p_m|_{z=L} \, rdr = 0$$

The corresponding finite element formulation is given as follows:

$$-\omega^2 \delta \mathbf{P}^T \mathbf{M}_F \mathbf{P} + \delta \mathbf{P}^T \mathbf{M}_F \mathbf{P} - \rho\omega^2 \delta \mathbf{P}^T \mathbf{L}^T \boldsymbol{\Lambda} = 0$$

for all $\delta \mathbf{P}$, with $\mathbf{P}$ assembling *all* pressure degrees-of-freedom and

$$\rho\omega^2 \delta \boldsymbol{\Lambda}^T \mathbf{LP} = 0$$

for all $\delta \boldsymbol{\Lambda}$.

$\mathbf{L}$ yields the finite element approximation of $\int_R^{R'} \delta \lambda p_m|_{z=L} \, rdr$: it has the same expression as the mass matrix $\mathbf{m}_F'$ associated with gravity, as per Equation (3.62), with $g = 1$.

The eigenvalue problem with Lagrange multipliers is formulated as follows:

$$\left(-\omega^2 \begin{bmatrix} \mathbf{M}_F & \rho \mathbf{L}^T \\ \rho \mathbf{L} & 0 \end{bmatrix} + \begin{bmatrix} \mathbf{K}_F & 0 \\ 0 & 0 \end{bmatrix} \right) \begin{Bmatrix} \mathbf{P} \\ \boldsymbol{\Lambda} \end{Bmatrix} = \begin{Bmatrix} 0 \\ 0 \end{Bmatrix}$$

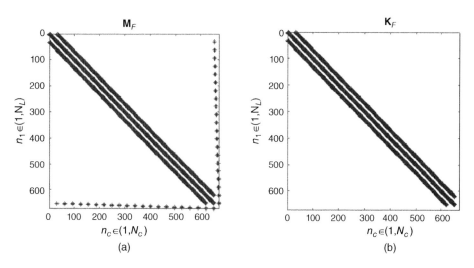

Figure 3.22   (a) Mass and (b) stiffness matrices of the acoustic cavity with opened end (formulation with Lagrange multipliers)

Table 3.12   Eigenfrequencies of the acoustic cavity with free surface: analytical solution and numerical computation

(n, l)	Frequency	Analytical (Hz)	Numerical
$(1, 0)$	f_1	208	209 Hz (0.2 %)
$(1, 1)$	f_2	547	549 Hz (0.3 %)
$(2, 0)$	f_3	552	554 Hz (0.4 %)
$(2, 1)$	f_4	748	751 Hz (0.5 %)

Assembling the elementary matrices according to the connectivity matrix proposed in Table 3.9 yields global matrices with limited bandwidth. Taking into account the zero pressure constraining relation at the opened end with Lagrange multipliers produces additional non-null terms, as illustrated in Figure 3.22.

3.6.3   Acoustic Modes

In what follows, an application is performed with the following data: $R = 1.25$ m, $R' = 2.5$ m, $L = 1.75$ m and $c = 1,250$ m/s and let the symmetric modes of order $m = 1$ be investigated. Finite element computations are carried out with $I \times J = 10 \times 15$ elements, which produces noticeable numerical results in accordance with the analytical solution, as highlighted in Table 3.12. The first four modes correspond to $n = 1, 2$ and $l = 0, 1$: the modulation of the pressure field in the radial and vertical directions is evidenced in Figure 3.23.

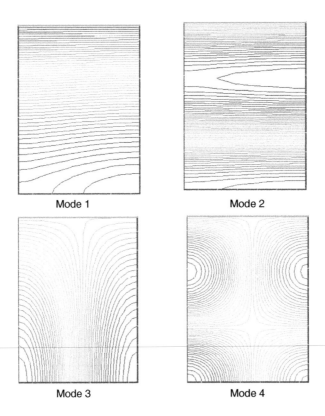

Figure 3.23 Acoustic mode shapes in terms of pressure (acoustic cavity with opened surface)

According to the analytical expression of the acoustic modes, the pressure gradient in the vertical direction is found to be:

$$\frac{\partial p}{\partial z} = -(l+1/2)\pi/L(J_m(q_{m,n}r/R)Y'_m(q_{m,n}) - Y_m(q_{m,n}r/R)J'_m(q_{m,n}))$$

$$\cos(m\theta)\sin((l+1/2)\pi z/L)$$

The expression of the vertical acceleration of the fluid at the free surface is thus:

$$\rho\omega^2\eta|_{z=L} =$$

$$(J_m(q_{m,n}r/R)Y'_m(q_{m,n}) - Y_m(q_{m,n}r/R)J'_m(q_{m,n}))\cos(m\theta)(-1)^l(l+1/2)\pi/L$$

The Lagrange multipliers associated with each eigenmode stand for the elevation of the fluid at $z = L$: as illustrated in Figure 3.24, the numerical computation of $\rho\omega^2\lambda$ with finite elements do compare with the expected analytical values.

3.7   Example: Acoustic Modes in a Closed Reservoir

3.7.1   Rectangular Acoustic Closed Cavity

As sketched in Figure 3.25, the acoustic modes of a closed rectangular cavity of length R and height H, filled with a compressible fluid of density ρ and speed of sound c are considered.

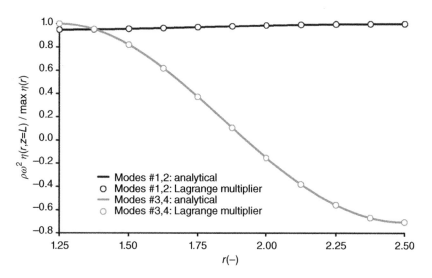

Figure 3.24 Normal pressure gradient at the opened surface: analytical solution and numerical computation with Lagrange multiplier

The wave equation is stated in terms of the pressure field in a Cartesian coordinate system as follows:

$$\frac{\omega^2}{c^2}p + \frac{\partial^2 p}{\partial r^2} + \frac{\partial^2 p}{\partial z^2} = 0 \text{ for } (r, z) \in [0, R] \times [0, L]$$

with boundary conditions standing for fixed walls:

$$\frac{\partial p}{\partial r}\bigg|_{r=0,R} = 0 \qquad \frac{\partial p}{\partial z}\bigg|_{z=0,L} = 0$$

The solution of the problem is written in a separated form $p(r, z) = p_R(r)p_Z(z)$, with $p_R(r)$ and $p_Z(z)$ expressed as a linear superposition of sine and cosine functions $\cos(qr)$, $\sin(qr)$ and $\cos(q'z)$, $\sin(q'z)$. Substituting the latter expression in the former equation and boundary conditions yields the admissible values for q, q':

$$q_n = \frac{n\pi}{R} \qquad q'_m = \frac{m\pi}{L} \qquad \text{with } n \geq 1, \ m \geq 1$$

Hence, the acoustic mode shapes formulated in terms of pressure:

$$p_{n,m}(r, z) = \cos\left(\frac{n\pi r}{R}\right)\cos\left(\frac{m\pi z}{L}\right) \tag{3.65}$$

with the associated pulsations:

$$\omega_{m,n} = c\sqrt{\left(\frac{n\pi}{R}\right)^2 + \left(\frac{m\pi}{L}\right)^2} \tag{3.66}$$

Note that $m = n = 0$ gives a zero-frequency mode with constant null pressure. All modes also have a pressure zero mean value, that is, they comply with the constraining relation:

$$\int_\Omega p \, d\Omega = 0$$

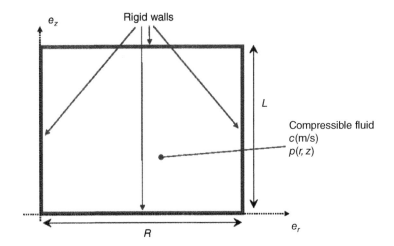

Figure 3.25 Compressible fluid filling a rectangular closed cavity

Enforcing the latter condition in the modal analysis eliminates the rigid pressure mode out of the modal basis.

3.7.2 2D Fluid Elements with Compressibility

The weighted integral formulation of the problem is stated as follows:

$$-\omega^2 \int_0^R \int_0^L \frac{1}{c^2}\, p\delta p \; drdz + \int_0^R \int_0^L \nabla p\, \nabla \delta p \; drdz = 0$$

where $\nabla p\, \nabla \delta p$ is calculated according to:

$$\nabla p\, \nabla \delta p = \frac{\partial p}{\partial r}\frac{\partial \delta p}{\partial r} + \frac{\partial p}{\partial z}\frac{\partial \delta p}{\partial z}$$

The finite element discretisation is performed with elements described in the previous section; see Figure 3.19. With linear shape functions, the analytical expressions of the mass and stiffness matrices are found to be:

$$\mathbf{m}_p(k, l) = \frac{ab}{4c^2}\left(1 + \frac{\xi_k\xi_l}{3}\right)\left(1 + \frac{\eta_k\eta_l}{3}\right)$$

and

$$\mathbf{k}_p(k, l) = \frac{ab}{4}\left(\frac{\xi_k\xi_l}{a^2}\left(1 + \frac{\eta_k\eta_l}{3}\right) + \frac{\eta_k\eta_l}{b^2}\left(1 + \frac{\xi_k\xi_l}{3}\right)\right)$$

Matrices are assembled as described above for the axisymmetric element. The eigenvalue problem (3.31) is solved with the constraining relation set here as $\int_\Omega p/c^2 \; d\Omega = 0$, which is equivalent to a pressure zero mean value if c does not vary over the fluid domain. This condition is enforced in a direct manner with Lagrange multipliers: following Equation (2.20), it is formulated as $\Lambda \mathbf{P} = 0$, where $\Lambda = \langle 1 \rangle \mathbf{M}_F$, $\langle 1 \rangle$ being a row vector with all components equal to 1.

Table 3.13 Eigenfrequencies of the acoustic cavity without free surface: analytical solution and numerical computation

(n, m)	Frequency	Analytical (Hz)	Numerical (Hz)
$(0, 1)$	f_1	600	603
$(1, 0)$	f_2	1,000	1,017
$(1, 1)$	f_3	1,166	1,182
$(0, 2)$	f_4	1,200	1,220
$(1, 2)$	f_5	1,562	1,588
$(0, 3)$	f_6	1,800	1,867
$(0, 3)$	f_7	2,000	2,126
$(1, 3)$	f_8	2,059	2,133

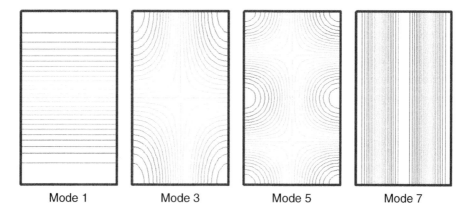

    Mode 1   Mode 3   Mode 5   Mode 7

Figure 3.26 Acoustic mode shapes in terms of pressure (acoustic cavity without free surface)

3.7.3 Acoustic Modes

An application is proposed hereafter with the following data: $R = 0.75$ m $L = 1.25$ m and $c = 1,500$ m/s. Numerical computations are performed with $I \times J = 10 \times 20$ elements. Table 3.13 compares the computed values of the frequencies to the analytical ones. A good agreement is observed for the first modes, whereas discrepancies inevitably arise with higher order modes, even with refined meshes.

Figure 3.26 gives a representation of the pressure field in the acoustic cavity for some particular modes; agreement with the analytical expression is remarkable.

3.8 Example: Acoustic Radiation in Infinite Fluid

3.8.1 Pulsating Ring in Infinite Acoustic Fluid

The propagation of an acoustic wave triggered by a pulsating ring in a bi-dimensional infinite fluid medium is considered in Figure 3.27. The fluid domain is $[R, \infty[\times[0, 2\pi]$, while ρ and c denote the fluid density and speed of sound, respectively.

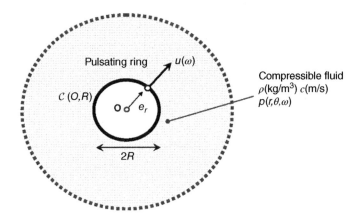

Figure 3.27   Pulsating ring in a 2D infinite acoustic fluid

The Helmholtz equation governs the evolution of the pressure field $p(r, \theta, \omega)$ in the frequency domain. p also complies with a non-reflexion condition at infinity and a coupling condition at $r = R$; as discussed in Chapter 5, it is formulated as follows:

$$\left.\frac{\partial p}{\partial r}\right|_{r=R} = \rho\omega^2 u(\omega)$$

where $-\omega^2 u(\omega)$ is the acceleration imposed on the fluid.

A Fourier expansion of u and p onto angular harmonic functions reads as follows:

$$u(\theta, \omega) = \sum_{m \geq 0} u_m(\omega)\cos(m\theta) \qquad p(r, \theta, \omega) = \sum_{m \geq 0} p_m(\omega)\cos(m\theta)$$

The projection of the Helmholtz equation and of the boundary conditions on the mth term yields the following equation:

$$\frac{\partial^2 p_m}{\partial r^2} + \frac{1}{r}\frac{\partial p_m}{\partial r} + \frac{1}{r^2}\left(\frac{\omega^2}{c^2} - m^2\right)p_m = 0$$

with the boundary condition

$$\left.\frac{\partial p_m}{\partial r}\right|_{r=R} = \rho\omega^2 u_m$$

The solution of the problem should also comply with a non-reflection condition at infinity. According to Axisa and Antunes (2007), the solution of the problem may be expressed as follows:

$$p_m(r, \omega) = \rho c\omega\frac{J_m(\omega r/c) - iY_m(\omega r/c)}{J'_m(\omega R/c) - iY'_m(\omega R/c)}u_r^m(\omega)$$

The pressure at $r = R$ takes the following form:

$$p_m(R, \omega) = -\rho R\mu_m(\omega R/c)\omega^2 u_r^m(\omega)$$

$\mu_m(\omega R/c)$ is termed here as the *vibro-acoustic coefficient*. It is a complex-valued and frequency-dependent scalar, which is calculated as follows:

$$\mu_m(\omega R/c) = -\frac{1}{\omega R/c}\frac{J_m(\omega R/c) - iY_m(\omega R/c)}{J'_m(\omega R/c) - iY'_m(\omega R/c)} \tag{3.67}$$

$-\rho R^2\omega^2\mu_m(\omega R/c)u_m(\omega)$ is the pressure force at the fluid interface; using the real and imaginary parts of μ_m gives the following:

$$-\rho R^2\omega^2\mu_m(\omega R/c)u_m(\omega) =$$

$$-\rho R^2(\omega^2\Re(\mu_m(\omega R/c))\,u_m(\omega) + i\omega(-\omega\Im(\mu_m(\omega R/c))\,u_m(\omega))$$

As further discussed in Chapter 5, the latter expression of the pressure force evidences both *inertial* and *acoustic damping* FSI effects: they are, respectively, represented by the real and imaginary parts of the vibro-acoustic coefficient, namely, by $\Re(\mu_m(\omega R/c))$ and $-\omega\Im(\mu_m(\omega R/c))$.

3.8.2   1D Axisymmetric Element with Radiation Condition

In order to solve the problem with finite elements, the fluid domain is truncated at $r = R'$. The radiation condition may be enforced using the plane or cylindrical wave approximation, and as an example the method is detailed here with the former condition. For the Fourier component of order m, the plane wave condition is stated as follows:

$$\left.\frac{\partial p_m}{\partial r}\right|_{r=R'} = -\frac{i\omega}{c}p_m|_{r=R'}$$

For any m, the weighted integral formulation of the problem reads as follows:

$$-\omega^2\int_R^{R'}\frac{1}{c^2}p_m\delta p_m\,rdr + \int_R^{R'}\frac{\partial p_m}{\partial r}\frac{\partial\delta p_m}{\partial r}\,rdr + m^2\int_R^{R'}p_m\delta p_m\,\frac{dr}{r}$$

$$+i\frac{\omega R'}{c}p_m|_{r=R'}\delta p_m|_{r=R'} = \rho R\omega^2 u_m\delta p_m|_{r=R}$$

Its discretisation is achieved using elements with linear shape functions and two nodes with one degree-of-freedom per node, as represented in Figure 3.28.

Denoting $\lambda_i = r_i/R_i$, $R_i = r_{i+1} - r_i$, the analytical expressions of the elementary matrices are

$$\mathbf{m}^m_{p,i} = \frac{\lambda_i^2}{12c^2}\begin{bmatrix} 4R_i + 1 & 2R_i + 1 \\ 2R_i + 1 & 4R_i + 3 \end{bmatrix}$$

for the 'mass' matrix and,

$$\mathbf{k}^i_{p,o} = (\lambda_i + 1/2)\begin{bmatrix} 1 & -1 \\ -1 & 1 \end{bmatrix}$$

$$\mathbf{k}^i_{p,m} = \begin{bmatrix} (\lambda_i + 1)^2\ln(1 + 1/\lambda_i) - \lambda_i - 3/2 & -\lambda_i(\lambda_i + 1)\ln(1 + 1/\lambda_i) + \lambda_i + 1/2 \\ -\lambda_i(\lambda_i + 1)\ln(1 + 1/\lambda_i) + \lambda_i + 1/2 & \lambda_i^2\ln(1 + 1/\lambda_i) - \lambda_i + 1/2 \end{bmatrix}$$

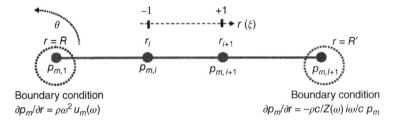

Figure 3.28   Linear fluid element with two nodes and one degree-of-freedom per node

for the 'stiffness' matrices. Matrices $\mathbf{k}_{p,o}^i$ and $\mathbf{k}_{p,m}^i$ yield the 'stiffness' matrix $\mathbf{k}_p^i$ according to $\mathbf{k}_p^i = \mathbf{k}_{p,o}^i + m^2 \mathbf{k}_{p,m}^i$.

The acoustic radiation is described by the *damping matrix*, with the sole contribution of node $I + 1$. Using the plane wave radiation condition, the corresponding elementary matrix is[19]:

$$\mathbf{c}_{p,m}^I = \begin{bmatrix} 0 & 0 \\ 0 & R'/c \end{bmatrix}$$

Finally, the coupling with the imposed acceleration is accounted for as follows:

$$\rho R \omega^2 u_m \delta p_m |_{r=R} = \delta \mathbf{P}_m^T |_{r=R} \mathbf{F}_m$$

where $\mathbf{F}_m^T = \langle \rho R \omega^2 u_m, \; \mathbf{0} \rangle$ is a vector of length $I + 1$ with $\mathbf{0}$ a vector of length I with components equal to zero.

Assembling the matrices yields the following equation:

$$(-\omega^2 \mathbf{M}_F + i\omega \mathbf{C}_F + \mathbf{K}_F) \mathbf{P}_m(\omega) = \mathbf{F}_m(\omega)$$

3.8.3   1D Boundary Elements

The integral representation of the wave propagation equation reads as follows:

$$\frac{1}{2} p(M_o) = \int_0^{2\pi} G(M_o, M) \frac{\partial p}{\partial r}(M) - p(M) \frac{\partial G}{\partial r}(M_o, M) \; R d\theta$$

---

[19] For the cylindrical wave radiation condition, the plane wave condition is represented with the damping matrix of element I for node $I + 1$ *and* an additional term in the elementary 'stiffness' matrix of element I for node $I + 1$:

$$\mathbf{k}_{p,m}^I = \begin{bmatrix} 0 & 0 \\ 0 & 1/2 \end{bmatrix}$$

It arises from the discretisation of the weighted integral formulation of the wave equation endowed with the radiation condition $\dfrac{\partial p}{\partial r} = -\left(\dfrac{1}{2r} + \dfrac{i\omega}{c} \right) p$ at $r = R'$, this integral representation being formulated as:

$$-\omega^2 \int_R^{R'} \frac{p_m \delta p_m}{c^2} r dr + \int_R^{R'} \frac{\partial p_m}{\partial r} \frac{\partial \delta p_m}{\partial r} r dr + m^2 \int_R^{R'} p_m \delta p_m \frac{dr}{r}$$

$$+ i \frac{\omega R'}{c} p_m|_{r=R'} \delta p_m|_{r=R'} + \frac{1}{2} p_m|_{r=R'} \delta p_m|_{r=R'} = \rho R \omega^2 u_m \delta p_m|_{r=R}$$

for any point M_o belonging to the circle centred on O with radius R. The Green function and its derivative are expressed in the cylindrical coordinate system as follows:

$$G(\mathbf{r}, \mathbf{r}_o; \omega) = -\frac{i}{4}H_0^-(\omega|\mathbf{r} - \mathbf{r}_o|/c) \qquad \frac{\partial G}{\partial n}(\mathbf{r}, \mathbf{r}_o; \omega) = \frac{-i\omega}{4c}H_1^-(i\omega|\mathbf{r} - \mathbf{r}_o|/c)\frac{\partial|\mathbf{r} - \mathbf{r}_o|}{\partial n}$$

with

$$|\mathbf{r} - \mathbf{r}_o| = \sqrt{(r\cos\theta - r_o\cos\theta_o)^2 + (r\sin\theta - r_o\sin\theta_o)^2}$$

The discretisation of the integral equation is performed using a direct method with curved boundary elements, as depicted in Figure 3.29, so that $\int_0^{2\pi}(\bullet)d\theta = \sum_{j=1}^{j=J}\int_{\theta_j}^{\theta_{j+1}}(\bullet)d\theta$. On each interval $[\theta_j, \theta_{j+1}]$, the approximation of p and $\frac{\partial p}{\partial r}$ is

$$p|_{[\theta_j,\theta_{j+1}]} = \mathbf{N}_j\mathbf{P}_j = \mathbf{N}_j\mathbf{\Lambda}_j\mathbf{P} \qquad \frac{\partial p}{\partial r}\bigg|_{[\theta_j,\theta_{j+1}]} = \mathbf{N}_j\frac{\partial\mathbf{P}}{\partial r}\bigg|_j = \mathbf{N}_j\mathbf{\Lambda}_j\frac{\partial\mathbf{P}}{\partial r}$$

where $\mathbf{N}_j$ is the element shape function, $\mathbf{P}_j$ and $\frac{\partial\mathbf{P}}{\partial r}\big|_j$ are the nodal values of the pressure and its gradient, and $\mathbf{\Lambda}_j$ is the localisation matrix associated with element j.

The discrete integral equation is expressed as follows:

$$\sum_{j=1,J}\left(\frac{\delta_{ij}}{2}\mathbf{N}_j(\theta_j) + \int_{\theta_j}^{\theta_{j+1}}\mathbf{N}_j(\theta)\frac{\partial G}{\partial r}(\theta_i, \theta)\, Rd\theta\right)\mathbf{\Lambda}_j\mathbf{P}$$

$$= \sum_{j=1,J}\int_{\theta_j}^{\theta_{j+1}}\mathbf{N}_j(\theta)G(\theta_i, \theta)\, Rd\theta\mathbf{\Lambda}_j\frac{\partial\mathbf{P}}{\partial r}$$

It is Equation (3.53), where the lines of $\mathbf{D}$ and $\mathbf{S}$ are assembled according to

$$\mathbf{D}_i = \sum_{j=1,J}\mathbf{d}_{ij}\mathbf{\Lambda}_j \qquad \mathbf{S}_i = \sum_{j=1,J}\mathbf{s}_{ij}\mathbf{\Lambda}_j$$

with

$$\mathbf{d}_{ij} = \frac{\delta_{ij}}{2}\mathbf{N}_j(\theta_j) + \int_{\theta_j}^{\theta_{j+1}}\mathbf{N}_j(\theta)\frac{\partial G}{\partial r}(\theta_i, \theta)\, Rd\theta$$

and

$$\mathbf{s}_{ij} = \int_{\theta_j}^{\theta_{j+1}}\mathbf{N}_j(\theta)G(\theta_i, \theta)\, Rd\theta$$

Using curved boundary elements is restricted to sufficiently regular surfaces, and it may not always be suited to all practical applications, for instance, when surfaces of less regularity are to be handled, and using straight elements as depicted in Figure 3.29 may be more adapted. In such cases, the integral formulation of the problem may be recast as follows:

$$\Pi(M_o)p(M_o) = \int_\Gamma G(M_o, M)\frac{\partial p}{\partial n}(M) - p(M)\frac{\partial G}{\partial n}(M_o, M)\, d\Gamma(M)$$

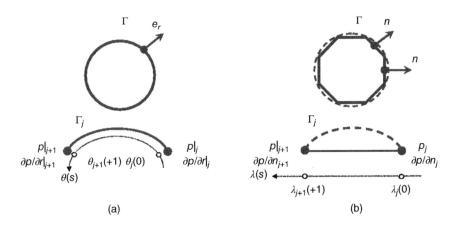

Figure 3.29 (a) Curved and (b) straight boundary elements

where $\Pi(M_o)$ is[20]:

$$\Pi(M_o) = \begin{cases} 1 & M_o \in \Omega \\ 1 - \int_\Gamma \frac{\partial H}{\partial n}(M_o, M)\, d\Gamma(M) & M_o \in \Gamma \\ 0 & M_o \notin \Omega \cup \Gamma \end{cases}$$

The discretisation of the integral formulation is performed according to

$$\sum_{j=1,J} \delta_{ij}\Pi(M_j)p(M_j) + \int_{\Gamma_j} p(M)\frac{\partial G}{\partial n}(M_i, \theta)\, d\Gamma_j(M)$$

$$= \sum_{j=1,J} \int_{\Gamma_j} G(M_i, M)\frac{\partial p}{\partial n}(M)\, d\Gamma_j(M) \qquad \forall i \in [1, J]$$

In the present case, Equation (3.53) is arrived at using boundary elements with two nodes and linear shape functions: as depicted in Figure 3.29, on element j, the nodal values are $p|_j, p|_{j+1}$ and $\left.\frac{\partial p}{\partial n}\right|_j, \left.\frac{\partial p}{\partial n}\right|_{j+1}$.

Following Niu *et al.* (2007) and Ozgener and Ozgener (2000), the computation of the matrices is performed with numerical schemes which circumvent the singularity of the integrals.

The pressure gradient is imposed by the acceleration of the deformable boundary $-\omega^2 U$ and the pressure is retrieved from Equation (3.53) according to

$$\mathbf{P} = \rho\omega^2 \mathbf{D}^{-1}\mathbf{SU}$$

---

[20] For bi-dimensional problems, as in this example, $H(M_o, M)$ is

$$H(M_o, M) = -\frac{1}{2\pi}\ln(|M_oM|)$$

The proposed formulation handles various configurations: in particular, $\Pi(M_o) = 1/2$ for a sufficiently regular surface Γ.

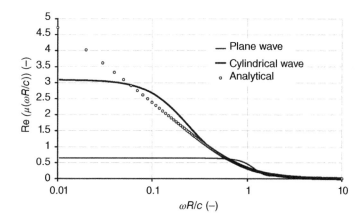

Figure 3.30   Axisymmetric pulsation and acoustic scattering: inertial effect

The projection of the pressure and displacement fields onto the harmonic function $\cos(m\theta)$ yields the vibro-acoustic coefficient μ_m. The latter is calculated according to Leblond and Sigrist (2010):

$$\mu_m(\omega R/c) \approx \frac{2\pi}{J}\frac{1}{\pi_n}\mathbf{C}(m\theta)^\mathsf{T}\mathbf{D}^{-1}(\theta,\omega)\mathbf{S}(\theta,\omega)\mathbf{C}(m\theta)/R \tag{3.68}$$

where $\mathbf{C}(m\theta)^\mathsf{T} = \langle\cos(m\theta_j)\rangle_{j\in[1,J]}$ and $\pi_0 = 2\pi$, $\pi_n = \pi$, $\forall n > 1$.

3.8.4   Acoustic Radiation

A numerical application is proposed with $R = 0.5$ m, $\rho = 1,000$ kg/m^3 and $c = 1,500$ m/s. The fluid domain is bounded at $R' = 3R$ and it is discretised with $I = 30$ elements. The forcing term corresponds to an axisymmetric unit pulsation throughout the frequency range of interest (0–5,000 Hz), that is, $u_0 = 1$ for all ω. The numerical values of the real and imaginary parts of the mass coefficient $\mu(\omega R/c)$ are compared with the analytical expressions for Fourier component $m = 0$ in Figures 3.30 and 3.31.

The plane wave and the cylindrical wave approximations are compared: as expected, the latter condition offers results which are in better agreement with the analytical solution. However, discrepancies are noted for low wave numbers $\omega R/c \ll 1$, that is, for the description of the inertial effect. Computations with a larger fluid domain ($R' = 10R$ with $I = 100$ and $R' = 30R$ with $I = 300$) yield better results, as emphasised in Figure 3.32, but convergence is slow.

The finite element description of acoustic radiation reaches in this case a limit for low wave numbers, and, as evidenced in Figure 3.33, it may be bypassed using boundary elements: the numerical computation of $\mu_0(\omega R/c)$ according to Equation (3.68) with a BEM reaches a noticeable accuracy over the whole frequency range, while requiring a few elements over the circumference (in the present case, the computation is performed with $I = 36$ elements).

Equation (3.68) indicates that the evaluation of the vibro-acoustic coefficient, and subsequently the computation of the pressure field, requires the inversion of matrix $\mathbf{D}$. As evoked in

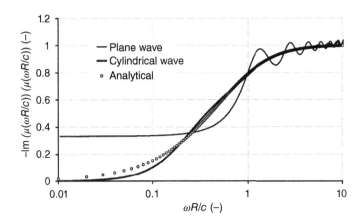

Figure 3.31 Axisymmetric pulsation and acoustic scattering: radiation effect

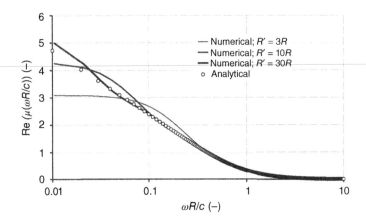

Figure 3.32 Acoustic scattering: influence of the fluid domain extension

Section 3.4.4, for exterior problems, **D** is singular at certain frequencies, namely, the eigenfrequencies of the corresponding interior problem with a zero pressure condition at the boundary.

The singularity of matrix **D** may be eluded with the use of CHIEF points: according to Seybert (1987), they are chosen randomly in the interior domain with the restriction that they should not be placed on or near a nodal surface of the interior modes. In the present case, the mode shapes of a few acoustic modes of the interior problem in terms of pressure are represented in Figure 3.34.

Table 3.14 compares the computed non-dimensional pressure on the ring surface to the analytical value $Rp(r = R, \omega)/\rho c^2 = -(R\omega/c)^2 \mu_0(\omega R/c)$, therefore, illustrating how the singularity of matrix **D** may influence the accuracy of the BEM computation. Computations are performed for a non-singular frequency ($\omega R/c = 1$) and for a singular frequency ($\omega R/c = 2.4048$) for various discretisations: in the latter case, the use of CHIEF points enhances the

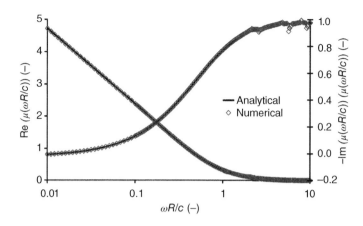

Figure 3.33   Acoustic scattering: BEM computation

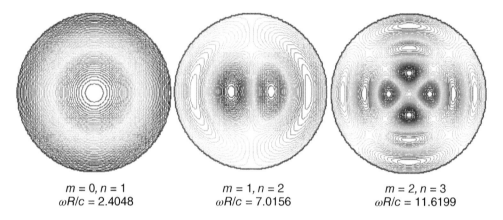

| $m = 0, n = 1$ | $m = 1, n = 2$ | $m = 2, n = 3$ |
| $\omega R/c = 2.4048$ | $\omega R/c = 7.0156$ | $\omega R/c = 11.6199$ |

Figure 3.34   Pressure mode shapes of an interior problem. The interior problem which is associated with the example depicted in Figure 3.27 is formulated in terms of pressure as $(\Delta + \frac{\omega^2}{c^2})p(r, \theta) = 0$ for $r \in [0, R]$ with the boundary condition $p(r = R, \theta) = 0$ for $\theta \in [0, 2\pi]$. The corresponding pressure field is found to be $p_{n,m} = J_m(\omega_{n,m}R/cr/R)\cos(m\theta)$ or $p_{n,m} = J_m(\omega_{n,m}R/cr/R)\sin(m\theta)$ and the associated pulsation $\omega_{n,m}$ are such that $J_m(\omega_{n,m}R/c) = 0$, where J_m is the Bessel function of the first kind and of order m. The figure features some of the acoustic mode shape in terms of pressure and thereby identifies the nodal lines

accuracy, especially for a coarse mesh, while in the former case, the numerical results are only dependent on discretisation.

CHIEF points are added in the integral formulation as the number of frequencies of interest increases, so that the conditioning of matrix $\mathbf{D}(\omega R/c)$ does not significantly overrate a specified threshold. Figure 3.35 indicates that without CHIEF points, $\mathbf{D}$ becomes ill-conditioned with respect to matrix inversion – it is even close to singular for the pulsations of the interior problem.

The conditioning number of matrix $\mathbf{D}$ derived from the direct boundary element discretisation of the problem represented in Figure 3.27 is plotted versus the wave number in Figure 3.35.

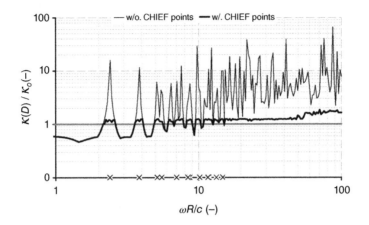

Figure 3.35 Conditioning of matrices obtained from BEM discretisation

Table 3.14 Non-dimensional wall pressure: analytical solution and numerical computation using BEM with and without CHIEF points

	$\omega R/c = 1$					
Analytical			$Rp/\rho c^2 = -0.3331 + 0.7919\,i$			
Numerical	$I = 36$		$I = 90$		$I = 180$	
	$\Re$	$\Im$	$\Re$	$\Im$	$\Re$	$\Im$
w/o. CHIEF points	−0.3321	+0.7932	−0.3328	+0.7922	−0.3329	+0.7920
w. CHIEF points	−0.3322	+0.7931	−0.3328	+0.7922	−0.3329	+0.7920
	$\omega R/c = 2.4048$					
Analytical			$Rp/\rho c^2 = -0.4498 + 2.2722\,i$			
Numerical	$I = 36$		$I = 90$		$I = 180$	
	$\Re$	$\Im$	$\Re$	$\Im$	$\Re$	$\Im$
w/o. CHIEF points	−0.0242	+16.980	−1.0853	+5.8604	−0.7286	+3.7153
w. CHIEF points	−0.4513	+2.2746	−0.4501	+2.2726	−0.4499	+2.2724

Crosses on the abscissae indicate the irregular frequencies and κ_o refers to a reference value of the conditioning number which ensures the accuracy of the matrix inversion. When the matrix is assembled without CHIEF points on the one hand, the conditioning number may overrate the desired threshold, particularly for large values of the wave number: as a consequence, the accuracy of the BEM computation becomes poor. For irregular frequencies, the accuracy is even questionable, as evidenced in Table 3.14. On the other hand, when the matrix is assembled with CHIEF points, the conditioning number remains within bounds which ensure the accuracy of the BEM computation, even at irregular frequencies.

References

Abramowitz M and Stegun IA 1970 *Handbook of Mathematical Functions*. Dover Publications.

Aliabadi MH 2001 *The Boundary Element Method*. John Wiley & Sons, Ltd.

Axisa F 2007 *Modelling of Mechanical Systems – Fluid-Structure Interaction*. Elsevier.

Bayliss A, Gunzburger M, and Turkel E 1982 Boundary conditions for the numerical solution of elliptic equations in exterior regions. *SIAM Journal on Applied Mathematics*, **42**, 430–451.

Beer G 2001 *Programming the Boundary Element Method – An Introduction for Engineers*. John Wiley & Sons, Inc.

Bermudez A, Duran R, Muschietti MA, Rodriguez R, and Solomin J 1995 Finite element vibration analysis of fluid-solid systems without spurious modes. *Journal of Numerical Analysis*, **32**, 1280–1295.

Bettess P 1992 *Infinite Elements*. Penshaw Press.

Brunner D, Junge M, and Gaul L 2009 A comparison of FE-BE coupling schemes for large-scale problems with fluid-structure interaction. *International Journal of Numerical Methods in Engineering*, **77**, 664–688.

Chassaing P 2000 *Mécanique des fluides*. Cépaduès-Éditions.

Colton D and Kress R 1983 *Integral Equation Methods in Scattering Theory*. John Wiley & Sons, Inc..

Ferziger JH and Peric M 2002 *Computational Methods for Fluid Dynamics*. Springer-Verlag.

Geers TL and Zhang P 1994a Doubly asymptotic approximation for submerged structures with internal fluids: formulation. *Journal of Applied Mechanics*, **61**, 893–899.

Geers TL and Zhang P 1994b Doubly asymptotic approximation for submerged structures with internal fluids: evaluation. *Journal of Applied Mechanics*, **61**, 900–906.

Gingold R and Monaghan JJ 1977 Smoothed particle hydrodynamics: theory and application to non-spherical stars. *Monthly Notices of the Royal Astronomical Society*, **181**, 375–389.

Harari I and Djellouli R 2004 Analytical study of the effect of wave number on the performance of local absorbing boundary conditions for acoustic scattering. *Applied Numerical Mathematics*, **50**, 15–47.

Ibrahim RA 2005 *Liquid Sloshing Dynamics: Theory and Applications*. Cambridge University Press.

Leblond C and Sigrist JF 2010 A versatile approach to the study of submerged two-dimensional thin shell transient response. *Journal of Sound and Vibration*, **329**, 56–71.

Lucy L 1977 A numerical approach to the testing of the fission hypothesis. *Astronomical Journal*, **82**, 1013–1024.

Manolis GD and Polyzos D 2009 *Recent Advances in Boundary Element Methods*. Springer-Verlag.

Mindlin RD and Bleich HH 1953 Response of an elastic cylindrical shell to a transverse step shock wave. *Journal of Applied Mechanics*, **20** 189–195.

Niu Z, Cheng C, Zhou H, and Hu Z 2007 Analytic formulations for calculating nearly singular integrals in two-dimensional BEM. *Engineering Analysis with Boundary Elements*, **31**, 949–964.

Ozgener B and Ozgener HA 2000 Gaussian quadratures for singular integrals in BEM with application to the 2D modified Helmholtz equation. *Engineering Analysis with Boundary Elements*, **24**, 259–269.

Qing W and Bo-Hou X 1990 On the uniqueness of boundary integral equation for exterior Helmholtz problem. *Applied Mathematics and Mechanics*, **11**, 965–971.

Rosen D and Cormack D 1993 Singular and near-singular integrals in the BEM: a global approach. *Journal of Applied Mathematics*, **53**, 340–357.

Sayhi MN, Ousset Y, and Verchery G 1981 Solution of radiation problems by collocation of integral formulation in terms of single and double layer potentials. *Journal of Sound and Vibration*, **74**, 187–204.

Schenck HA 1968 Improved integral formulation for acoustic radiation problems. *Journal of the Acoustical Society of America*, **44**, 41–58.

Scuderi A 2008 On the computation of nearly singular integrals in 3D BEM collocation. *International Journal for Numerical Methods in Engineering*, **74**, 1733–1770.

Selfridge A 1985 Approximate material properties in isotropic materials. *Transaction on Sonics and Ultrasonics*, **32**, 381–394.

Seybert AF 1987 The use of CHIEF to obtain solutions for acoustic radiation using boundary integral equations. *Journal of the Acoustical Society of America*, **81**, 1299–1306.

Soize C and Ohayon R 2014 *Advanced Computational Vibroacoustics: Reduced-Order Models and Uncertainty Quantification*. Cambridge University Press.

Stewart I 2013 *Seventeen Equations that Changed the World*. Profile Books.

Zhang YO, Zhang T, Ouyong H, and Li TY 2014 SPH simulations of sound propagation and interference. In *Proceedings of the 5th International Conference on Computational Methods (ICCM 2014)*.

4

Inertial Coupling

The objective of this chapter is to study the vibrations of an elastic structure containing an incompressible and non-viscous fluid. If in addition gravity effects are discarded, fluid–structure interaction (FSI) is found of inertial nature. As far as the discretised formulation of the problem is concerned, either using a finite element or a boundary element method, it can be described in terms of the sole nodal vector of the structure acceleration, acting on the so-called *added mass matrix*. The latter accounts for the unsteady pressure force exerted on the walls. Some aspects of the inertial coupling are investigated on a few examples together with some important points regarding the computation of the added mass matrix.

Figure 4.1 Fluid–structure coupling. The vibrations of an elastic structure are influenced by pressure forces induced at the interface by the fluid motion. The inertial effect is one of the manifestations of FSI which reveals of major importance in the case of dense and nearly incompressible fluids. Except in a few geometries of academic interest, the problem must be solved using numerical techniques, the finite element method (FEM) being by far the most commonly used in the industrial context. *Source*: © Jean-François Sigrist

Fluid–Structure Interaction: An Introduction to Finite Element Coupling, First Edition. Jean-François Sigrist.
© 2015 John Wiley & Sons, Ltd. Published 2015 by John Wiley & Sons, Ltd.
Companion Website: www.wiley.com/go/sigrist

4.1 Mathematical Modelling

As already outlined in Chapter 1, FSI reduces to inertial coupling provided the fluid is assumed to be non-viscous and incompressible, gravity being discarded. Although not generally valid, such a simplified model is found particularly convenient in the case of dense and weakly compressible fluids, typically liquids.

The analytical formulation of the problem is based on a set of partial differential equations expressed in terms of displacement as the structure is concerned and in terms of pressure as the fluid is concerned. Restricting the analysis to the case of small displacements and strains, the displacement and pressure fields may be considered indifferently Lagrangian or Eulerian, while the governing equations and associated boundary conditions are linearised about the static equilibrium of both the solid and fluid parts of the coupled problem.

Referring to the notations of Figure 4.2, let Σ be the structure domain of boundary $\partial\Sigma = \Gamma_o \cup \Gamma_\sigma \cup \Gamma$, with outer normal $\boldsymbol{\nu}$, and let Ω stand for the fluid domain. $\Gamma = \partial\Omega$ is the fluid boundary as well as the interface with the structure with outer normal $\mathbf{n}$. The material properties of the structure are Young's modulus E and Poisson's coefficient v, while the density of the structure or the fluid is denoted ρ – without further distinction for the sake of simplicity: in what follows, ρ featured in the structure (or fluid) equation implicitly refers to the structure (or fluid) density.

As already described in Chapter 2, the harmonic vibrations of the structure are accounted for using the Navier equation, formulated here for a tri-dimensional problem, and associated with clamped/free end boundary conditions – though other support conditions could also be considered. The equations of motion are repeated below for convenience:

$$-\omega^2 \rho u_i - \frac{\partial \sigma_{ij}(\mathbf{u})}{\partial x_j} = 0 \text{ in } \Sigma \qquad u_i = 0 \text{ on } \Gamma_o \qquad \sigma_{ij}(\mathbf{u})v_j = 0 \text{ on } \Gamma_\sigma$$

where ρ is the structure density.

The motion of the structure is influenced by the fluid through the unsteady pressure force exerted at the interface:

$$\sigma_{ij}(\mathbf{u})n_j = pn_i \text{ on } \Gamma$$

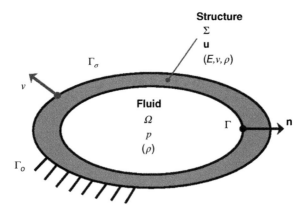

Figure 4.2 Incompressible perfect fluid embedded within an elastic structure

On the other hand, as discussed in the previous chapter, for incompressible flows, the pressure satisfies the Laplace equation throughout the fluid domain:

$$\frac{\partial^2 p}{\partial x_i \partial x_i} = 0 \text{ in } \Omega$$

The coupling condition with the deformable structure may be expressed as follows:

$$\frac{\partial p}{\partial x_i} n_i = \rho \omega^2 u_i n_i \text{ on } \Gamma$$

where ρ is the fluid density. It states that, at the interface, the fluid and the structure have the same acceleration in the normal direction.

In order for the problem to be well posed, the fluid equations are complemented with the following conditions:

- Condition of *uniqueness*: when it exists, the solution to the Laplace equation with the coupling condition formulated above is not unique. When $p(\mathbf{x})$ is a solution of the problem, $p'(\mathbf{x}) = p(\mathbf{x}) + P_o$ with any constant P_o is also a solution. An additional condition is therefore needed for the pressure field to be non-equivocally determined, and this condition may take various forms. From the physical point of view, the situation depicted on Figure 4.2 is a mere abstract representation of real problems. A more realistic condition arises from the consideration that pressure is imposed by an external device (e.g. through any orifice in the volume). There exists a location in the fluid domain where the pressure is P_o: a condition such as $p(\mathbf{x}_o) = 0$, $\mathbf{x}_o \in \Omega$ suffices as a uniqueness condition. Sticking to the ideal case of a totally enclosed fluid volume, and considering that the fluid state is described by pressure fluctuations, it is mathematically sound to assume a zero-mean value condition as a model condition for the uniqueness of the pressure field. Thus,

$$\int_\Omega p \, d\Omega = 0 \tag{4.1}$$

- Condition of *existence*: for a solution of $\Delta p = 0$ on Ω with $\frac{\partial p}{\partial n} = \rho \omega^2 \mathbf{u} \cdot \mathbf{n}$ on Γ to exist, a condition is to be verified by the pressure normal gradient imposed on the boundary. This condition is derived from the divergence theorem:

$$\int_\Omega \Delta p \, d\Omega = 0 = \int_\Gamma \frac{\partial p}{\partial n} \, d\Gamma = \rho \omega^2 \int_\Gamma \mathbf{u} \cdot \mathbf{n} \, d\Gamma$$

Formulated in terms of displacement, it is referred to in what follows as a *compatibility condition* and it reads as follows:

$$\int_\Gamma u_i n_i \, d\Gamma = 0 \tag{4.2}$$

These additional conditions ensure that the pressure and displacement formulated fluid problems are equivalent with the following meaning:

$$\frac{\partial^2 p}{\partial x_i \partial x_i} = 0 \Longleftrightarrow \frac{\partial \xi_k}{\partial x_k} = 0 \text{ with } \rho \omega^2 \xi_i = \frac{\partial p}{\partial x_i} \text{ and } \omega \neq 0$$

Equation (4.1) warrants a coupled formulation which is free of zero-frequency mode, while Equation (4.2) is interpreted as a condition of volume conservation for the fluid domain, when it is subjected to deformations of its boundaries.

Turning back to the problem outlined by Figure 4.2, the following weighted integral formulation is arrived at.

Find $\mathbf{u} \in \mathcal{U}, p \in \mathcal{P}$ and ω such that:

$$-\omega^2 \int_\Sigma \rho u_i \, \delta u_i \, d\Sigma + \int_\Sigma \sigma_{ij}(\mathbf{u}) \, \varepsilon_{ij}(\delta \mathbf{u}) \, d\Sigma - \int_\Gamma p n_i \delta u_i \, d\Gamma = 0$$

for all $\delta \mathbf{u}$ in $\mathcal{U}$, where:

$$\varepsilon_{ij}(\mathbf{u}) = \frac{1}{2}\left(\frac{\partial u_i}{\partial x_j} + \frac{\partial u_j}{\partial x_i}\right)$$

$$\sigma_{ij}(\mathbf{u}) = \frac{E}{1+v}\left(\frac{v}{1-2v}\varepsilon_{kk}(\mathbf{u})\delta_{ij} + \varepsilon_{ij}(\mathbf{u})\right)$$

and:

$$\int_\Omega \frac{\partial p}{\partial x_i}\frac{\partial \delta p}{\partial x_i} = \rho\omega^2 \int_\Gamma u_i n_i \delta p \, d\Gamma$$

for all $\delta p \in \mathcal{P}$, with:

$$\mathcal{U} = \left\{\mathbf{u} \in H^1(\Sigma), \mathbf{u}|_{\Gamma_o} = \mathbf{0}, \int_\Gamma u_i n_i \, d\Gamma = 0\right\}$$

and $\mathcal{P} = \{p \in H^1(\Omega), \int_\Omega p \, d\Omega = 0\}$.

4.2 Added Mass Matrix

4.2.1 Coupling Matrix

As already established in Chapters 2 and 3, the discretisation of Σ and Ω produces a mesh composed of fluid and structure finite elements:

$$\Sigma \simeq \bigcup_e \Sigma_e \qquad \Omega \simeq \bigcup_e \Omega_e$$

The discretisation of elementary mass and stiffness terms of the weighted integral formulation is obtained with the following matrices:

$$\int_\Sigma \rho u_i \, \delta u_i \, d\Sigma \rightarrow \delta \mathbf{u}^\mathsf{T} \mathbf{M}_S \mathbf{U} \tag{4.3}$$

$$\int_\Sigma \sigma_{ij}(\mathbf{u}) \, \varepsilon_{ij}(\delta \mathbf{u}) \, d\Sigma \rightarrow \delta \mathbf{U}^\mathsf{T} \mathbf{K}_S \mathbf{U} \tag{4.4}$$

$$\int_\Omega \frac{\partial p}{\partial x_i}\frac{\partial \delta p}{\partial x_i} \, d\Omega \rightarrow \delta \mathbf{P}^\mathsf{T} \mathbf{K}_F \mathbf{P} \tag{4.5}$$

The approximation of the fluid–structure interface with surface elements is represented by

$$\Gamma \simeq \bigcup_{e=1}^{e=E} \Gamma_e$$

Without loss of generality regarding the definition of the coupling, an ideal pairing of fluid and structure elements at the interface is considered, as represented in Figure 4.3, so that $\Gamma_e = \Sigma_e \cap \Omega_e$.

In the weighted integral formulation, the coupling terms

$$\int_\Gamma p n_i \, \delta u_i \, d\Gamma$$

and

$$-\rho \omega^2 \int_\Gamma u_i \, n_i \, \delta p \, d\Gamma$$

describe the mechanical exchanges between the fluid and the structure.

The first coupling term stands for the virtual work of the pressure force at the interface; it is calculated as follows:

$$\int_\Gamma p n_i \, \delta u_i \, d\Gamma \simeq \sum_{e=1}^{e=E} \int_{\Gamma_e} p n_i \, \delta u_i \, d\Gamma_e$$

With $\mathbf{N}_S$ and $\mathbf{N}_F$ denoting the shape functions associated with the structure and fluid elements, respectively, the approximated displacement and the pressure fields are given by $\mathbf{u} = \mathbf{N}_S^e \mathbf{U}_e$

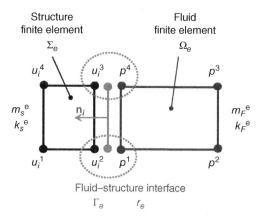

Structure finite element
Σ_e

Fluid finite element
Ω_e

u_i^4 u_i^3 p^4 p^3

m_S^e n_i m_F^e
k_S^e k_F^e

u_i^1 u_i^2 p^1 p^2

Fluid–structure interface
Γ_e Γ_e

Figure 4.3 Finite element coupling. An ideal matching between fluid and structure elements is considered here: it allows for a straightforward definition of the elementary fluid–structure coupling matrix $\mathbf{r}_e$, according to Equation (4.6). In practice, achieving such a simple coupling is not always possible: a more general configuration occurs when one structure (or fluid) element faces several fluid (or structure) elements. In such a case, the computation of the elementary coupling matrix $\mathbf{r}_e$ may be performed with the aid of an interpolation scheme, but the accuracy is usually better with a one-to-one element coupling. When assembled, the elementary matrices produce the fluid–structure coupling matrix $\mathbf{R}$ which accounts for the mechanical energy exchanges between the structure and the fluid on the whole interface Γ

and $p = \mathbf{N}_F^e \mathbf{P}_e$. The finite element discretisation of the elementary coupling term reads as follows:

$$\int_{\Gamma_e} p n_i \, \delta u_i \, d\Gamma_e = \delta \mathbf{U}_e^{\mathsf{T}} (\int_{\Gamma_e} \mathbf{N}_S^{e\mathsf{T}} \, \mathbf{n}_e \, \mathbf{N}_F^e \, d\Gamma_e) \mathbf{P}_e$$

$\mathbf{r}_e$ denotes the *elementary fluid–structure coupling matrix*; it is evaluated at the interface element according to:

$$\mathbf{r}_e = \int_{\Gamma_e} \mathbf{N}_S^{e\mathsf{T}} \, \mathbf{n}_e \, \mathbf{N}_F^e \, d\Gamma_e \tag{4.6}$$

In agreement with the notations of Figures 4.2 and 4.3, $\mathbf{n}_e$ is the unit normal on Γ_e. The discretisation of the coupling term is written as follows:

$$\int_{\Gamma} p n_i \, \delta u_i \, d\Gamma \to \delta \mathbf{U}^{\mathsf{T}} \mathbf{R} \mathbf{P}$$

$\mathbf{R}$ denotes the *fluid–structure coupling matrix* which is assembled according to the standard formula:

$$\mathbf{R} = \sum_{e=1}^{e=E} \mathbf{\Lambda}_S^{e\mathsf{T}} \mathbf{r}_e \mathbf{\Lambda}_F^e \tag{4.7}$$

where $\mathbf{\Lambda}_S^e$ and $\mathbf{\Lambda}_F^e$ denote the structure and fluid localisation matrices associated with the finite elements Σ_e and Γ_e.

The second coupling term immediately follows:

$$-\rho\omega^2 \int_{\Gamma} u_i \, n_i \delta p \, d\Gamma \to -\rho\omega^2 \delta \mathbf{P}^{\mathsf{T}} \mathbf{R}^{\mathsf{T}} \mathbf{U}$$

4.2.2   Added Mass Matrix

The additional constraints (4.1) and (4.2) are discretised as follows:[1]

$$\int_{\Gamma} u_i n_i \, d\Gamma = \mathbf{L}_S \mathbf{U} \qquad \int_{\Omega} p \, d\Omega = \mathbf{L}_F \mathbf{P}$$

with $[\mathbf{L}_S] = \langle \mathbf{1} \rangle [\mathbf{R}]^{\mathsf{T}}$ and $[\mathbf{L}_F] = \langle \mathbf{1} \rangle [\mathbf{M}_F^*]$, where the 'mass' matrix is non-dimensional and is evaluated assuming $c = 1$ m/s.

Gathering all the partial results obtained so far, it is possible to formulate the matrix equation governing the discretised form of the fluid–structure problem. As a first step, the structure and the fluid parts are described using two distinct dynamical equations, where the structure is loaded by the pressure at the interface and the fluid is loaded by the normal acceleration of the interface:

$$\left(-\omega^2 \begin{bmatrix} \mathbf{M}_S & \mathbf{0} \\ \mathbf{0} & \mathbf{0} \end{bmatrix} + \begin{bmatrix} \mathbf{K}_S & -\mathbf{L}_S^{\mathsf{T}} \\ -\mathbf{L}_S & \mathbf{0} \end{bmatrix} \right) \left\{ \begin{matrix} \mathbf{U} \\ \mathbf{\Lambda}_u \end{matrix} \right\} = \left\{ \begin{matrix} \mathbf{R}\mathbf{P} \\ \mathbf{0} \end{matrix} \right\}$$

and:

$$\begin{bmatrix} \mathbf{K}_F & -\mathbf{L}_F^{\mathsf{T}} \\ -\mathbf{L}_F & \mathbf{0} \end{bmatrix} \left\{ \begin{matrix} \mathbf{P} \\ \mathbf{\Lambda}_p \end{matrix} \right\} = \rho\omega^2 \left\{ \begin{matrix} \mathbf{R}^{\mathsf{T}}\mathbf{U} \\ \mathbf{0} \end{matrix} \right\}$$

---

[1] The constraint relations representing the clamped boundary condition for the structure may also be accounted for using Lagrange multipliers, but their contribution is not made explicit here for the sake of clarity.

With $\mathbb{U} = \begin{Bmatrix} \mathbf{U} \\ \Lambda_u \end{Bmatrix}$ and $\mathbb{P} = \begin{Bmatrix} \mathbf{P} \\ \Lambda_p \end{Bmatrix}$ standing for nodal unknowns of the system, the fluid–structure coupling matrix is expressed as follows:

$$\mathbb{R} = \begin{bmatrix} \mathbf{R} & \mathbf{0} \\ \mathbf{0} & \mathbf{0} \end{bmatrix}$$

Hence, the coupled formulation using two distinct matrix equations expressed in terms of both the structure and fluid unknowns may be recast in the standard form:

$$(-\omega^2 \mathbb{M}_S + \mathbb{K}_S)\mathbb{U} = \mathbb{R}\mathbb{P} \qquad \mathbb{K}_F \mathbb{P} = \rho\omega^2 \mathbb{R}^\mathsf{T} \mathbb{U}$$

Furthermore, the uniqueness condition ensures that $\mathbb{K}_F$ is non-singular, and as a consequence, the pressure may be related to the displacement of the structure according to

$$\mathbb{P} = \rho\omega^2 \mathbb{K}_F^{-1} \mathbb{R}^\mathsf{T} \mathbb{U}$$

The pressure may therefore be treated as an auxiliary variable which can be eliminated from the structural dynamic equation, thereby yielding the standard form of the eigenvalue problem:

$$(-\omega^2(\mathbb{M}_S + \mathbb{M}_A) + \mathbb{K}_S)\mathbb{U} = \mathbf{0}$$

Accordingly, it is found that the incompressible fluid does not bring any additional degree-of-freedom and that the force acting on the structure is of *inertial nature*. It is represented by the so-called *added mass matrix*:

$$\mathbb{M}_A = \rho\mathbb{R}\mathbb{K}_F^{-1}\mathbb{R}^\mathsf{T}$$

In order to alleviate the notations, the preceding equations are also stated without making reference to the Lagrange multipliers:

$$(-\omega^2(\mathbf{M}_S + \mathbf{M}_A) + \mathbf{K}_S)\mathbf{U} = \mathbf{0} \qquad (4.8)$$

with the added mass matrix expressed as follows:

$$\mathbf{M}_A = \rho\mathbf{R}\mathbf{K}_F^{-1}\mathbf{R}^\mathsf{T} \qquad (4.9)$$

4.2.3   Inertial Effect

As made explicit in Equation (4.9), the inertial effect combines the influence of the fluid density, the geometry of the interface with the structure (accounted for by the fluid–structure coupling matrix) and the incompressibility of the fluid volume (represented by the inversion of the fluid 'stiffness' matrix).

The added mass matrix is *fully populated* and obviously symmetric, and it may also be shown to be definite positive by virtue of the relation giving the kinetic energy of the fluid. Indeed, the vibration of the structure sets the fluid into motion, which therefore gains kinetic energy and for harmonic vibrations, the latter is calculated as follows:

$$\mathcal{E} \propto -\frac{1}{2\rho\omega^2}\mathbf{P}^\mathsf{T}\mathbf{K}_F\mathbf{P}$$

Since at the interface the pressure may be derived from the displacement as $\mathbf{P} = \rho\omega^2 \mathbf{K}_F^{-1} \mathbf{R}^\mathsf{T} \mathbf{U}$, the fluid kinetic energy may also be expressed as follows:

$$\mathcal{E} \propto -\frac{\omega^2}{2} \mathbf{U}^\mathsf{T} \mathbf{M}_A \mathbf{U}$$

which is the multi-degrees-of-freedom equivalent of the relation outlined in Chapter 1.

The fluid motion which is generated by a constant displacement imposed at the boundary is now considered. It may be described using the potential displacement φ, such that

$$\Delta\varphi = 0 \text{ in } \Omega, \quad \frac{\partial\varphi}{\partial n} = \boldsymbol{\delta} \cdot \mathbf{n} \text{ on } \Gamma \text{ and } \int_\Omega \varphi \, d\Omega = 0$$

where $\boldsymbol{\delta}$ designates here a unit vector in a given direction. It follows that

$$\varphi = \boldsymbol{\delta} \cdot \mathbf{x} - \frac{1}{|\Omega|} \int_\Omega \boldsymbol{\delta} \cdot \mathbf{x} \, d\Omega$$

This corresponds to a bulk motion of the fluid, $\boldsymbol{\xi} \equiv \boldsymbol{\delta}$ throughout Ω, so that for a harmonic motion the kinetic energy is given as follows:

$$\mathcal{E} = -\frac{\omega^2}{2} \int_\Omega \rho\xi^2 \, d\Omega = -\frac{\omega^2}{2} m_F = -\frac{\omega^2}{2} \boldsymbol{\Delta}^\mathsf{T} \mathbf{M}_A \boldsymbol{\Delta}$$

where $\boldsymbol{\Delta}$ is a vector with components equal to 1 in the considered direction and equal to 0 in the others.

The total mass of fluid embedded within the structure is therefore retrieved from the added mass matrix according to the following equation:

$$\boldsymbol{\Delta}^\mathsf{T} \mathbf{M}_A \boldsymbol{\Delta} = m_F \tag{4.10}$$

Equation (4.8) is of the standard form valid for any discrete mechanical system which is linear and *conservative* in nature. The mechanical energy of the free vibrations of the coupled system is composed of the fluid and structure kinetic energy, on the one hand, and the structure (elastic) potential energy, on the other hand. Each of these components oscillates at the frequency of vibrations, the sum remaining constant.

The eigenvectors of the structure coupled with the fluid are denoted $(\mathbf{U}_n^h)_{n \in [1,N]}$, with N being the size of the problem. They are real and constitute an orthogonal basis to expand the solution of any forced problem as a modal series; the corresponding eigenpulsations, which are real and non-negative, are denoted $(\omega_n^h)_{n \in [1,N]}$.

Let $\mathbb{U}_o = (\mathbf{U}_n^o)_{n \in [1,N]}$ be the eigenvectors of the structure in vacuo, associated with eigenpulsations $(\omega_n^o)_{n \in [1,N]}$ written from the lowest to the highest value. Using this set of eigenmode shape so defined as an expansion basis for the solutions of Equation (4.8), namely $\mathbf{U} = \sum_n q_n^o \mathbf{U}_n^o$, yields an eigenvalue problem stated in terms of coordinates q_n^o:

$$\left(-\omega^2 \left(m_n + \sum_{n'} m_{n,n'} \right) + k_n \right) q_n^o = 0 \qquad \forall n \in [1, N] \tag{4.11}$$

with $m_{n,n'} = \mathbf{U}_{n'}^{o \, \mathsf{T}} \mathbf{M}_A \mathbf{U}_n^o$.

The projection of the structural eigenmodes on the added mass matrix, as defined in the coordinate system of the finite element model, produces a new added mass matrix defined in the modal coordinates of the structure in vacuo, and this matrix is generally non-diagonal.

As an immediate consequence, the mode shapes are found to differ according to whether fluid inertia is accounted for or not in the finite element model.

However, there exist simpler configurations where the mode shapes are not affected by the presence of the fluid. If is the case, the added mass matrix is diagonal in the eigenvector basis and fluid inertia may be accounted for by a positive scalar, namely, the coefficient of the so-called *modal added mass*, as further outlined in the example proposed in Section 4.4.

When $m_{n,n'} = 0$, Equation (4.11) yields

$$\omega_n^h = \sqrt{\frac{k_n}{m_n + \mu_n}} \leq \sqrt{\frac{k_n}{m_n}} = \omega_n^o$$

where $m_{n,n} = \mu_n > 0$, for all n since $\mathbf{M}_A$ is definite and positive.

The latter relation may be extended to the cases where the added mass matrix is *not* diagonal in the modal basis of the structure in vacuo. This relation can be stated using a characterisation of a system eigenpulsations with the *energy quotient*; see Remark 2.4 and Equation (2.22):

$$\forall n \in [1, N] \qquad \frac{1}{\omega_n^2} = \min_{\mathcal{W}_n} \max_{\mathbf{w}_n \in \mathcal{W}_n} (q(\mathbf{w}_n)) \text{ with } q(\mathbf{w}_n) = \frac{\mathbf{w}_n^T \mathbf{M}_n \mathbf{w}_n}{\mathbf{w}_n^T \mathbf{K}_n \mathbf{w}_n}$$

$q(\bullet)$ is the energy quotient associated with matrices $\mathbf{M}$ and $\mathbf{K}$ of dimension $N \times N$. $\mathbf{w}_n$ is a vector with n degrees-of-freedom and belongs to a space $\mathcal{W}_n$ of dimension n, while $\mathbf{M}_n$, $\mathbf{K}_n$ are square sub-matrices of dimension $n \times n$ (with $n \leq N$).

Let $q_o(\bullet)$ and $q_h(\bullet)$ be the energy quotient for the structure in vacuo and for the structure coupled to the fluid; since $\mathbf{M}_S$ and $\mathbf{M}_A$ are definite positive matrices, the following inequality is established:

$$\forall \mathcal{W}_n, \ \forall \mathbf{w}_n \in \mathcal{W}_n \qquad q_o(\mathbf{w}_n) \leq q(\mathbf{w}_n)$$

Thus,

$$\forall n \in [1, N] \qquad \omega_n^h \leq \omega_n^o \tag{4.12}$$

An illustration of the inertial effect is proposed on the simple system depicted in Figure 4.4. Two rigid cylinders of radius R and R' form a two degrees-of-freedom spring/mass system, with physical parameters m, m' and k, k' and are coupled through an incompressible fluid of density ρ, embedded in the annular space $R < r < R'$. Let u and u' denote the displacement of the cylinders, assumed to be small with respect to the typical dimension of the problem $R' - R$.

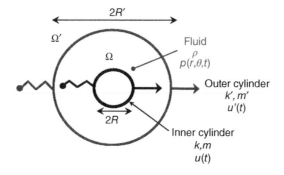

Figure 4.4   Inertial effect: 2D degrees-of-freedom system

The system equations of motion are as follows:

$$\begin{cases} m\ddot{u}(t) + ku(t) = \varphi(t) \\ m'\ddot{u}'(t) + k'u'(t) = \varphi'(t) \end{cases}$$

where $\varphi(t)$ and $\varphi'(t)$ are the fluid forces on the inner and outer cylinders, respectively. As detailed by Fritz (1972), they are calculated as follows:

$$\begin{Bmatrix} \varphi(t) \\ \varphi'(t) \end{Bmatrix} = - \begin{bmatrix} m_a & -(m_a + m_d) \\ -(m_a + m_d) & m_a + m_d + m'_d \end{bmatrix} \begin{Bmatrix} \ddot{u}(t) \\ \ddot{u}'(t) \end{Bmatrix} \tag{4.13}$$

with:

$$m_a = \rho\pi R^2 \frac{R'^2 + R^2}{R'^2 - R^2} \qquad m_d = \rho\pi R^2 \qquad m'_d = \rho\pi R'^2$$

m_a is the added mass on the inner cylinder, while m_d and m'_d represent the mass of fluid which would be contained in Ω and Ω'. $m'_d - m_d$ is the mass of fluid confined in the annular space $\Omega' - \Omega$. This mass is retrieved from the added mass matrix according to:

$$\boldsymbol{\Delta}^{\mathsf{T}}\mathbf{M}_A\boldsymbol{\Delta} = \begin{Bmatrix} 1 \\ 1 \end{Bmatrix}^{\mathsf{T}} \begin{bmatrix} m_a & -(m_a + m_d) \\ -(m_a + m_d) & m_a + m_d + m'_d \end{bmatrix} \begin{Bmatrix} 1 \\ 1 \end{Bmatrix} = m'_d - m_d$$

For the sake of simplicity, the application is proposed here for $k' = k = 1$ N/m and $m' = m = 1$ kg. The eigenmodes of the coupled system are solutions of

$$\left(-\omega^2 \begin{bmatrix} 1 + \rho\pi R^2 \dfrac{\alpha^2 + 1}{\alpha^2 - 1} & -\rho\pi R^2 \dfrac{2\alpha^2}{\alpha^2 - 1} \\ -\rho\pi R^2 \dfrac{2\alpha^2}{\alpha^2 - 1} & 1 + \rho\pi R^2 \dfrac{\alpha^2(\alpha^2 + 1)}{\alpha^2 - 1} \end{bmatrix} + \begin{bmatrix} 1 & 0 \\ 0 & 1 \end{bmatrix} \right) \begin{Bmatrix} u \\ u' \end{Bmatrix} = \begin{Bmatrix} 0 \\ 0 \end{Bmatrix}$$

with $\alpha = R'/R$.

The natural pulsations of the cylinders without fluid coupling are $\omega = \omega' = 1$ rad/s with orthogonal modes $\langle 1\ 0\rangle$ and $\langle 0\ 1\rangle$ and effective mass $\mu = \mu' = 1$ kg. When coupled to the fluid, two eigenmodes of pulsations $\beta_1 \leq \beta_2$ are observed, as depicted in Figure 4.5, where β is the ratio between the natural frequency of the cylinders with fluid and without fluid.

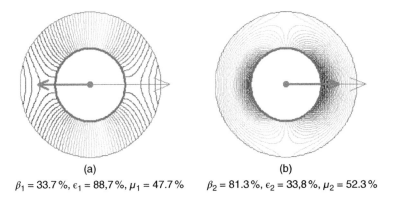

(a)   (b)

$\beta_1 = 33.7\%,\ \epsilon_1 = 88{,}7\%,\ \mu_1 = 47.7\%$ $\beta_2 = 81.3\%,\ \epsilon_2 = 33{,}8\%,\ \mu_2 = 52.3\%$

Figure 4.5   (b) Out-of-phase and (a) in-phase modes (the application is proposed here with $\alpha = 2$ and $\rho\pi R^2 = 1$; for the definition of the reduced quantities μ and ϵ, see just below)

The mode with the lower eigenpulsation corresponds to an *out-of-phase* motion of the cylinders, while the mode with the higher eigenpulsation corresponds to an *in-phase* motion of the cylinders. The amount of kinetic energy imparted to the springs is the same in both modes. However, the amount of kinetic energy imparted to the fluid is significantly higher when the system vibrates according to the out-of-phase mode than when it vibrates according to the in-phase mode, as evidenced by the evaluation of the following quantities:

- ε is the fluid kinetic energy compared with total (fluid and structure) kinetic energy:

$$\varepsilon = \frac{\begin{Bmatrix} u \\ u' \end{Bmatrix}^{\mathsf{T}} \begin{bmatrix} \rho\pi R^2 \dfrac{\alpha^2+1}{\alpha^2-1} & -\rho\pi R^2 \dfrac{2\alpha^2}{\alpha^2-1} \\ -\rho\pi R^2 \dfrac{2\alpha^2}{\alpha^2-1} & \rho\pi R^2 \dfrac{\alpha^2(\alpha^2+1)}{\alpha^2-1} \end{bmatrix} \begin{Bmatrix} u \\ u' \end{Bmatrix}}{\begin{Bmatrix} u \\ u' \end{Bmatrix}^{\mathsf{T}} \begin{bmatrix} 1+\rho\pi R^2 \dfrac{\alpha^2+1}{\alpha^2-1} & -\rho\pi R^2 \dfrac{2\alpha^2}{\alpha^2-1} \\ -\rho\pi R^2 \dfrac{2\alpha^2}{\alpha^2-1} & 1+\rho\pi R^2 \dfrac{\alpha^2(\alpha^2+1)}{\alpha^2-1} \end{bmatrix} \begin{Bmatrix} u \\ u' \end{Bmatrix}}$$

As outlined in Figure 4.5, ε is roughly 90% for the out-of-phase mode, while it is only 30% for the in-phase mode.

- μ is the effective modal mass, compared with the total mass:

$$\mu = \frac{1}{2+\rho\pi R^2(\alpha^2-1)} \frac{\left(\begin{Bmatrix} 1 \\ 1 \end{Bmatrix}^{\mathsf{T}} \begin{bmatrix} 1+\dfrac{\alpha^2+1}{\alpha^2-1} & -\dfrac{2\alpha^2}{\alpha^2-1} \\ -\dfrac{2\alpha^2}{\alpha^2-1} & 1+\dfrac{\alpha^2(\alpha^2+1)}{\alpha^2-1} \end{bmatrix} \begin{Bmatrix} u \\ u' \end{Bmatrix} \right)^2}{\begin{Bmatrix} u \\ u' \end{Bmatrix}^{\mathsf{T}} \begin{bmatrix} 1+\dfrac{\alpha^2+1}{\alpha^2-1} & -\dfrac{2\alpha^2}{\alpha^2-1} \\ -\dfrac{2\alpha^2}{\alpha^2-1} & 1+\dfrac{\alpha^2(\alpha^2+1)}{\alpha^2-1} \end{bmatrix} \begin{Bmatrix} u \\ u' \end{Bmatrix}}$$

As made conspicuous in Figure 4.5, the total mass of the system is evenly distributed between the two modes (48% for the out-of-phase mode and 52% for the in-phase mode).

Remark 4.1 Inertial coupling with unbounded fluid – infinite elements *When considering a structure which is fully submerged in a fluid that occupies an unbounded domain, the infinity may be regarded as a separate boundary with a condition that makes the problem well posed. For the Laplace equation, a condition of regularity at infinity is proved to be sufficient to ensure the uniqueness of the solution. The regularity condition is stated as follows:*

$$p = \mathcal{O}(1/r) \qquad r \to \infty \tag{4.14}$$

for tri-dimensional problems, and

$$p = \mathcal{O}(1) \qquad r \to \infty \tag{4.15}$$

for bi-dimensional problems. It is inferred from Equation (4.14) that:

$$\lim_{r \to \infty} p(r) = 0$$

in the 3D case, while for the 2D case the condition is $p(r \to \infty) < \infty$.

For bi-dimensional problems, the pressure is bounded at infinity, yet the bound is not known; in this case, however, the pressure field is shown to verify

$$\lim_{r \to \infty} \frac{\partial p}{\partial r} = 0$$

In the context of the finite element method, the natural idea would be to enforce the above boundary conditions (whether $p = 0$ or $\frac{\partial p}{\partial r} = 0$) at a finite distance suitably chosen. However, it proves to be a too crude approximation of the problem at stake. Consistent approximation of condition (4.14) or (4.15) is the core of the Infinite element method (IEM).

Element-based representations of Equations (4.14) or (4.15) give rise to the development of two kinds of infinite elements, namely, the decay element and the mapped element (Bettess, 1977, 1980). A short description of these is presented in what follows:

Decay infinite elements *describe the physics of the problem at infinity with a decay function, which is used in conjunction with the element shape function. On the element, approximation of the pressure is given as follows:*

$$p \approx \langle N_k(\xi) \rangle \{ p_k \}$$

where ξ denotes the local variable on the element and p_k the nodal value of p, as sketched in Figure 4.6. The shape function N_k associated with node k is given by

$$N_k(\xi) = P_k(\xi) f_k(\xi)$$

P_k is the parent shape function while f_k denotes the decay function, as specified in Table 4.1. Many decay functions may be used, as long as they are equal to unity at their nodes ($f_k(\xi_k) = 1$) and as they ensure that $N_k(\xi)$ tends to the desired far-field value, when ξ tends to ∞. The following decay functions were originally used in the design of the IEM, while various improvements of the method have been proposed, using different functions.

- *Exponential decay function:*

$$f_k(\xi) = \exp \left(\frac{\xi_k - \xi}{\delta_k} \right)$$

where δ_k is a length which determines the rate of the decay.

- *Reciprocal decay function:*

$$f_k(\xi) = \left(\frac{\xi_k - \xi_o}{\xi - \xi_o} \right)^{\nu_k}$$

where the value of the exponent ν_k governs the rate of decay and ξ_o stands for an origin point placed outside the element (for instance, $\xi_o < -1$ for a unidimensional quadratic element).

Mapped elements *are defined by a mapping of a parent element from a finite to an infinite domain. The space variable r, ranging from a finite value to infinity, is approximated according to*

$$r \approx \langle M_k(\xi) \rangle \{ r_k \}$$

where M_k is the mapping function associated with node k. The approximation of the pressure field on the element is written as follows:

$$p \approx \langle P_k(\xi) \rangle \{ p_k \}$$

where P_k is the shape function of node k on the parent element (as sketched in Figure 4.6 and specified in Table 4.1).

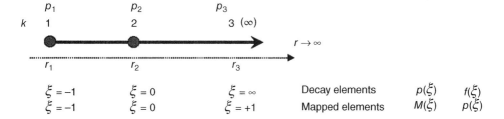

Figure 4.6   *Unidimensional quadratic (infinite) element provided with three nodes*

Table 4.1   Example of mapping and parent functions: unidimensional quadratic infinite element (Bettess, 1992)

k	ξ_k	$P_k(\xi)$	$M_k(\xi)$	$\dfrac{\partial P_k}{\partial \xi}(\xi)$	$\dfrac{\partial M_k}{\partial \xi}(\xi)$
1	-1	$-\dfrac{\xi(1-\xi)}{2}$	$-\dfrac{2}{1-\xi}$	$\xi - \dfrac{1}{2}$	$-\dfrac{2}{(1-\xi)^2}$
2	0	$(1+\xi)(1-\xi)$	$\dfrac{1+\xi}{1-\xi}$	-2ξ	$\dfrac{2}{(1-\xi)^2}$
3	$+1$	$-\dfrac{\xi(1+\xi)}{2}$	$-$	$\xi + \dfrac{1}{2}$	$-$

The integration of infinite elements may be performed without major difficulty either analytically for simple enough geometries or numerically using the Gauss scheme, for instance. The elementary integral to formulate a mapped element can be computed using Equation (2.21), while those needed to formulate a decay element are of the kind $\int_{-1}^{\infty} \psi(x)dx$ or $\int_{0}^{\infty} \psi(x)dx$. Furthermore, a change of variable is appropriate to recast the integration within the interval $[-1, +1]$, enabling Equation (2.21) to be applied, for instance:

$$\int_{-1}^{\infty} \psi(x)dx = 2 \int_{-1}^{+1} \psi\left(\frac{2\xi}{1-\xi}\right) \frac{d\xi}{1-\xi^2}$$

More on the construction and the application of infinite elements may be found in Bettess (1992) and in Gerdes (2000). The method is further described in Section 4.5, where it is applied to a problem of academic interest aimed at providing the reader with an illustration of this topic.

Actually, the technique of combining the structure finite element with fluid finite and infinite elements is of broad use in many industrial applications, ranging from civil engineering to offshore and naval engineering, as illustrated by Figure 2.9.[2]   ■

4.3   Modelling Inertial Coupling for Complex Systems: Example of Tube Bundle

While the analytical calculation of the added mass is often possible for simple geometries, see for instance Section 4.4.1, its numerical computation encounters some practical difficulties in most cases of practical interest.

On account of data storage capacity and numerical efficiency of matrix inversion algorithms, computing the added mass matrix according to the formula $\mathbf{M}_A = \rho \mathbf{R} \mathbf{K}_F^{-1} \mathbf{R}^\mathsf{T}$ – whether stated in a direct or indirect manner – becomes often unworkable in the case of very large matrices as those produced by the finite element mesh of complex structures coupled to a large extent of fluids.

A typical example of such difficulties is encountered in power nuclear engineering when building a finite element model of the tube bundle of steam generators, as represented in Figure 4.7. In order to produce a numerical model of the tubes and fluid system in the component, various techniques of increasing complexity are required, which depend on the specific objective of the study.

4.3.1   Analytical Models for Added Mass

Blevins (1979) gathers added mass coefficients of simple systems, among which rigid or deformable bodies (such as beam, plates or shells). The added mass of a moving tube in an array of fixed tubes is given in Figure 4.8: this provides the diagonal component of the added mass matrix, but coupling between the tubes is not accounted for with this model.

4.3.2   'Term-to-Term' Computation of the Added Mass Matrix

A 'term-to-term' computation of the added mass matrix is based on the interaction of any individual tube with all the surrounding tubes. According to the following principles, it is an alternative to the direct computation of $\mathbf{M}_A$.

The tube bundle is modelled as a set of L identical tubes as represented in Figure 4.9.

---

[2] In Figure 2.9, the numerical model is based on structural finite elements with fluid finite and infinite elements. FSI is accounted for with a coupling matrix which requires the fluid–structure interface to be discretised. Infinite elements are used to account for the inertial effect in an unbounded fluid domain: in the present case, the boundary condition corresponds to a cylindrical wave propagation.

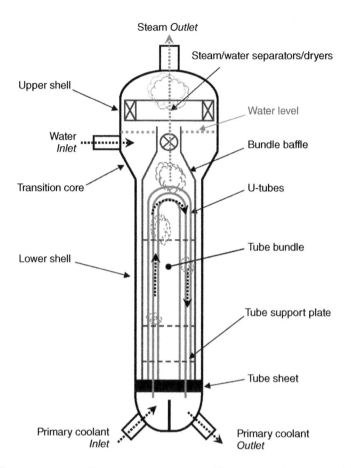

Figure 4.7   Steam generator. Steam generators are used in power nuclear energy for heat transfer between the primary and secondary coolant loops. A steam generator is a complex structure: describing inertial effect for the tube bundle with finite element coupling is not straightforward. In the context of a seismic event, how does FSI influence the dynamic response of the tubes? Can inertial effect be represented with a simple model (for instance, through an added mass coefficient for each tube), or is it necessary to use a more complex description (for example, by taking the interaction between tubes into account)? In the latter case, how can FSI be rendered with a model which combines physical consistency and numerical efficiency? Some of these issues are briefly discussed hereafter

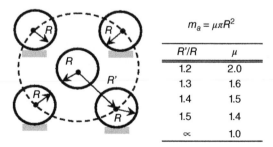

Figure 4.8   Analytical calculation of added mass: example of cylinder in array of fixed cylinders (Blevins, 1979)

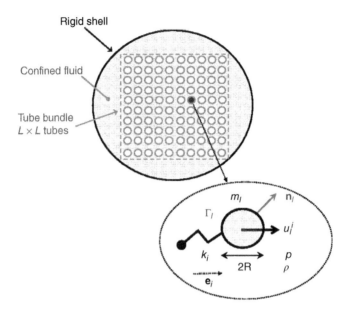

Figure 4.9   Dynamics of an individual tube in a confined and immersed array of tubes. A 2D tube bundle composed of $L \times L$ tubes immersed in a fluid confined by a rigid shell is considered here. The tubes are free to move in the two directions and their motion along $\mathbf{e}_i$ is represented by u_l^i. m_l and k_l are the mass and stiffness characteristic of tube l. The tubes are assumed to be of circular shape, with radius R; $\mathbf{n}_l$ stands for the outer normal on the tube boundary Γ_l in contact with the fluid. p stands for the pressure in the fluid, which is assumed incompressible, ρ denoting its density

The motion of any tube l in direction $\mathbf{e}_i$ is described by the following equation:

$$m_l \frac{\partial^2 u_l^i}{\partial t^2} + k_l u_l^i = - \int_{\Gamma_l} p|_{\Gamma_l} \mathbf{n}_l \cdot \mathbf{e}_i d\Gamma_l$$

where the pressure field is solution to $\Delta p = 0$ with the coupling condition on each tube expressed as $\dfrac{\partial p}{\partial n_{l'}} = -\rho \dfrac{\partial^2 u_{l'}^i}{\partial t^2} \mathbf{n}_{l'} \cdot \mathbf{e}_i$, for all $l' \in [0, L]$.

Assuming all tubes fixed, except tube l' which is supposed to undergo a unit acceleration in direction $\mathbf{e}_i$, the pressure field which results from this motion is denoted $p_{l'}^i$. It is the solution of $\Delta p_{l'}^i = 0$ with boundary conditions $\dfrac{\partial p_{l'}^i}{\partial n_{l'}} = -\mathbf{n}_{l'} \cdot \mathbf{e}_i \delta_{ll'}$ for all $l \in [0, L]$. As represented in Figure 4.10, $p_{l'}^i$ may be computed with a finite element technique.

The pressure field p induced by all tubes in motion is obtained by a linear superposition of the elementary solutions $p_{l'}^i$:

$$p = \rho \sum_{l'} p_{l'}^i \frac{\partial^2 u_{l'}^i}{\partial t^2}$$

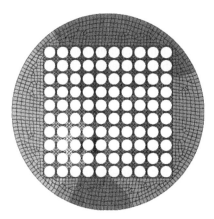

Figure 4.10 Elementary computation. In this example of a 10×10 tube bundle, the pressure field generated by the acceleration of a tube in a given direction may be evaluated with the aid of any finite element program which solves the Laplace equation

The fluid force on tube l is therefore

$$\varphi_l = \rho \sum_{l'} \int_{\Gamma_l} p^i_{l'}|_{\Gamma_l} \frac{\partial^2 u^i_{l'}}{\partial t^2} \mathbf{n}'_l \cdot \mathbf{e}_i d\Gamma_l$$

$\rho \int_{\Gamma_l} p^i_{l'}|_{\Gamma_l} \mathbf{n}'_l \cdot \mathbf{e}_i d\Gamma_l$ stands for the fluid force on tube l when tube l' is subjected to a 1 m/ s^2 acceleration in direction $\mathbf{e}_i$.

The equation of motion of tube l reads as follows:

$$m_l \frac{\partial^2 u^i_l}{\partial t^2} + \rho \sum_{l'} \mu_{ll'} \frac{\partial^2 u^i_{l'}}{\partial t^2} + k_l u^i_l = 0$$

and the equation of motion of all tubes is $(\mathbf{M} + \rho\boldsymbol{\mu})\ddot{\mathbf{U}} + \mathbf{KU} = \mathbf{0}$, where $\mathbf{M}$ and $\mathbf{K}$ stand for the mass and stiffness matrices of the tubes. $\rho\boldsymbol{\mu}$ is the added mass matrix, which has the following components:

$$\mu^i_{ll'} = \int_{\Gamma_l} p^i_{l'}|_{\Gamma_l} \mathbf{n}_l \cdot \mathbf{e}_i d\Gamma_l$$

The computation of the added mass matrix can be complete or partial: in the latter case, the interaction of a tube to some neighbouring tubes is considered, whereas in the former, all interactions are accounted for. From the practical point of view, such an approach is tedious and its feasibility is in fact directly related to the storage capacity of the computer used. Furthermore, the numerical manipulation of the added mass matrix in the time or frequency domain also raises practical difficulties because all the algorithms designed to solve linear systems or eigenvalue problems tend to lose their numerical efficiency as the size of densely – or fully – populated matrices is concerned.

Several techniques of model reduction are available to significantly reduce the size of the problem at hand. One of them is of particular interest to deal with periodic fluid–structure systems as the tube bundle of a heat exchanger, for instance, and is known as the homogenisation method. It consists of replacing the heterogeneous and periodic system by an equivalent continuous medium. The procedure generates sparse matrices of low size, thereby avoiding the tedious task of meshing the whole real structure and fluid domains.

4.3.3   A Homogenisation Technique

4.3.3.1   Homogenisation of Periodic Fluid–Structure Systems

Homogenisation methods stand for a powerful class of mathematical tools which are suited to convert a heterogeneous medium into an equivalent homogeneous one. If reviewing the variety of such approaches is beyond the scope of this book, it is however worthwhile to outline here the mains steps of the procedure, which remains the same independently of the particularities of the specific method proposed here.

- Double scales asymptotic development: when the problem exhibits a geometrical periodicity, it is convenient to divide the domain in identical cells, with a typical length period ε, which is supposed to be small in comparison to a characteristic length of the system L, as detailed in Figure 4.11. The problem unknowns are supposed to be ε dependent and are expanded in terms of a series:

$$\psi_\varepsilon = \psi_o + \sum_{n \geq 1} \psi_n(X, x) \tag{4.16}$$

where the functions ψ_n depend on the *local* and *global* scale variables x and X, with $x = X/\varepsilon$.
- Term-to-term identification: substituting the former expression of ψ into the partial differential equations which govern the dynamics of the heterogeneous medium, that is, the fluid–structure coupled system in the present context, and performing a term-to-term identification give a set of multi-scale problems with ψ_n as unknown. In addition, an averaging process is required to formulate the equations for the mean problem ψ_o.
- Closure condition: for periodic systems, the *multi-scale* and the *mean* problems are to be solved in an elementary cell, which results in the so-called *homogenised model*.

In power nuclear engineering, such approaches were applied to several components immersed in water, namely, reactor cores (Zhang *et al.*, 2001), reactor internals (Brochard *et al.*, 1987), fuel assemblies (Planchard, 1985) or tube bundles (Planchard, 1987; Zhang, 1999). These studies were carried out considering bi-dimensional and tri-dimensional homogenised models.

An alternative homogenisation technique has been proposed by Sigrist and Broc (2008a), which is based on the simplified elementary cell analysis as detailed below. The averaging process to be performed on the elementary cell is depicted in Figure 4.12. Despite the inevitable loss of accuracy, which arises from the ground hypotheses of the homogenisation model (Broc and Sigrist, 2014), such an approach has been found to be a relevant alternative to classical finite element coupling when it comes to some engineering applications – see for instance Sigrist and Broc (2008b).

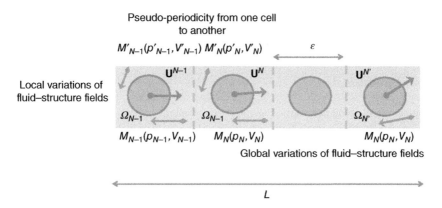

Figure 4.11 Pseudo-periodicity, global and local variations. A unidimensional periodic fluid–structure system is considered here: it is composed of identical elementary cells $(\Omega_N)_{N\geq 1}$, each cell containing one tube and being filled with the fluid. The global-local description is based on an asymptotic description which is valid when the system is composed of a significant number of subsystems, that is, when $\varepsilon \ll L$ (ε standing for the characteristic length of the elementary cells and L standing for that of the whole system). On cell Ω_N, the displacement of the tube is denoted $\mathbf{u}_N$, while p_N and $\mathbf{v}_N$ stand for the pressure and velocity at a given location M_N. *Local variations* refer to the evolution of p and $\mathbf{v}$ from point M_N to point M'_N in Ω. *Pseudo-periodicity* indicates that the fluid and structure fields are similar (if not identical) from one cell (Ω_{N-1}) to the next (Ω_N). While the distance increases, the fluid flow and the tube motion gradually evolve so that *global variations* are observed between distant cells Ω_N and $\Omega_{N'}$. These evolutions are accounted for using a double-scale representation of the structure or the fluid field ψ, expressed by Equation (4.16), using x as the *local scale variable* and X as the *global scale variable*

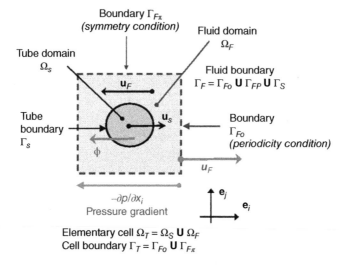

Figure 4.12 Elementary cell: single tube in array confinement

4.3.3.2 Elementary Cell Analysis

When the fluid–structure system exhibits a repetitive pattern, that is, when the accelerations of the tubes are expected to change by a small amount from one cell to another, it may be assumed that the tube and fluid motions are described at the local scale by a two-degrees-of-freedom system, namely, with

- the displacement of the tube $\mathbf{u}_S$;
- the displacement of the outer boundary of the elementary cell $\mathbf{U}_F$, in agreement with the notations of Figure 4.12.

The fluid flow within the elementary cell is assumed incompressible; the pressure field verifies the Laplace equation $\Delta p = 0$ in Ω_F with boundary conditions:

$$\left.\frac{\partial p}{\partial n}\right|_{\Gamma_S} = -\rho\ddot{\mathbf{u}}_S \cdot \mathbf{n} \qquad \left.\frac{\partial p}{\partial n}\right|_{\Gamma_T} = -\rho\ddot{\mathbf{U}}_F \cdot \mathbf{n}$$

The weighted integral formulation of this problem reads as follows:

$$\int_{\Omega_F} \nabla p \cdot \nabla \delta p \, d\Omega_F = \int_{\Gamma_F} \frac{\partial p}{\partial n} \delta p \, d\Gamma_F$$

for all virtual pressure field δp. The displacement field in the elementary cell is denoted $\mathbf{u}_F$; it derives from the pressure field according to $\nabla p = -\rho\ddot{\mathbf{u}}_F$ so that the integral formulation may also be recast as:

$$\int_{\Omega_F} -\rho\ddot{\mathbf{u}}_F \cdot \nabla \delta p \, d\Omega_F = \int_{\Gamma_F} -\rho\ddot{\mathbf{u}}_F \cdot \mathbf{n} p \, d\Gamma_F$$

As $\Gamma_F = \Gamma_S \cup \Gamma_T$ with the notation of Figure 4.12, the preceding relation may also be written as follows:

$$-\int_{\Omega_F} \rho\ddot{\mathbf{u}}_F \cdot \nabla \delta p \, d\Omega_F = -\int_{\Gamma_T} \rho\ddot{\mathbf{U}}_F \cdot \mathbf{n} p \, d\Gamma_T - \int_{\Gamma_S} \rho\ddot{\mathbf{u}}_S \cdot \mathbf{n} p \, d\Gamma_S$$

Applying the divergence theorem on Ω_S, with normal $\mathbf{n}$ pointing outwards, yields the following:

$$\int_{\Gamma_S} \mathbf{u}_S \cdot \mathbf{n} \delta p \, d\Gamma_S = \int_{\Omega_S} \nabla \cdot (\mathbf{u}_S \delta p) \, d\Omega_S$$

Developing the divergence of $\mathbf{u}_S \delta p$ with $\mathbf{u}_S$ constant on Ω_S gives

$$\int_{\Gamma_S} \rho\ddot{\mathbf{u}}_S \cdot \mathbf{n} p \, d\Gamma_S = -\int_{\Omega_S} \rho\ddot{\mathbf{u}}_S \cdot \nabla \delta p \, d\Omega_S$$

In the same way, the following relation is arrived at on Ω_T:

$$\int_{\Gamma_T} \rho\ddot{\mathbf{U}}_F \cdot \mathbf{n} p \, d\Gamma_T = \int_{\Omega_T} \rho\ddot{\mathbf{U}}_F \cdot \nabla \delta p \, d\Omega_T$$

Using the previous relations, the following integral formulation is obtained for all δp:

$$\int_{\Omega_F} \rho\ddot{\mathbf{u}}_F \cdot \nabla \delta p \, d\Omega_F + \int_{\Omega_S} \rho\ddot{\mathbf{u}}_S \cdot \nabla \delta p \, d\Omega_S = \int_{\Omega_T} \ddot{\mathbf{U}}_F \cdot \nabla \delta p \, d\Omega_T$$

With $\delta p = \mathbf{e}_i \cdot \mathbf{r}$, the former expression reads as follows:

$$\int_{\Omega_F} \rho \ddot{\mathbf{u}}_F \cdot \mathbf{e}_i \, d\Omega_F + \int_{\Omega_S} \rho \ddot{\mathbf{u}}_S \cdot \mathbf{e}_i \, d\Omega_S = \int_{\Omega_T} \rho \ddot{\mathbf{U}}_F \cdot \mathbf{e}_i \, d\Omega_T$$

Since $\mathbf{u}_S$ and $\mathbf{U}_F$ are constant on Ω_S and Ω_T, the former equation can also be formulated as follows:

$$\int_{\Omega_F} \rho \ddot{\mathbf{u}}_F \cdot \mathbf{e}_i \, d\Omega_F + \rho |\Omega_S| \ddot{\mathbf{u}}_S \cdot \mathbf{e}_i = \rho |\Omega_T| \ddot{\mathbf{U}}_F \cdot \mathbf{e}_i$$

This relation holds in any direction $\mathbf{e}_i$, thus:

$$\ddot{\mathbf{U}}_F = \frac{|\Omega_S|}{|\Omega_T|} \ddot{\mathbf{u}}_S + \frac{1}{|\Omega_T|} \int_{\Omega_F} \ddot{\mathbf{u}}_F \, d\Omega_F \tag{4.17}$$

φ_S and φ_F are the pressure force on the tube and on the fluid cell in direction i; they are calculated as follows:

$$\varphi_S = \int_{\Gamma_S} p\mathbf{n} \cdot \mathbf{e}_i \, d\Gamma_S \qquad \varphi_F = \int_{\Gamma_T} p\mathbf{n} \cdot \mathbf{e}_i \, d\Gamma_T$$

with the notations of Figure 4.12.

Inertial coupling by the fluid is described by the following added mass matrix:

$$\left\{ \begin{array}{c} \varphi_S(t) \\ \varphi_F(t) \end{array} \right\} = - \left[\begin{array}{cc} m_a & -(m_a + \rho|\Omega_S|) \\ -(m_a + \rho|\Omega_S|) & m_a + \rho(|\Omega_S| + |\Omega_T|) \end{array} \right] \left\{ \begin{array}{c} \ddot{u}_S(t) \\ \ddot{U}_F(t) \end{array} \right\} \tag{4.18}$$

with $\ddot{u}_S = \ddot{\mathbf{u}}_S \cdot \mathbf{e}_i$ and $\ddot{U}_F = \ddot{\mathbf{U}}_F \cdot \mathbf{e}_i$.

The former expression has the same canonical form as in Equation (4.13): m_a is the added mass of the tube confined in the cell, $\rho|\Omega_S|$ is the mass of fluid displaced by the tube, while $\rho|\Omega_T|$ is the mass of fluid displaced by the enclosed cell.

Let $\mathbf{\Phi}$ be the pressure force on the moving tube; according to Equation (4.18), it is:

$$\mathbf{\Phi} = -m_A \ddot{u}_S + (m_A + \rho|\Omega_S|)\ddot{U}_F \tag{4.19}$$

The fluid dynamics within the elementary cell is described by

$$\int_{\Omega_F} \rho \ddot{u}_F \, d\Omega_F = \mathbf{\Pi} - \mathbf{\Phi} \tag{4.20}$$

where $-\mathbf{\Phi}$ is the force exerted by the tube on the fluid and $\mathbf{\Pi}$ is the force on the fluid volume. The latter is calculated from the pressure gradient according to $\mathbf{\Pi} = -|\Omega_T|\nabla p$, where $|\Omega_T|$ stands for the volume of the elementary cell. Combining the latter equation with Equations (4.17) and (4.19), the following relations are arrived at:

$$\int_{\Omega_F} \rho \ddot{u}_F = \rho|\Omega_T|\ddot{U}_F - \rho|\Omega_S|\ddot{u}_S$$

and:

$$\int_{\Omega_F} \rho \ddot{u}_F = -|\Omega_T|\nabla p + m_A \ddot{u}_S - (m_A + \rho|\Omega_S|)\ddot{U}_F$$

The homogenised equation of motion for the tube and fluid system is finally obtained (Sigrist and Broc, 2008a):

$$\rho\ddot{U}_F = (1 - B_S)\nabla p + \rho B_S \ddot{u}_S \tag{4.21}$$

where B_S is the *confinement ratio* of the tube in the elementary cell:

$$B_S = \frac{m_a + \rho|\Omega_S|}{m_a + \rho(|\Omega_S| + |\Omega_T|)} \tag{4.22}$$

According to Equation (4.17), the displacement flux through the elementary cell boundary (U_F) may also be interpreted as the mean displacement of the tube/fluid system on the elementary cell volume. Equation (4.21) indicates that this flux originates from the acceleration of the tube and from the pressure gradient. These sources of motion have an opposite action and are balanced by the confinement ratio: the proposed description is similar to a fluid flow model in porous media.

The confinement is extreme when the elementary cell is filled by the solid only. Thus, $B_S = 0$, so that Equation (4.22) reads $\ddot{U}_F = \ddot{u}_S$, the physical meaning of which is obvious. Conversely, $B_S = 1$ indicates that the elementary cell is filled by the fluid only: in such a case, Equation (4.22) reduces to the fluid momentum equation $\rho\ddot{U}_F = -\nabla p$.

With $\ddot{u}_S$ and $\ddot{U}_F$ as degrees-of-freedom for the tubes and fluid system, any situation is a linear superposition of the two elementary cases depicted in Figure 4.13.

- The fluid flows around the tube which is at rest ($\ddot{u}_S = 0$), the resulting acceleration of the fluid being $\ddot{U}_F = \gamma$. The fluid forces on the cell boundary derive from Equation (4.18): $\varphi_S = (m_a + \rho|\Omega_S|)\gamma$ and $\varphi_F = -(m_a + \rho|\Omega_S| + \rho|\Omega_T|)\gamma$.
- The tube is accelerated ($\ddot{u}_S = \gamma > 0$) and generates fluid motion within the cell: at the local scale, the fluid gains acceleration ($\ddot{u}_F \neq 0$). The mean fluid acceleration is negative, whereas the tube acceleration is positive: as a result, the mean acceleration given by Equation (4.17) cancels to zero. The global flux between two adjacent cells is null, whereas locally, some fluid flows from one cell to the next. In this second elementary case, the fluid pressure forces are $\varphi_S = -m_a\gamma$ and $\varphi_F = (m_a + \rho|\Omega_S|)\gamma$. The added mass of the tube in the cell is derived from a finite element computation of the fluid forces, in one of the former configurations. The confinement ratio B_S is then calculated according to Equation (4.22).

The continuity condition between two adjacent cells is enforced in a simple manner using the flux of the fluid passing at the cell boundary: as sketched in Figure 4.14, the method is able to cope with various configurations of tube arrays of industrial interest.

4.3.3.3   FSI Modelling for a Tube Bundle Confined within a Structure

FSI for an additional structure coupled to a tube bundle through a fluid may be accounted for to the general framework depicted in Figure 4.15.

The structure equation is stated as $\rho_\sigma\ddot{u}_\Sigma - \nabla \cdot \sigma(u_\Sigma) = 0$ in the absence of external loads. Equation (4.21) describes the homogenised fluid-tube dynamics. The pressure forces on the tubes are described through the density load ϕ, the latter being derived from the elementary cell description $\phi = \varphi_S/|\Omega_T|$. The equation for the tubes motion is then

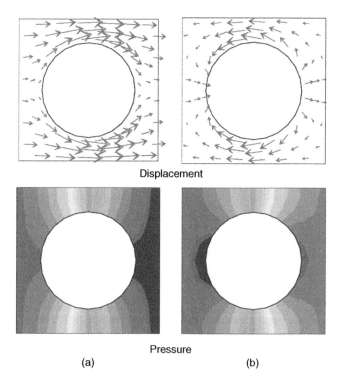

Displacement

Pressure

(a)   (b)

Figure 4.13   Elementary cell: a two-degrees-of-freedom system. On the elementary cell, the inertial effect is accounted for using a two degrees-of-freedom model of the system dynamics. Two elementary cases (a, b) account for the behaviour of the tubes in fluid, as illustrated above, in terms of fluid displacement and pressure

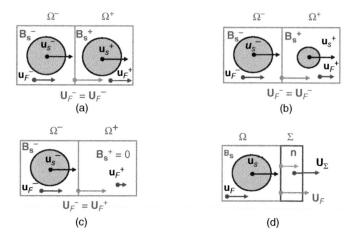

Figure 4.14   Various possible configurations of tubes are encountered in industrial applications: all of them may be handled by the homogenisation method proposed by Broc and Sigrist (2014). (a) Two identical adjacent cells. (b) Two different adjacent cells. (c) Two adjacent cells with/without tube. (d) Single cell coupled with an elastic structure

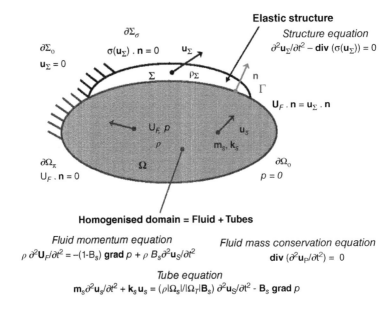

Figure 4.15   Tube array coupled to another structure through fluid. According to the homogeni-sation approach proposed by Sigrist and Broc (2008a), the tubes and fluid are described as a single continuous medium in which the fluid pressure p and the tube displacement $\mathbf{u}_S$ stand for scalar and vec-tor fields, which are assumed to vary continuously within the homogenised domain Ω. $\ddot{\mathbf{U}}_F$ denotes the mean fluid acceleration, according to Equation (4.17), while $\mathbf{u}_\Sigma$ denotes the structure displacement. $\partial\Omega_o$ and $\partial\Omega_\pi$ stands for the fluid boundaries with imposed pressure and with imposed displacement. Σ is the structure domain, $\partial\Sigma_o$ and $\partial\Sigma_\sigma$ denote the structure boundaries with prescribed displacements and with imposed forces. The fluid–structure interface is Γ, with unit normal $\mathbf{n}$. ρ and ρ_Σ are the fluid and structure density, m_S and k_S are the mass and stiffness of the tubes

$m_S \ddot{\mathbf{u}}_S + k_S \mathbf{u}_S = \rho |\Omega_S| / |\Omega_T| - B_S \ddot{\mathbf{u}}_S - B_S \nabla p$ (Sigrist and Broc, 2008a). The boundary and coupling conditions associated with this set of equations are specified in Figure 4.15; they read as:

$$\mathbf{u}_\Sigma = 0 \text{ on } \partial\Sigma_o \qquad \sigma(\mathbf{u}_\Sigma) \cdot \mathbf{n} = 0 \text{ on } \partial\Sigma_\sigma$$

$$\mathbf{U}_F \cdot \mathbf{n} = 0 \text{ on } \partial\Omega_\pi \qquad \mathbf{U}_F \cdot \mathbf{n} = \mathbf{u}_\Sigma \cdot \mathbf{n} \text{ on } \Gamma$$

and:

$$p = 0 \text{ on } \partial\Omega_o$$

Following Sigrist and Broc (2008a), the weighted integral formulation of the coupled prob-lem is stated as follows:

$$\int_\Omega \rho(1 - B_S)\nabla p \cdot \nabla \delta p \, d\Omega + \int_\Gamma \rho \frac{\partial^2 \mathbf{u}_\Sigma}{\partial t^2} \cdot \mathbf{n}\delta p \, d\Gamma$$

$$= \int_\Omega \rho B_S \frac{\partial^2 \mathbf{u}_S}{\partial t^2} \nabla \delta p \, d\Omega \qquad (4.23)$$

for all δp and:

$$\int_\Sigma \rho_\Sigma \frac{\partial^2 \mathbf{u}_\Sigma}{\partial t^2} \cdot \delta \mathbf{u}_\Sigma \, d\Sigma + \int_\Sigma \sigma(\mathbf{u}_\Sigma) : \varepsilon(\delta \mathbf{u}_\Sigma) \, d\Sigma + \int_\Omega m_S \frac{\partial^2 \mathbf{u}_S}{\partial t^2} \cdot \delta \mathbf{u}_S \, d\Omega + \int_\Omega k_S \mathbf{u}_S \cdot \delta \mathbf{u}_S \, d\Omega$$

$$= \int_\Gamma p\mathbf{n} \cdot \delta \mathbf{u}_\Sigma \, d\Gamma + \int_\Omega \rho(|\Omega_S|/|\Omega_T| - \mathrm{B}_S)\frac{\partial^2 \mathbf{u}_\Sigma}{\partial t^2} \cdot \delta \mathbf{u}_\Sigma \, d\Omega - \int_\Omega \mathrm{B}_S \nabla p \cdot \delta \mathbf{u}_S \, d\Omega \qquad (4.24)$$

for all $\delta \mathbf{u}_\Sigma$ and $\delta \mathbf{u}_S$.

The finite element discretisation of the preceding equations yields the following matrix system:

$$\begin{bmatrix} \mathbf{M}_S + \mathbf{M}_S^* & 0 & 0 \\ 0 & \mathbf{M}_\Sigma & 0 \\ -\rho\mathbf{B}_S\mathbf{C}^\mathsf{T} & \rho\mathbf{R}^\mathsf{T} & 0 \end{bmatrix} \begin{Bmatrix} \ddot{\mathbf{U}}_S(t) \\ \ddot{\mathbf{U}}_\Sigma(t) \\ \ddot{\mathbf{P}}(t) \end{Bmatrix}$$

$$+ \begin{bmatrix} \mathbf{K}_S & 0 & \mathbf{B}_S\mathbf{C} \\ 0 & \mathbf{K}_\Sigma & -\mathbf{R} \\ 0 & 0 & (1-\mathbf{B}_S)\mathbf{K}_F \end{bmatrix} \begin{Bmatrix} \mathbf{U}_S(t) \\ \mathbf{U}_\Sigma(t) \\ \mathbf{P}(t) \end{Bmatrix} = \begin{Bmatrix} 0 \\ 0 \\ 0 \end{Bmatrix} \qquad (4.25)$$

While $\mathbf{M}_S$, $\mathbf{K}_F$, $\mathbf{K}_S$ and $\mathbf{R}$ are the standard FSI matrices, additional operators arise from the homogenisation approach, namely:

- the tubes-fluid coupling matrix $\mathbf{C}$, which describes the interaction between the fluid and the tubes:

$$\int_\Omega \nabla p \cdot \delta \mathbf{u}_S \, d\Omega \rightarrow \delta \mathbf{U}_S \mathbf{C}\mathbf{P} \qquad \int_\Omega \frac{\partial^2 \mathbf{u}_S}{\partial t^2} \cdot \nabla \delta p \, d\Omega \rightarrow \delta \mathbf{P}^\mathsf{T} \mathbf{C}^\mathsf{T} \mathbf{U}_S \qquad (4.26)$$

- the added mass matrix $\mathbf{M}_S^*$, which accounts for an additional inertial effect on tubes, arising from the confinement of the fluid in the cell:

$$\int_\Omega \rho(\mathbf{B}_S - |\Omega_S|/|\Omega_T|)\frac{\partial^2 \mathbf{u}_S}{\partial t^2} \cdot \delta \mathbf{u}_S \rightarrow \delta \mathbf{U}_S^\mathsf{T} \mathbf{M}_S^* \ddot{\mathbf{U}}_S \qquad (4.27)$$

Condensing the pressure degrees-of-freedom on the displacement degree-of-freedom in Equation (4.25) yields:

$$\begin{bmatrix} \mathbf{M}_S + \mathbf{M}_S^* + \dfrac{\rho\mathbf{B}_S^{\,2}}{1-\mathbf{B}_S}\mathbf{C}\mathbf{K}_F^{-1}\mathbf{C}^\mathsf{T} & -\dfrac{\rho\mathbf{B}_S}{1-\mathbf{B}_S}\mathbf{C}\mathbf{K}_F^{-1}\mathbf{R}^\mathsf{T} \\ -\dfrac{\rho\mathbf{B}_S}{1-\mathbf{B}_S}\mathbf{R}\mathbf{K}_F^{-1}\mathbf{C}^\mathsf{T} & \mathbf{M}_\Sigma + \dfrac{\rho}{1-\mathbf{B}_S}\mathbf{R}\mathbf{K}_F^{-1}\mathbf{R}^\mathsf{T} \end{bmatrix} \begin{Bmatrix} \ddot{\mathbf{U}}_S(t) \\ \ddot{\mathbf{U}}_\Sigma(t) \end{Bmatrix}$$

$$+ \begin{bmatrix} \mathbf{K}_S & 0 \\ 0 & \mathbf{K}_\Sigma \end{bmatrix} \begin{Bmatrix} \mathbf{U}_S(t) \\ \mathbf{U}_\Sigma(t) \end{Bmatrix} = \begin{Bmatrix} 0 \\ 0 \end{Bmatrix} \qquad (4.28)$$

The inertial coupling is accounted for with the *homogenised added mass matrix*:

$$
\mathbf{M}_A^* =
\begin{bmatrix}
\mathbf{M}_S^* + \dfrac{\rho B_S^{\,2}}{1 - B_S}\mathbf{C}\mathbf{K}_F^{-1}\mathbf{C}^\mathsf{T} & -\dfrac{\rho B_S}{1 - B_S}\mathbf{C}\mathbf{K}_F^{-1}\mathbf{R}^\mathsf{T} \\[2ex]
-\dfrac{\rho B_S}{1 - B_S}\mathbf{R}\mathbf{K}_F^{-1}\mathbf{C}^\mathsf{T} & \dfrac{\rho}{1 - B_S}\mathbf{R}\mathbf{K}_F^{-1}\mathbf{R}^\mathsf{T}
\end{bmatrix}
$$

where $\mathbf{M}_S^* + \rho B_S^{\,2}/(1 - B_S)\mathbf{C}\mathbf{K}_F^{-1}\mathbf{C}^\mathsf{T}$ is the added mass on tubes, $\rho/(1 - B_S)\mathbf{R}\mathbf{K}_F^{-1}\mathbf{R}^\mathsf{T}$ is the added mass on the structure and $-\rho B_S/(1 - B_S)\mathbf{R}\mathbf{K}_F^{-1}\mathbf{C}^\mathsf{T}$ stands for the coupling between the tubes and the structure.

It is shown that the *homogenised model* complies with the condition of fluid mass conservation (Sigrist and Broc, 2008a), *i.e* that Equation (4.10) is verified with $\mathbf{M}_A^*$.

An application of the method is proposed by Veron *et al.* (2014) concerning a 3D tube array subjected to seismic loading, as depicted in Figure 4.16. The response of the system is computed in the time domain, using an implicit time integration of Equation (4.28), as discussed in Chapter 6.

Figure 4.17 and Table 4.2 propose a comparison of the classical and homogenised approaches. If the classical method is selected, a detailed finite element model must be elaborated in order to represent the fluid between all tubes, whereas the homogenised method allows for a drastic simplification of the numerical model.

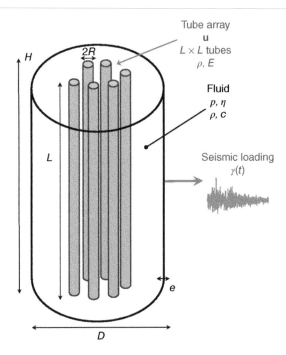

Figure 4.16 Tube array immersed in a fluid with free surface and subjected to seismic loading. u stands for the tube displacement, p and η for the fluid pressure and free level elevation; the structure and fluid material properties are denoted E (Young's modulus) and ρ (density) and c (speed of sound) and ρ (density), respectively

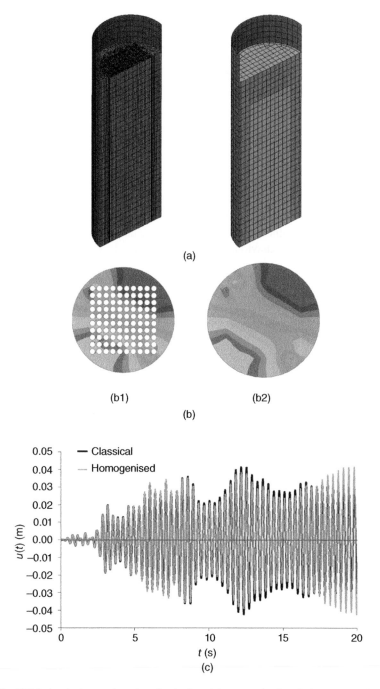

Figure 4.17 FEM simulations using the classical and homogenised techniques. (a) Finite element model. (b) Computation of the pressure field in the fluid, (b1) Classical technique, (b2) Homogenised technique. (c) Computation of the displacement of one tube in the bundle

Table 4.2 Comparison of the classical and homogenised FEM simulations in terms of problem size and computational cost

	Classical model	Homogenised model
Number of nodes	~165,000	~10,000
Number of degrees-of-freedom	~800,000	~50,000
Computational cost	~900,000 s	~50,000 s
Database	~35 Go	~5 Go

As expected, a better numerical efficiency is achieved with the homogenisation method, while preserving the physical consistency of the numerical model. The computational cost is cut down by almost 80%, while data storage requirements are considerably reduced. Furthermore, the homogenisation method is able to capture some detailed pattern of the pressure field in the fluid, while the global behaviour of the tubes is reproduced with a remarkable degree of accuracy.

4.4 Examples: Inertial Effect in Bounded Domain

4.4.1 Analytical Calculation of the Added Mass Matrix

A straight tube of circular cross section (with radius R, height L and thickness h) modelled as an elastic beam is coupled to an incompressible fluid filling a co-axial cylindrical cavity of

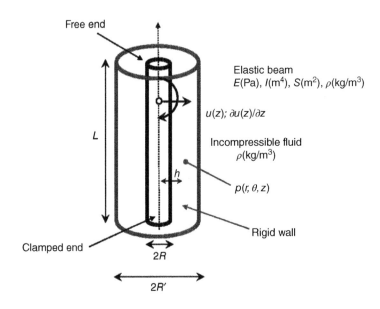

Figure 4.18 Straight cylindrical tube coupled to a fluid filling a cylindrical cavity

radius R', as represented in Figure 4.18. The material properties of the structure and fluid are denoted E and ρ, without explicit reference to structure or fluid to alleviate the notations of the density. The beam is clamped on one end, the other end being free and walls of the cavity are assumed fixed.

The bending motion of the tube immersed in fluid is described by

$$-\omega^2 \rho S u + EI \frac{\partial^4 u}{\partial z^4} = -\int_0^{2\pi} p|_{r=R} \cos\theta R d\theta$$

with clamped/free ends boundary conditions, as expressed in Section 2.3.1. The term on the right-hand side of the equation of motion is the fluid force per unit tube length exerted on section $z \in [0, H]$; the pressure field is solution to the Laplace equation:

$$\frac{\partial^2 p}{\partial r^2} + \frac{1}{r}\frac{\partial p}{\partial r} + \frac{1}{r^2}\frac{\partial^2 p}{\partial \theta^2} + \frac{\partial^2 p}{\partial z^2} = 0 \text{ for } (r,\theta,z) \in [R, R'] \times [0, 2\pi] \times [0, H]$$

with the fixed wall boundary conditions:

$$\frac{\partial p}{\partial z}\bigg|_{z=0} = 0 \qquad \frac{\partial p}{\partial z}\bigg|_{z=L} = 0 \qquad \frac{\partial p}{\partial r}\bigg|_{r=R'} = 0$$

and the coupling condition with the tube immersed wall:

$$\frac{\partial p}{\partial r}\bigg|_{r=R} = \rho\omega^2 u \cos\theta$$

Owing to the simplicity of the cylindrical geometry, an analytical solution of the pressure field may be written in terms of separated variables as $p(r, \theta, z) = p_R(r)p_Z(z)\cos\theta$. p_Z is shown to be a linear superposition of cosine and sine functions and to comply with the fixed wall conditions; only specific cosine terms are admissible, so that

$$p_Z(z) = \cos(q_l z) \text{ with } q_l = \frac{l\pi}{L}, \forall l \geq 1$$

p_R is found to be the solution of

$$\frac{d^2 p_R}{dr^2} + \frac{1}{r}\frac{dp_R}{dr} - \left(q_l^2 + \frac{1}{r^2}\right)p_R = 0$$

- $l = 0$ yields the solution of a bi-dimensional flow: $p_R(r) = \alpha_o r + \frac{\beta_o}{r}$;
- tri-dimensional effects are accounted for with $l \geq 1$. p_R is a linear combination of Bessel functions: $p_R(r) = \alpha_l I(q_l r) + \beta_l K(q_l r)$.

$p(r, \theta, z)$ may therefore be calculated as follows:

$$p(r, \theta, z) = \left(\alpha_o r + \frac{\beta_o}{r}\right)\cos\theta + \sum_{l=1}^{l=+\infty}(\alpha_l I(q_l r) + \beta_l K(q_l r))\cos(q_l z)\cos\theta \qquad (4.29)$$

α_o, β_o, α_l and β_l derive from the boundary and coupling conditions. After some mathematical manipulations,[3] the following expression of the pressure field is arrived at:

$$p(r,\theta,z) = -\rho\omega^2 \frac{R^2}{R'^2 - R^2}(r + \frac{R'^2}{r})\frac{1}{L}\int_0^L u(\zeta)d\zeta \cos\theta$$

$$+\rho\omega^2 \sum_{l\geq 1} \frac{1}{h_l} \frac{I'(q_lR')K(q_lr) - K'(q_lR')I(q_lr)}{I'(q_lR')K'(q_lR) - K'(q_lR')I'(q_lR)} \int_0^L u(\zeta)\cos(q_l\zeta)d\zeta \cos(q_lz)\cos\theta$$

with $h_l = \frac{l\pi}{2}$ for all $l \geq 1$.

The fluid force exerted on the cross section of the tube at z is given by

$$\phi = \rho\omega^2 \pi R^2 \sum_{l\geq 0} \pi_l \int_0^L u(\zeta)\cos(q_l\zeta)d\zeta \cos(q_lz)$$

with:

$$\pi_o = \frac{1}{L}\frac{R'^2 + R^2}{R'^2 - R^2} \qquad \pi_l = -\frac{1}{Rh_l}\frac{I'(q_lR')K(q_lR) - K'(q_lR')I(q_lR)}{I'(q_lR')K'(q_lR) - K'(q_lR')I'(q_lR)}$$

The weighted integral formulation of the structure problem then reads as follows:

$$-\omega^2 \int_0^L \rho S u\, \delta u\, dz + \int_0^L EI\frac{\partial^2 u}{\partial z^2}\frac{\partial^2 \delta u}{\partial z^2}\, dz$$

$$= \omega^2 \rho\pi R^2 \sum_{l\geq 0} \pi_l \int_0^L u\cos(q_lz)\, dz \int_0^L \delta u\cos(q_lz)\, dz \qquad (4.30)$$

The finite element discretisation of this expression yields an eigenvalue problem of the standard form, Equation (4.8), where the added mass matrix is the operator associated with the sum of the products of two integrals:

$$\rho\pi R^2 \sum_{l\geq 0} \pi_l \int_0^L u(z)\cos(q_lz)dz \int_0^L \delta u(z)\cos(q_lz)dz$$

---

[3] The main steps of the calculation are presented hereafter. According to Equation (4.29), the derivative of p with respect to r is

$$\frac{\partial p}{\partial r}(r,\theta,z) = \left(\alpha_o - \frac{\beta_o}{r^2}\right)\cos\theta + \sum_{l=1}^{l=+\infty} q_l(\alpha_lI'(q_lr) + \beta_lK'(q_lr))\cos(q_lz)\cos\theta$$

The projection of the pressure derivative onto the cosine function $\cos(q_lz)$ yields

$$\int_0^{2\pi}\int_0^L \frac{\partial p}{\partial r}(r,\theta,z)\cos\theta\cos(q_lz)\, d\theta dz = q_l(\alpha_lI'(q_lr) + \beta_lK'(q_lr))\int_0^{2\pi}\int_0^L \cos^2\theta\cos^2(q_lz)\, d\theta dz$$

Using the coupling condition on $r = R$ gives

$$\rho\omega^2 \int_0^L u(z)\cos(q_lz)\, dz = \frac{l\pi}{2}(\alpha_lI'(q_lR) + \beta_lK'(q_lR))$$

and the boundary condition on $r = R'$ gives $0 = \frac{l\pi}{2}(\alpha_lI'(q_lR') + \beta_lK'(q_lR'))$. The expressions of α_l and β_l ensue from the preceding relations. α_o and β_o are obtained in a similar manner, with the projection of $\frac{\partial p}{\partial r}$ on the component describing the 2D-flow expressed as follows:

$$\int_0^L \frac{\partial p}{\partial r}(r,\theta,z)\, dz = \left(\alpha_o - \frac{\beta_o}{r^2}\right)$$

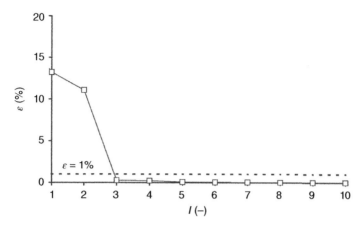

Figure 4.19   2D and 3D contributions of the fluid flow. The graph illustrates the convergence of the analytical solution of the fourth mode frequency. The influence of the number of terms in the series expressed by Equation (4.30) is evidenced (with $I = 40$ finite elements). In the present case, the slenderness ratio R/L is close to unity: the fluid flow has a marked 3D aspect, so that taking only the 2D component of the fluid flow is not sufficient to account for the inertial effect

The above expression indicates that the fluid flow, described by the pressure as a linear combination of cosine functions in the axial direction, may induce a coupling of the structural modes. The added mass matrix is obviously symmetric, and it assembles the contribution of the 2D and 3D flows.

For each term of the series $\rho\pi R^2 \sum_{l=0}^{l=+\infty} \pi_l(\bullet)$, an elementary added mass matrix $\mathbf{m}_a^{i,i'}(l)$ is calculated according to

$$\mathbf{m}_a^{i,i'}(l) = \int_{z_{i'}}^{z_{i'+1}} \mathbf{N}_{i'}^{\mathsf{T}}(z)\cos(q_l z)\, dz \int_{z_i}^{z_{i+1}} \mathbf{N}_i(z)\cos(q_l z)\, dz$$

$\mathbf{N}_i$ is the shape function for finite element i: in this case, elements as described in Section 2.3.2 are used.

A numerical application is proposed with the following parameters: $L = 1$ m, $R = 0.1$ m, $h = 0.01$ m, $\rho = 7,800$ kg/m^3 (for the structure), $E = 2.1 \cdot 10^{11}$ Pa, $R' = 0.2$ m and $\rho = 1,000$ kg/m^3 (for the fluid). Calculations are performed with $I = 40$ finite elements and $l = 10$ terms of the series in Equation (4.30), ensuring sufficiently accurate results as the shapes and the frequencies of the first modes are concerned, as illustrated in Figure 4.19.

The influence of the fluid on the structure behaviour is twofold.

On eigenmode shapes The fluid does couple eigenmodes of the structure as defined in vacuo (the added mass matrix is fully populated), so that mode shapes of the structure in the fluid may differ more or less significantly from the modes observed in vacuo. Comparison of the modal bases with and without fluid can be performed using the so-called *modal assurance criterion* (MAC), which is traditionally used to quantify the correlation between two sets of eigenmodes (Allemang, 2003). Let $(\mathbf{U}_n^o)_{n\geq 1}$ and $(\mathbf{U}_n^h)_{n\geq 1}$ the modes computed for the structure with and without fluid; the coefficients of the MAC matrix $\mathbf{\Psi}$ associated with these modal bases are defined as:

$$\mathbf{\Psi}_{n,n'} = \frac{(\mathbf{U}_n^o, \mathbf{U}_{n'}^h)^2}{\|\mathbf{U}_n^o\|^2 \|\mathbf{U}_{n'}^h\|^2}$$

In this relation, the scalar product is $(\mathbf{U}_n^o, \mathbf{U}_{n'}^h) = \bar{\mathbf{U}}_n^{o\mathsf{T}} \mathbf{U}_{n'}^h$, and the associated norm is $\|\mathbf{U}_n^o\|^2 = (\mathbf{U}_n^o, \mathbf{U}_n^o)$, where $\bar{\bullet}$ and $\bullet^\mathsf{T}$ denote the complex conjugate and the transpose of $\bullet$.

$0 \leq \mathbf{\Psi}_{n,n'} \leq 1$ identifies the correlation between $\mathbf{U}_n^o$ and $\mathbf{U}_{n'}^h$: $\mathbf{\Psi}_{n,n'} = 1$ indicates a perfect correlation, whereas $\mathbf{\Psi}_{n,n'} = 0$ means no correlation. Figure 4.20 gives the MAC matrix for two sets of 20 eigenmodes computed with and without fluid coupling: the diagonal-dominant property of the MAC matrix indicates that for $n = n'$, modes with and without fluid are roughly identical. Modes of distinct rank are significantly less correlated to one another, indicating that mode coupling by FSI remains limited to a small amount in this example.

On eigenfrequencies The frequencies of the structure coupled with fluid are *lower* compared with the natural frequencies without fluid, as indicated by Equation (4.12). As highlighted above, the mode shapes of the structure remain unchanged, whether coupled to the fluid or not. Quantifying the inertial effect for each mode is straightforward if, as an additional property, the structure mode shapes are orthogonal with respect to the mass matrix. In such cases, with m_n and k_n denoting the modal mass and stiffness of mode n, the pulsation of the mode without fluid is k_n/m_n and the pulsation of the mode with fluid is $k_n/(m_n(1 + \mu_n))$. μ_n is a modal added mass coefficient, calculated as $\mu_n = \mathbf{U}_n \mathbf{M}_a \mathbf{U}_n$. Of course, the validity of

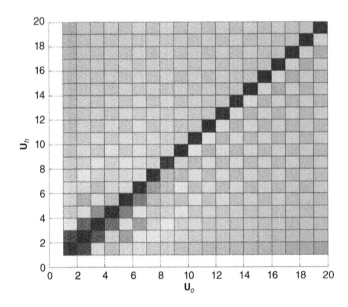

Figure 4.20 MAC matrix of modal bases for the structure with and without fluid coupling; comparison for the first 20 modes (for better readability, the MAC matrix is represented in a grey scale, with black standing for 1 and light grey standing for 0)

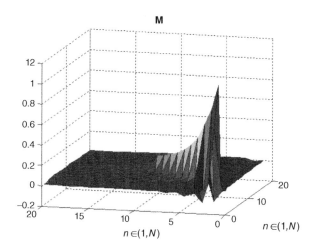

Figure 4.21   Projection of the added mass matrix onto the first N structure eigenmode shape in vacuo

Table 4.3   Inertial effect: frequencies of the structure computed with and without fluid coupling

Frequency	Structure w/o. fluid (Hz)	Structure w. fluid (Hz)	μ (%)
f_1	195	136	105
f_2	1,224	910	80
f_3	3,427	2,697	60
f_4	6,716	5,542	45

such results is restricted to the case where the following orthogonality holds $\mathbf{U}_n \mathbf{M}_a \mathbf{U}_{n'} = 0$ for $n \neq n'$, which is the case here, as evidenced in Figure 4.21.

Table 4.3 gives the frequencies of the beam with and without fluid coupling, as well as the corresponding added mass coefficient. Inertial effects are more pronounced for the first bending modes and decrease as the rank of the mode is increased, because as the wavelength of the mode shape is increased, the associated fluid flow tends to fade out more rapidly. In the present example, the fluid inertia may be safely neglected for modes of rank equal to and greater than 10, as highlighted in Figure 4.21.

In this case, the major properties of the inertial coupling are easily evidenced, partly as a consequence of the simple geometry at hand. These properties can largely differ depending on the particularities of the system, and the same analysis process may however be applied.

4.4.2   Numerical Computation of the Added Mass Matrix

A rectangular cavity of width R and height L, filled with an incompressible fluid of density ρ is considered, as represented in Figure 4.22. The upper and lower walls, as well as the lateral

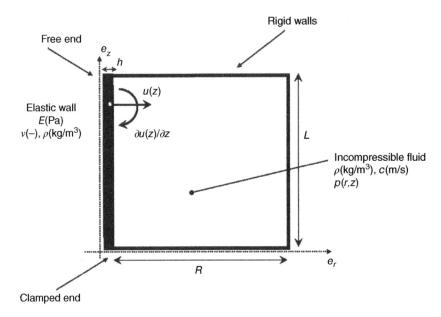

Figure 4.22   Rectangular cavity filled with a liquid and provided with one elastic wall, the others being assumed rigid

right wall, are rigid, whereas the lateral left wall is assumed elastic. The latter is modelled as a plate of thickness h, density ρ, Young's modulus E and Poisson's coefficient v; clamped/free conditions are assumed at the lower and upper ends.

$u(z)$ is the plate displacement in direction r along the length of the plate, $z \in [0, L]$; using a unidimensional model, the bending motion of the plate is governed, in fact, by the following beam equation:

$$-\omega^2 \rho h u + \frac{Eh^3}{12(1 - v^2)} \frac{\partial^4 u}{\partial z^4} = -p|_{r=0} \text{ for } z \in [0, L]$$

with fluid loading represented by the pressure force $-p|_{r=0}$ at the wall. Clamped/free ends imply the following boundary conditions:

$$u|_{z=0} = 0 \qquad \frac{\partial u}{\partial z}\bigg|_{z=0} = 0 \qquad \frac{\partial^2 u}{\partial z^2}\bigg|_{z=L} = 0 \qquad \frac{\partial^3 u}{\partial z^3}\bigg|_{z=L} = 0$$

The fluid pressure field $p(r, z)$ within the water tank $(r, z) \in [0, R] \times [0, L]$ satisfies the Laplace equation:

$$\frac{\partial^2 p}{\partial r^2} + \frac{\partial^2 p}{\partial z^2} = 0 \text{ for } (r, z) \in [0, R] \times [0, L]$$

together with the coupling conditions at the moving and fixed walls:

$$\frac{\partial p}{\partial r}\bigg|_{r=0} = \rho \omega^2 u \qquad \frac{\partial p}{\partial r}\bigg|_{r=R} = 0 \qquad \frac{\partial p}{\partial z}\bigg|_{z=0,L} = 0$$

It is recalled that the problem is well posed when the following conditions are verified.

- Uniqueness condition of the pressure field:

$$\int_0^R \int_0^L p(r,z)\, drdz = 0 \qquad (4.31)$$

- Compatibility condition of the structure displacement:

$$\int_0^L u(z)\, dz = 0 \qquad (4.32)$$

The weighted integral formulation of the problem is expressed as follows:

$$-\omega^2 \int_0^L \rho h u \delta u \, dz + \int_0^L \frac{Eh^3}{12(1-v^2)} \frac{\partial^2 u}{\partial z^2} \frac{\partial^2 \delta u}{\partial z^2}\, dz - \lambda \delta u|_{z=0} - \lambda' \frac{\partial \delta u}{\partial z}\Big|_{z=0}$$

$$-\lambda_u \int_0^L u\, dz = -\int_0^L p|_{r=0} \delta u\, dz \qquad \forall \delta u$$

for the structure part, and

$$\int_0^R \int_0^L \left(\frac{\partial p}{\partial r} \frac{\partial \delta p}{\partial r} + \frac{\partial p}{\partial r} \frac{\partial \delta p}{\partial r} \right)\, drdz - \lambda_p \int_0^R \int_0^L \delta p\, drdz = \rho\omega^2 \int_0^L u \delta p\, dz \qquad \forall \delta p$$

for the fluid part.

λ and λ' are the Lagrange multipliers associated with the boundary conditions at $z = 0, L$; λ_p and λ_u are associated with conditions (4.31) and (4.32), respectively.

The finite element discretisation couples 1D beam-type elements as described in Section 2.3.2, with the moment of inertia $I = \dfrac{h^3}{12(1-v^2)}$, to 2D acoustic-type fluid elements presented in Section 3.7.2. As the potential energy of the fluid is discarded here, the 'mass' matrix of the pressure-formulated element is not needed.

The geometry of the present example is so simple that a one-to-one element coupling, as depicted in Figure 4.23, is easily achieved. The finite element discretisation of the coupling term in the weighted integral formulation is as follows:

$$\int_0^L p|_{r=R} \, \delta u\, dz = \delta \mathbf{U}^{\mathsf{T}} \sum_{i=1}^{i=I} \mathbf{\Lambda}_S^{i\,\mathsf{T}} \int_{z_i}^{z_{i+1}} \mathbf{N}_S^{i\,\mathsf{T}} \mathbf{N}_F^{i,1}\, dz\, \mathbf{\Lambda}_F^{i,1}\, \mathbf{P}$$

With cubic shape functions $\mathbf{N}_S$ for the structure and linear shape functions $\mathbf{N}_F$ for the fluid, the elementary fluid–structure coupling matrix is found as follows:

$$\mathbf{r}_i = \pi R \begin{bmatrix} \dfrac{7l_i}{20} & 0 & 0 & \dfrac{3l_i}{20} \\[2mm] \dfrac{l_i^2}{20} & 0 & 0 & \dfrac{l_i^2}{30} \\[2mm] \dfrac{3l_i}{20} & 0 & 0 & \dfrac{7l_i}{20} \\[2mm] \dfrac{-l_i^2}{30} & 0 & 0 & \dfrac{-l_i^2}{20} \end{bmatrix}$$

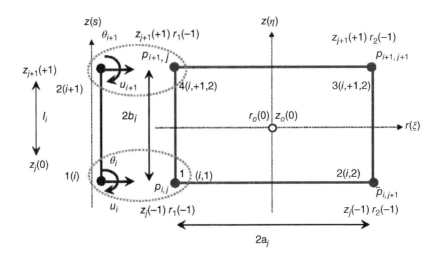

Figure 4.23   1D beam element coupled to 2D acoustic element

Equation (4.8) is finally arrived at: in the present case, the compatibility condition is accounted for with matrix L_S, which is calculated as $[L_S] = \langle 1 \rangle [M_S]/(\mu h)$.

The eigenmodes of the structure without fluid are computed adopting the following numerical data: $L = 1$ m, $R = 0.5$ m, $h = 0.1$ m, $\rho = 7,800$ kg/ m^3, $E = 2.1 \cdot 10^{11}$ Pa, $v = 0.3$. Figure 4.24 gives the shapes of the first four bending modes obtained with and without the compatibility condition, Equation (4.32). $I = 40$ beam elements are sufficient to achieve good accuracy as far as the computed data of the first four eigenmodes in the absence of the fluid are concerned.

The first bending modes of the non-constrained structure obviously do not comply with the compatibility condition. The other modes do not comply with this condition either; however, their mode shapes are roughly similar to those of the constrained beam, as made conspicuous in Figure 4.24: the mode shapes indexed 2,3 and 4 of the constrained beam are similar to those of the constrained beam indexed 1, 2 and 3.

Fluid–structure coupling is described by the added mass matrix, which is determined according to Equation (4.9). The computation is performed using a refined fluid mesh of $I \times J = 20 \times 40$ fluid elements for achieving good accuracy.

The first four eigenmodes of the fluid–structure coupled system are depicted in Figure 4.25, where the beam lateral displacement of the flexible wall is superposed on a set of isobars in the fluid rectangular domain.[4] The compatibility condition is enforced on the structure: as evidenced in Table 4.4, when performed without the compatibility condition, the computation does not yield mode shapes which are consistent with the physics of the FSI.

The inertial effect is further highlighted in Figure 4.26, which illustrates the influence of the fluid confinement on inertial effect, using the coefficient μ introduced in the preceding

---

[4] The pressure field in the fluid is deduced from the structure acceleration according to $\mathbf{P} = \rho\omega^2 \mathbf{K}_F^{-1} \mathbf{R}^\mathsf{T} \mathbf{U}$.

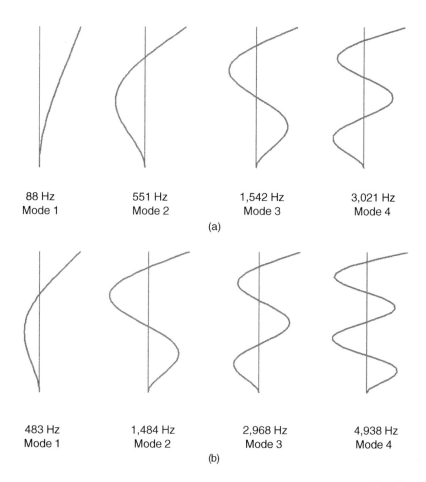

Figure 4.24 Structure bending modes (b) with and (a) without compatibility condition $\int_0^L u(z)dz = 0$ and without fluid coupling

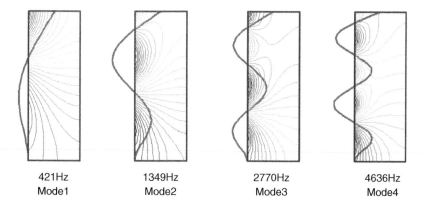

Figure 4.25 Mode shapes of the flexible wall coupled to the incompressible fluid within the rectangular tank

Table 4.4 Natural frequencies of a few coupled modes with and without the compatibility condition

Frequency	Modes w. compatibility (Hz)	Modes w/o. compatibility (Hz)
f_0	–	77
f_1	421	493
f_2	1,349	1,411
f_3	2,770	2,834
f_4	4,636	4,748

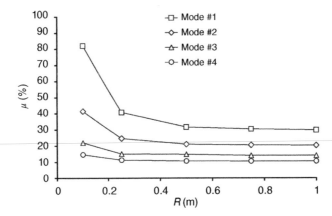

Figure 4.26 Inertial effect: influence of fluid confinement

example. The inertial effect is related to the kinetic energy of the fluid set into motion by the structure:

- the low-order modes, which generate more important motions of the fluid than the high-order modes, are much affected by the presence of the fluid;
- for all modes, the fluid oscillations are essentially limited within a rather small area next to the vibrating wall of a typical thickness, which decreases as the rank of the mode is increased. Therefore, the fluid inertia remains almost constant when the thickness of the fluid exceeds the characteristic value.

Remark 4.2 Retrieval of the fluid mass from the added mass matrix *It is recalled that the mass of fluid m_F contained within the tank can be retrieved from the added mass matrix according to*

$$\mathbf{\Delta}^T \mathbf{M}_A \mathbf{\Delta} = m_F$$

with $\mathbf{\Delta}$ a unit vector in a given direction. This relation holds when $\mathbf{R}$ is calculated by taking into account the appropriate boundary conditions at all the fluid boundaries (in the present case,

$\Delta \equiv \mathbf{e}_r$). *In order to build the finite element model, the matrix* $\mathbf{R}$ *is assembled using elementary matrices* $\mathbf{r}_e$ *and* $\mathbf{r}_{e'}$, *which account for inertial coupling of the fluid to the deformable wall at* $r = 0$ *and to the rigid and fixed wall at* $r = R$, *as sketched in Figure 4.27.*

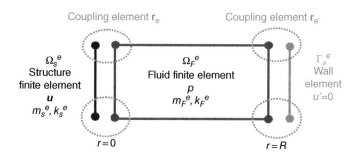

Figure 4.27 *FSI matrix* $\mathbf{R}$ *with coupling elements* $\mathbf{r}_e$ *and* $\mathbf{r}_{e'}$ *on* $r = 0$ *and* $r = R$

Considering a rectangular tank of dimension $R = 0.75$ m, $L = 1.25$ m *filled with a fluid of density* $\rho = 875$ kg/ m³, *the mass of fluid contained in the volume is* $m_F = \rho R L = 830$ kg/m, *while a numerical computation with the added mass matrix* $\mathbf{M}_A$ *assembled with* $I \times J = 15 \times 25$ *elements gives the value* $\Delta^T \mathbf{M}_A \Delta = 831$ kg/m. ∎

4.5   Example: Inertial Effect in Unbounded Domain

4.5.1   Elastic Ring Immersed in a Fluid

The example is concerned with a bi-dimensional model aimed at describing the natural modes of vibration of a thin cylindrical shell of revolution immersed in an infinite extent of fluid. Provided that the shell height $2L$ is large enough and that interest is focussed on the shell modes – in contrast with the beamlike bending modes – the problem may be conveniently simplified by restricting the geometry into a single plane normal to the shell axis, as represented in Figure 4.28.

The in-plane eigenmodes of a circular ring have already been discussed in Chapter 2, Section 2.4. As highlighted in Figure 2.29, which plots the eigenfrequencies of the ring versus the rank of the mode, two distinct branches are obtained on account of the coupling between bending and membrane strains. Each branch characterises distinct mode shapes, dominated by either a radial or ortho-radial displacement. Depending on the rank of the mode, the eigenfrequency of the first type of mode is lower, or higher, than that of the second type of mode. The study is extended here to the case where the ring is coupled to an infinite extent of the fluid. As a preliminary to the quantitative analysis of the problem, it may be anticipated that the modes dominated by bending, hence by radial displacement, will be found to be more affected by the fluid than the modes dominated by membrane, hence by ortho-radial displacement.

Let $u_r(\theta)$ and $u_\theta(\theta)$ denote the radial and ortho-radial displacements of the structure, respectively, and $p(r, \theta)$ denote the pressure in the fluid domain, which is considered unbounded. The incompressible flow is again governed by $\Delta p = 0$ with a coupling condition, $\dfrac{\partial p}{\partial r} = \rho \omega^2 u_r$ at $r = R$, and a condition at infinity, of the type $p < \infty$ for $r \to \infty$.

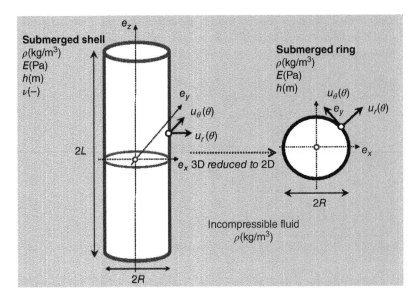

Figure 4.28   3D shell and 2D ring submerged in a fluid. For slenderness ratio $2L/R$ larger than 10 typically, the 3D shell may be described with a 2D model. In the process, the shell equations are reduced to that of a thin circular ring in its own plane, while the fluid fills an annular space extending from infinity.

Equations of bending/membrane modes are detailed in Section 2.4; with fluid coupling, Equation (2.30) is modified to take into account the pressure loading:

$$-\omega^2 \rho h u_r + \frac{Eh}{R^2}\left[u_r + \frac{\partial u_\theta}{\partial \theta} + \frac{h^2}{12R^2}\left(\frac{\partial^4 u_r}{\partial \theta^4} - \frac{\partial^3 u_\theta}{\partial \theta^3}\right)\right] = -p|_{r=R}$$

while Equation (2.31) remains unchanged.

Projection of the Laplace equation onto the Fourier component of order m for pressure is readily found to be

$$\frac{\partial^2 p_m}{\partial r^2} + \frac{1}{r}\frac{\partial p_m}{\partial r} - \frac{m^2}{r^2}p_m = 0$$

with boundary conditions at $r = R$ and for $r \to \infty$:

$$\left.\frac{\partial p_m}{\partial r}\right|_{r=R} = \rho\omega^2 u_r^m \qquad p_m(r \to \infty) < \infty$$

The pressure field is calculated as follows:

$$p_m(r, \omega) = -\rho\omega^2\frac{R^{m+1}}{mr^m}u_r^m(\omega)$$

so that $p_m(R, \omega) = -\rho R/m\omega^2 u_r^m$. According to the preceding relation, the pressure at infinity is null, which means that being far away enough from the perturbation source, the fluid may safely be considered at rest.

For the mode $m = 0$, the pressure field is singular, which indicates that no structure motion is possible – in other words, the axisymmetric breathing mode of the ring does not comply with the compatibility condition associated with the Laplace equation.

The discretised equations governing the motion of the ring reads as the canonical form $(-\omega^2(\mathbf{M}_S^m + \mathbf{M}_A^m) + \mathbf{K}_S^m)\mathbf{U}_m = \mathbf{0}$, where $\mathbf{M}_S^m$ and $\mathbf{K}_S^m$ have been introduced in Section 2.4.2, and $\mathbf{M}_A^m$ is the added mass matrix of the ring:

$$\mathbf{M}_A^m = \begin{bmatrix} \rho R \mu_m^\infty & 0 \\ 0 & 0 \end{bmatrix}$$

μ_m^∞ is the added mass coefficient associated with the Fourier θ−harmonic of order m:

$$\mu_m^\infty = \frac{1}{m} \qquad \forall m \neq 0$$

In the present problem, physical effects induced by FSI are particularly simple because no inertial coupling occurs between any Fourier components $m \neq m'$, nor between radial and ortho-radial displacements for a given Fourier component.

The pressure field is fading out to zero with the radial distance, all the more so as the rank of the mode is increased: this means that the fluid flow is unperturbed away from the structure. From the numerical point of view, it is therefore possible to take into account only a finite extent of the fluid, which paves way for finite element discretisation.

For a fluid domain bounded at $r = R'$, the pressure field satisfies the Laplace equation together with the coupling condition with the structure at $r = R$. At $r = R'$ away enough from the structure, it is sound to impose the condition $p = 0$. The Fourier component of order m for the pressure is found to be

$$p_m(r, \omega) = -\rho\omega^2 \frac{1}{m} u_r^m \frac{R^{m+1}}{R'^{2m} - R^{2m}} \left(r^m - \frac{R'^{2m}}{r^m} \right)$$

The wall pressure is $p_m(R, \omega) = -\rho R \mu_m^\infty(R, R')\omega^2 u_r^m$, with the added mass coefficient for mode m expressed as follows:

$$\mu_m^\infty(R, R') = \frac{1}{m} \frac{R'^{2m} + R^{2m}}{R'^{2m} - R^{2m}} \qquad \forall m \neq 0$$

Figure 4.29 compares the added mass coefficients with bounded and unbounded fluid domains, for the first Fourier components. The inertial effect is represented with good accuracy for all modes $m \geq 2$ with $R' = 3R$; a larger fluid domain $(R' > 5R)$ is however required for mode $m = 1$.

4.5.2 Finite Element Coupling with Infinite Element

The weighted integral formulation of the problem reads as

$$\int_R^\infty \frac{\partial p_m}{\partial r} \frac{\partial \delta p_m}{\partial r} r dr + m^2 \int_R^\infty p_m \delta p_m \frac{dr}{r} = \rho R\omega^2 u_m \delta p_m|_{r=R}$$

The finite element discretisation requires to couple one 0D structure element as described in Section 2.4.2 to a finite element mesh of the fluid domain composed of:

- a set of I 1D finite elements for $[R, R']$, as shown in Figure 4.30. Each fluid element has two nodes with linear shape functions, as detailed in Section 3.8.2. FSI is entirely described by

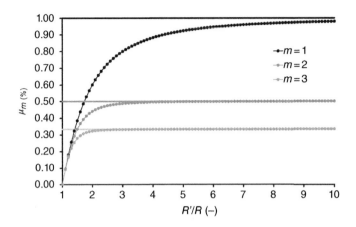

Figure 4.29 Added mass coefficient for Fourier component m: bounded and unbounded fluid domains

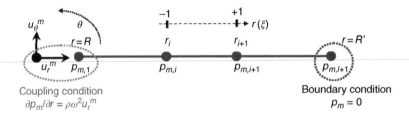

Figure 4.30 0D structure coupled to 1D fluid axisymmetric elements

the functional term $-\rho R\omega^2 u_r^m \delta p_m|_{r=R}$, which is also expressed as $-\rho\omega^2 \delta\mathbf{P}_m\mathbf{R}^\mathsf{T}\mathbf{U}_m$, making use of the fluid–structure coupling matrix $\mathbf{R}$:

$$\mathbf{R} = \begin{array}{c} \\ u_r^m \\ u_\theta^m \end{array} \overset{\begin{array}{ccccccc} p_1^m & p_2^m & \cdots & p_i^m & p_{i+1}^m & \cdots & p_I^m & p_{I+1}^m \end{array}}{\left(\begin{array}{cccccccc} R & 0 & \cdots & 0 & 0 & \cdots & 0 & 0 \\ 0 & 0 & \cdots & 0 & 0 & \cdots & 0 & 0 \end{array}\right)}$$

- One 1D infinite element for the interval $[R', \infty[$. In the present case, a mapped infinite element is used, as sketched in Figure 4.31. It is a two-node element with $r_I = R'$ and $r_{I+1} = \infty$ and linear shape functions. The local to global mapping is defined as follows:

$$r(\xi) = \frac{2R'}{1-\xi} \leftrightarrow \xi(r) = 1 - \frac{2R'}{r}$$

so that $\xi = -1$ corresponds to $r = R'$ and $\xi = +1$ to $r = \infty$. The fluid pressure for $r > R'$ is then calculated as follows:

$$p_m(\xi) = p_m' \frac{1-\xi}{2} + p_m^\infty \frac{1+\xi}{2} \qquad \xi \in [-1,+1]$$

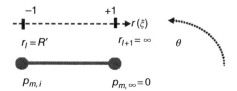

Figure 4.31   1D fluid axisymmetric infinite element

with p'_m and p^∞_m the pressure at $r = R'$ and $r \to \infty$, respectively; since the latter is null, $p_m(\xi) = p'_m/2(1 - \xi)$.

The contribution of domain $[R', \infty]$ to the weighted integral formulation is

$$\int_{R'}^{\infty} \frac{\partial p_m}{\partial r} \frac{\partial \delta p_m}{\partial r} r dr + m^2 \int_{R'}^{\infty} p_m \delta p_m \frac{dr}{r}$$

$$= \int_{-1}^{+1} \frac{\partial p_m(\xi)}{\partial \xi} \frac{\partial \delta p_m(\xi)}{\partial \xi} \left(\frac{\partial \xi}{\partial r} \right)^2 r(\xi) dr(\xi) + m^2 \int_{-1}^{+1} p_m(\xi) \delta p_m(\xi) \frac{dr(\xi)}{r(\xi)}$$

$$= (1 + m^2) \frac{p'_m \delta p'_m}{4} \int_{-1}^{+1} (1 - \xi) d\xi = (1 + m^2) \frac{p'_m \delta p'_m}{2}$$

This contribution is represented by an elementary 'stiffness' matrix associated with the infinite element:

$$\mathbf{k}^{\infty,I}_{F,m} = \begin{bmatrix} (1 + m^2)/2 & 0 \\ 0 & 0 \end{bmatrix}$$

In order to check the interest of using a layer of infinite fluid elements as an external boundary to account for the appropriate condition on pressure at infinity, the computations of the added mass matrix are carried out using two distinct types of models successively:

- the first type comprises finite elements solely with the condition $p = 0$ as the artificial boundary condition $r = R'$. Several simulations are performed where the fluid domain is extended progressively from $R' = 2R$ to $R' = 16R$;
- in the second type of model, which combines a sub-domain discretised with finite elements and an external infinite element, the fluid domain is radially limited to $R' = 2R$.

The numerical application presented here refers to the following input data: $R = 0.5$ m, $h = 0.05$ m, $E = 2.1 \times 10^{11}$ Pa, $\rho = 7,800$ kg/ m³ for the structure and $\rho = 1,000$ kg/ m³ for the fluid. The values of the added mass relative to the first three eigenmodes are gathered in Table 4.5.

It may be concluded that the technique of combining finite and infinite elements proves efficient: accurate numerical values of the added mass coefficient are obtained with a limited fluid domain ($R' = 2R$) discretised by using a few elements ($I = 10$), plus a single infinite element. To achieve similar accuracy with finite elements exclusively, it is necessary here to discretise a much larger fluid domain, involving a substantial increase in the computational cost. As could be anticipated, this is especially true as far as the Fourier component $m = 1$ is concerned, as highlighted in Figure 4.32.

Table 4.5 Added mass computation with finite elements with and without infinite element

	Analytical solution	Finite elements infinite element	Numerical computation Finite elements Boundary condition $p = 0$ at $r = R'$			
m	$\rho\pi R^2 \mu_m^\infty$ kg/m	$R' = 2 \times R$ kg/m	$R' = 2 \times R$ kg/m	$R' = 4 \times R$ kg/m	$R' = 8 \times R$ kg/m	$R' = 16 \times R$ kg/m
1	785	784	471	627	760	778
2	392	386	345	388	391	391
3	262	259	252	260	260	261

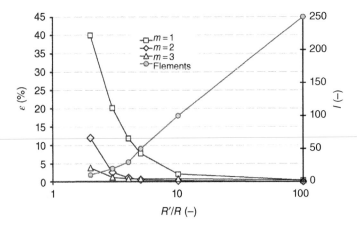

Figure 4.32 Convergence of added mass with increasing size of the fluid domain (ε refers to the relative error between the numerical computation and the analytical solution of the added mass; I is the number of fluid finite elements which are used to discretise the fluid domain $[R, R']$)

The mode shape of the in-plane natural modes of vibration with Fourier components $m = 1$, 2 and 3 of the elastic ring coupled to an in-plane domain of incompressible fluid are depicted in Figure 4.33 in a similar way to that in the previous example. The displacement field of the ring is normalised to a maximum amplitude equal to a fraction of its radius and it is superposed to the non-deformed configuration. The pressure field within the fluid domain of the finite element model is computed with the aid of the relation $\mathbf{P} = -\rho\omega^2 \mathbf{K}_F^{-1} \mathbf{R}^T \mathbf{U}$ and pictured as a discrete set of isobars.

The in-plane modes of the ring are coupling bending and membrane strains. To each value of the Fourier component m is related a pair of distinct eigenmodes which differs as far as the frequency and mode shape are concerned: one is dominated by radial displacement and the other is by ortho-radial displacement. Depending on the rank of the Fourier component, the eigenfrequency of the first type of mode is lower or higher than the eigenfrequency of the second type of mode.

The zero frequency mode $m = 1$ stands for a free non-accelerated translation $\omega = 0$, so no pressure perturbation is induced in the fluid. The second mode is an ortho-radial-dominant

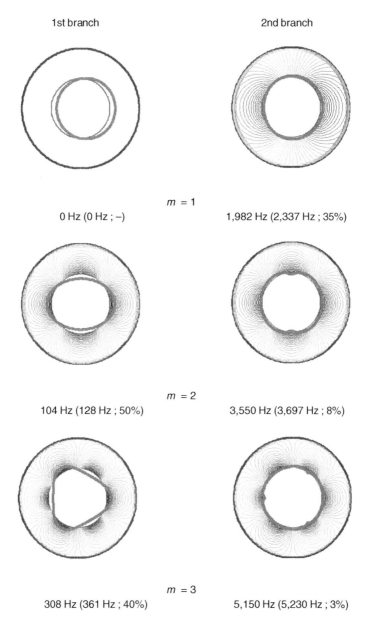

1st branch   2nd branch

$m = 1$

0 Hz (0 Hz ; –)   1,982 Hz (2,337 Hz ; 35%)

$m = 2$

104 Hz (128 Hz ; 50%)   3,550 Hz (3,697 Hz ; 8%)

$m = 3$

308 Hz (361 Hz ; 40%)   5,150 Hz (5,230 Hz ; 3%)

Figure 4.33   Mode shapes for an elastic fluid ring immersed in an incompressible fluid. The numerical computations are performed with finite elements complemented by one infinite element at radial distance $R' = 2R$. The frequency of the structure without fluid is indicated in brackets. The corresponding added mass coefficient is also indicated: it is evaluated according to $\mu = 1/\beta^2 - 1$ with $\beta = f_{w.fluid}/f_{w/o.fluid}$

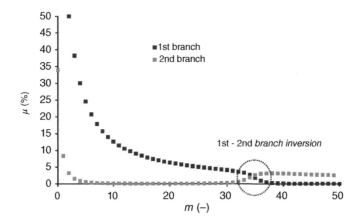

Figure 4.34   Added mass coefficient of the elastic ring modes

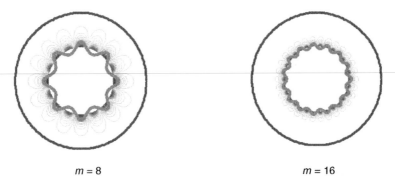

$m = 8$ $m = 16$

Figure 4.35   Mode shapes of (fairly) a higher rank for an elastic fluid ring immersed in an incompressible fluid

mode at frequency $1,982$ Hz instead of $2,337$ Hz in vacuo, leading to the added mass coefficient $\mu \approx 40\%$. For $m = 2$ and 3, the inertial effect is marked for the first branch (which is radial dominant), while it is much less significant for the second branch (which is ortho-radial dominant).

As m is increased, the inertial effect is less pronounced for both radial- and ortho-radial-dominant modes, as marked in Figure 4.34.

The fluid flow induced by the structure deformations remains limited for Fourier components with high index, as underlined in Figure 4.35: low kinetic energy in the fluid is therefore low added mass on the structure. The first/second mode inversion is also apparent with fluid coupling: for Fourier component $m \geq 35$, the added mass coefficient is higher for the second branch mode than for the first branch.

The added mass computation may also be carried out using an integral formulation of the Laplace equation. It is a convenient alternative to artificial boundary conditions implemented

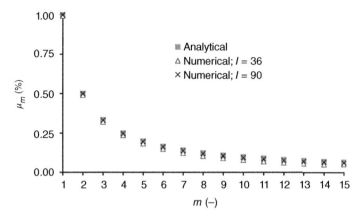

Figure 4.36   BEM computation of the added mass coefficient μ_m^∞

with finite elements, since for exterior problems, the Green function complies with the condition at infinity – as expressed by Equation (4.15) for bi-dimensional problems.

An application of the boundary element technique developed in Section 3.8.3 yields the added mass coefficient projected on each Fourier component, according to Equation (3.68). Figure 4.36 compares the analytical value of $\mu_m^\infty = 1/m$ and the computed values obtained, respectively, with $I = 36$ and $I = 90$ boundary elements. In this case, the accuracy of the BEM approach is satisfactory, while the computational effort remains limited, so that the BEM offers an efficient alternative to finite element method (FEM) and IEM for dealing with inertial coupling – generally, in many applications of engineering interest, both methods seem to be possible options to the designer.

References

Allemang RJ 2003 The modal assurance criterion – twenty years of use and abuse. *Sound and Vibration Magazine*, **37**, 14–20.

Bettess P 1977 Infinite elements. *International Journal for Numerical Methods in Engineering*, **11**, 53–64.

Bettess P 1980 More on infinite elements. *International Journal for Numerical Methods in Engineering*, **15**, 1613–1626.

Bettess P 1992 *Infinite Elements*. Penshaw Press.

Blevins RD 1979 *Formulas for Natural Frequency and Mode Shape*. Krieger Publishing Company.

Broc D and Sigrist JF 2014 Modelling inertial effects in periodic fluid-structure systems with an homogenization approach: application to seismic analysis of tube bundles. *Journal of Fluids and Structures*, **49**, 73–90.

Brochard D, Gantenbein F, and Gibert RJ 1987 Modelling of the dynamic behaviour of LWR internals. In *Proceedings of the 9th International Conference on Structural Mechanic in Reactor Technology (SMiRT 87)*.

Fritz RJ 1972 The effect of liquids on the dynamic motion of immersed solids. *Journal of Engineering for Industry*, **91**, 167–173.

Gerdes K 2000 A review of infinite element method for exterior Helmholtz problems. *Journal of Computational Acoustics*, **8**, 43–62.

Planchard J 1985 Modelling the dynamic behaviour of nuclear reactor fuel assemblies. *Nuclear Engineering and Design*, **90**, 331–339.

Planchard J 1987 Global behaviour of large elastic tube bundle immersed in a fluid. *Computational Mechanics*, **2**, 105–118.

Sigrist JF and Broc D 2008a Dynamic analysis of a tube bundle with fluid-structure interaction modelling using a homogenisation Method. *Computer Methods in Applied Mechanics and Engineering*, **197**, 1080–1099.

Sigrist JF and Broc D 2008b Homogenisation method for the dynamic analysis of a complete nuclear steam generator with fluid-structure interaction. *Nuclear Engineering and Design*, **238**, 2261–2271.

Veron E, Sigrist JF, and Broc D 2014 Implementation of a structural-acoustic homogenized method for the dynamic analysis of a tube bundle with fluid-structure interaction modeling within ABAQUS: formulation and applications. In *Proceedings of ASME Pressure Vessels & Piping Conference (PVP 2014)*.

Zhang RJ 1999 A beam bundle in a compressible inviscid fluid. *Journal of Applied Mechanics*, **66**, 546–548.

Zhang RJ, Wang WQ, Hou SH, and Chan CK 2001 Seismic analysis of reactor core. *Computers and Structures*, **76**, 1395–1404.

5

Fluid–Structure Coupling

Various coupled formulations are presented in this chapter, when compressibility and gravity are present in the fluid modelling. The finite element models resulting from the discretisation of the coupled equations involve matrices which have been already introduced in the preceding chapters. Symmetric or non-symmetric formulations are described and illustrated on simple configurations suited for analytical solutions. The numerical technique of coupling together finite elements describing the solid part of the system and boundary elements describing the fluid part of it is then introduced and compared to the more standard procedure according to which both the solid and fluid parts of the system are discretised using the finite element technique, taking an example case of a vibro-acoustic problem of academic and practical interest.

Figure 5.1 **Vibro-acoustic** FSI formulations provide a general framework for modelling vibro-acoustic problems, which are of concern in many industrial applications. Vibro-acoustics is also of theoretical and practical interest in the design of musical instruments – as discussed for instance in Béchade *et al.* (2005). *Source*: ©Jean-François SIGRIST

Fluid–Structure Interaction: An Introduction to Finite Element Coupling, First Edition. Jean-François SIGRIST.
© 2015 John Wiley & Sons, Ltd. Published 2015 by John Wiley & Sons, Ltd.
Companion Website: www.wiley.com/go/sigrist

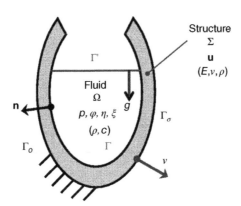

Figure 5.2   Fluid–structure coupling: interior problems

5.1   Modelling Assumption

Coupling between an elastic structure and a fluid is considered in the following. The potential energy associated with compressibility and gravity is now accounted for, so that the coupled problem involves degrees-of-freedom of both media.

The equation of motion for the structure is formulated in terms of displacement, while various formulations are possible for the fluid, as described by Boujot (1987). The non-symmetric or symmetric nature of the formulation arising from the coupled equations is discussed in both cases for two classes of problems.

Interior problems The structure contains an inviscid fluid, as in the configuration depicted in Figure 5.2. Gravity and compressibility waves in the fluid are accounted for either in a unified or separate manner. Following, for instance, Morand and Ohayon (1995), *vibro-acoustic coupling* defines FSI when modelling fluid compressibility, while *hydro-elastic sloshing* refers to FSI when gravity effects in the fluid are considered.

The structure domain is Σ, while the liquid occupies the volume Ω; the wetted interface is denoted Γ, while Γ' refers to the fluid-free surface. E and v denote the structure material properties, ρ designates the fluid or the structure density. The speed of sound in the fluid is c, while the gravity acceleration is g. Γ_o and Γ_σ stand for the structure boundaries with imposed displacements and forces, respectively; for the sake of convenience, homogeneous boundary conditions are considered. $\mathbf{n}$ denotes the normal on Γ, pointing outside the fluid (inside the structure), and v is the normal on $\Gamma_o \cup \Gamma_\sigma$, oriented outwardly.

Exterior problem The elastic structure is totally submerged in a compressible fluid, as considered in Figure 5.3. In this case, only vibro-acoustic coupling is accounted for, with compressibility waves propagating in an unbounded domain.

5.2   Interior Problems: Vibro-Acoustic and Hydro-Elastic Coupling

5.2.1   Non-Symmetric Formulation

FSI is formulated using a mixed *displacement/pressure*-based formulation in what follows. As presented in Section 3.3.1, the pressure formulation of the fluid problem can take

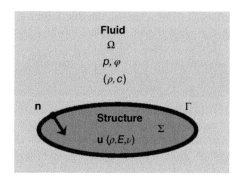

Figure 5.3   Fluid–structure coupling: exterior problem

into account gravity and compressibility effects. In order to simplify the presentation, the vibro-acoustic coupling and the hydro-elastic sloshing are presented in what follows using a unified model – bearing in mind that they are usually assessed in a separate manner because in most applications, acoustic and sloshing modes involve quite distinct frequency ranges.

The vibration of the elastic structure is described using the displacement field **u**, which verifies:

$$-\omega^2 \rho u_i - \frac{\partial \sigma_{ij}(\mathbf{u})}{\partial x_j} = 0 \text{ in } \Sigma \tag{5.1}$$

with boundary conditions:

$$u_i = 0 \text{ on } \Gamma_o \tag{5.2}$$

$$\sigma_{ij}(\mathbf{u})v_j = 0 \text{ on } \Gamma_\sigma \tag{5.3}$$

The fluid imposes pressure forces at the interface with the structure:

$$\sigma_{ij}(\mathbf{u})n_j = p\, n_i \text{ on } \Gamma \tag{5.4}$$

The perturbation of the pressure field p in the fluid domain on account of compressibility is governed by the acoustic wave equation:

$$-\frac{\omega^2}{c^2}p - \frac{\partial^2 p}{\partial x_i^2} = 0 \text{ in } \Omega \tag{5.5}$$

while gravity may be accounted for by the free level condition:

$$\frac{\partial p}{\partial x_j}n_j = \frac{\omega^2}{g}p \text{ on } \Gamma' \tag{5.6}$$

The normal acceleration of the fluid is imposed by the vibration of the structure at the wetted wall:

$$\frac{\partial p}{\partial x_j}n_j = \omega^2 \rho u_j n_j \text{ on } \Gamma \tag{5.7}$$

In addition, the displacement and the pressure comply with the *compatibility condition*. Gathering compressibility *and* gravity effects, this condition is stated as follows:

$$\int_\Omega \frac{1}{\rho c^2} p \, d\Omega + \int_{\Gamma'} \frac{1}{\rho g} p \, d\Gamma' + \int_\Gamma u_i n_i \, d\Gamma = 0 \tag{5.8}$$

Remark 5.1 Compatibility condition *Equation (5.8) ensures that the* $(\mathbf{u}, p)$ *formulation yields a well-posed problem with the meaning that the zero-frequency behaviour is retrieved in the coupled model. The compatibility condition corresponds to the global mass conservation – as pointed out by Ohayon (2001), this condition is often omitted in most presentations of FSI formulations.*

- *For vibro-acoustic coupling, the compatibility condition is*

$$\int_\Omega \frac{1}{\rho c^2} p \, d\Omega + \int_\Gamma u_i n_i \, d\Gamma = 0 \tag{5.9}$$

It ensures that the pressure formulation of the fluid problem is equivalent to the original displacement/pressure formulation with compressibility effects, with the following meaning:

$$-\frac{\omega^2}{c^2} p - \frac{\partial^2 p}{\partial x_i^2} = 0 \iff \frac{\partial p}{\partial x_i} = \rho \omega^2 \xi_i \ \text{and} \ p = -\rho c^2 \frac{\partial \xi_k}{\partial x_k}$$

With the compatibility condition, the fluid problem is well posed for $\omega = 0$*: in the static case, it yields the constant pressure field induced by any prescribed motions at the interface* Γ*. Denoted* p^o*, this pressure field is calculated as follows:*

$$p^o = -\frac{\rho c^2}{|\Omega|} \int_\Gamma u_i n_i \, d\Gamma$$

where $|\Omega|$ *is the volume of the fluid domain.*
- *For hydro-elastic sloshing, the compatibility condition is*

$$\int_{\Gamma'} \frac{1}{\rho g} p \, d\Gamma' + \int_\Gamma u_i n_i \, d\Gamma = 0 \tag{5.10}$$

It ensures that the pressure formulation of the fluid problem is equivalent to the original pressure-displacement formulation with gravity effects, with the following meaning:

$$\frac{\partial^2 p}{\partial x_i^2} = 0 \ \text{and} \ \frac{\partial p}{\partial x_j} n_j = \frac{\omega^2}{g} p$$

$$\iff \frac{\partial p}{\partial x_i} = \rho \omega^2 \xi_i \ \text{with} \ \frac{\partial \xi_k}{\partial x_k} = 0 \ \text{and} \ p = \rho g \xi_k n_k$$

The fluid problem is well posed for $\omega = 0$ *so that in the static case, the constant pressure field induced by any prescribed motions at the interface is calculated as follows:*

$$p^o = -\frac{\rho g}{|\Gamma'|} \int_\Gamma u_i n_i \, d\Gamma$$

where $|\Gamma'|$ *is the area of the fluid-free surface.*   ■

The weighted integral formulation of the problem is expressed as follows, where FSI is entirely described by the coupling terms $\int_\Gamma pn_i \delta u_i\ d\Gamma$ and $-\rho\omega^2 \int_\Gamma u_i n_i\ \delta p d\Gamma$.

Find $\mathbf{u} \in \mathcal{U}, p \in \mathcal{P}$ and ω such that

$$-\omega^2 \int_\Sigma \rho u_i\ \delta u_i\ d\Sigma + \int_\Sigma \sigma_{ij}(\mathbf{u})\ \varepsilon_{ij}(\delta\mathbf{u})\ d\Sigma - \int_\Gamma pn_i \delta u_i\ d\Gamma = 0$$

for all $\delta\mathbf{u} \in \mathcal{U}$, and

$$-\omega^2 \int_\Omega \frac{1}{c^2} p\ \delta p\ d\Omega - \omega^2 \int_{\Gamma'} \frac{1}{g} p\ \delta p\ d\Gamma' + \int_\Omega \frac{\partial p}{\partial x_i}\frac{\partial \delta p}{\partial x_i}\ d\Omega - \rho\omega^2 \int_\Gamma u_i n_i\ \delta p d\Gamma = 0$$

for all $\delta p \in \mathcal{P}$, with the constraint

$$\int_\Omega \frac{1}{\rho c^2} p\ d\Omega + \int_{\Gamma'} \frac{1}{\rho g} p\ d\Gamma' + \int_\Gamma u_i n_i\ d\Gamma = 0$$

and with $\mathcal{U} = \{\mathbf{u} \in H^1(\Sigma), \mathbf{u}|_{\Gamma_o} = \mathbf{0}\}$ and $\mathcal{P} = \{p \in H^1(\Omega)\}$.

The finite element discretisation produces the following matrices:

- The structure mass and stiffness matrices are calculated on Σ, as exposed in Chapter 2:

$$\int_\Sigma \rho u_i\ \delta u_i\ d\Sigma \rightarrow \delta\mathbf{U}^\mathsf{T}\mathbf{M}_S\mathbf{U}$$

$$\int_\Sigma \sigma_{ij}(\mathbf{u})\varepsilon_{ij}(\delta\mathbf{u})\ d\Sigma \rightarrow \delta\mathbf{U}^\mathsf{T}\mathbf{K}_S\mathbf{U}$$

- The fluid 'mass' and 'stiffness' matrices are calculated on Ω, as detailed in Chapter 3:

$$\int_\Omega \frac{1}{c^2} p\ \delta p\ d\Omega \rightarrow \delta\mathbf{P}^\mathsf{T}\mathbf{M}_F\mathbf{P} \qquad \int_{\Gamma'} \frac{1}{g} p\ \delta p\ d\Gamma' \rightarrow \delta\mathbf{P}^\mathsf{T}\mathbf{M}'_F\mathbf{P}$$

$$\int_\Omega \frac{\partial p}{\partial x_i}\frac{\partial \delta p}{\partial x_i}\ d\Omega \rightarrow \delta\mathbf{P}^\mathsf{T}\mathbf{K}_F\mathbf{P}$$

- The FSI matrix is calculated on Γ, by coupling structure and fluid elements, as developed in the preceding chapter:

$$\int_\Gamma p\ n_i\delta u_i\ d\Gamma \rightarrow \delta\mathbf{U}^\mathsf{T}\mathbf{RP} \qquad \int_\Gamma u_i n_i\ \delta p\ d\Gamma \rightarrow \delta\mathbf{P}^\mathsf{T}\mathbf{R}^\mathsf{T}\mathbf{U}$$

The discretisation of Equation (5.8) is

$$\int_\Omega \frac{1}{\rho c^2} p\ d\Omega + \int_{\Gamma'} \frac{1}{\rho g} p\ d\Gamma' + \int_\Gamma u_i n_i\ d\Gamma = 0 \rightarrow \mathbf{L}_F\mathbf{P} + \mathbf{L}_S\mathbf{U} = \mathbf{0}$$

$\mathbf{L}_F$ and $\mathbf{L}_S$ are built as:

$$\mathbf{L}_F = \langle\mathbf{1}\rangle(\mathbf{M}_F + \mathbf{M}'_F)/\rho \qquad \mathbf{L}_S = \langle\mathbf{1}\rangle\mathbf{R}^\mathsf{T}$$

where $\langle\mathbf{1}\rangle$ is a row vector with all components equal to 1.

Using the displacement $\mathbf{U}$ and the pressure $\mathbf{P}$ degrees-of-freedom, together with the Lagrange multiplier $\boldsymbol{\Lambda}$ associated with the compatibility condition, the following eigenvalue problem is arrived at:

$$-\omega^2 \begin{bmatrix} \mathbf{M}_S & 0 & 0 \\ \rho \mathbf{R}^{\mathsf{T}} & \mathbf{M}_F + \mathbf{M}'_F & 0 \\ 0 & 0 & 0 \end{bmatrix} \begin{Bmatrix} \mathbf{U} \\ \mathbf{P} \\ \boldsymbol{\Lambda} \end{Bmatrix}$$

$$+ \begin{bmatrix} \mathbf{K}_S & -\mathbf{R} & -\mathbf{L}_S^{\mathsf{T}} \\ 0 & \mathbf{K}_F & -\mathbf{L}_F^{\mathsf{T}} \\ -\mathbf{L}_S & -\mathbf{L}_F & 0 \end{bmatrix} \begin{Bmatrix} \mathbf{U} \\ \mathbf{P} \\ \boldsymbol{\Lambda} \end{Bmatrix} = \begin{Bmatrix} 0 \\ 0 \\ 0 \end{Bmatrix} \qquad (5.11)$$

In most finite element codes, however, the compatibility condition is not taken into account; in such cases, the $(\mathbf{u}, p)$ coupled problem is simply stated as follows:

$$\left(-\omega^2 \begin{bmatrix} \mathbf{M}_S & 0 \\ \rho \mathbf{R}^{\mathsf{T}} & \mathbf{M}_F \end{bmatrix} + \begin{bmatrix} \mathbf{K}_S & -\mathbf{R} \\ 0 & \mathbf{K}_F \end{bmatrix} \right) \begin{Bmatrix} \mathbf{U} \\ \mathbf{P} \end{Bmatrix} = \begin{Bmatrix} 0 \\ 0 \end{Bmatrix} \qquad (5.12)$$

The pressure/displacement formulation yields *non-symmetric* matrices despite the *conservative* nature of the mechanical coupling, since no energy dissipation nor energy source is taken into account in the coupled model. Although the coupled eigenmodes and associated eigenpulsations are *real* (Stammberger and Voss, 2010), their computation generally requires algorithms which are designed to handle complex-valued and/or non-symmetric matrices (Rajakumar and Rogers, 1991).

5.2.2 Symmetric Formulation

Although the numerical methods involved in structural dynamics are applicable equally to symmetric and non-symmetric matrices, with specificities for each case, symmetric formulations are often preferred to non-symmetric ones. In fact, the efficiency of the algorithms implemented in finite element codes for matrix inversion or eigenvalue computation is generally optimal for symmetric matrices, while the application of modal projection techniques is straightforward for problems involving eigenmodes associated with symmetric matrices.

Symmetric formulations of FSI problems have been extensively developed and exposed, see for instance Morand and Ohayon (1995). In the following, they are recalled and discussed from an engineering standpoint. Vibro-acoustic coupling and hydro-elastic sloshing are considered separately in order to point out the essential features of the mathematical models.

5.2.2.1 Vibro-Acoustic Coupling

A symmetric formulation describing the vibro-acoustic coupling is obtained by introducing the so-called displacement potential (Morand and Ohayon, 1995). As exposed in Remark 3.4, the displacement potential is denoted φ and is related to the pressure according to

$$p = \omega^2 \rho \varphi$$

for $\omega \neq 0$.

The vibrations of the structure are described by Equation (5.1) with boundary conditions (5.2) and (5.3). The coupling condition with the fluid is given as follows:

$$\sigma_{ij}(\mathbf{u})n_j = \omega^2 \rho \varphi\, n_i \text{ on } \Gamma \tag{5.13}$$

The compressibility wave propagation in the fluid is described in terms of the variables p and φ:

$$\frac{p}{c^2} + \rho \frac{\partial^2 \varphi}{\partial x_i^2} = 0 \text{ in } \Omega \tag{5.14}$$

$$\frac{p}{\rho c^2} - \frac{\omega^2}{c^2} \varphi = 0 \text{ in } \Omega \tag{5.15}$$

Gravity being discarded here for convenience, the free surface condition formulated in terms of the displacement potential reads as follows:

$$\varphi = 0 \text{ on } \Gamma' \tag{5.16}$$

which is equivalent to the acoustic-free surface condition stated as $p = 0$ when the pressure formulation is used.

The coupling condition at the fluid–structure interface is expressed in terms of displacements as follows:

$$\frac{\partial \varphi}{\partial x_j} n_j = u_j n_j \text{ on } \Gamma \tag{5.17}$$

The weighted integral formulation of the problem is stated as follows.

Find $\mathbf{u} \in \mathcal{U}, p \in \mathcal{P}, \varphi \in \mathcal{F}$ and ω such that:

$$-\omega^2 \int_\Sigma \rho u_i\, \delta u_i\, d\Sigma + \int_\Sigma \sigma_{ij}(\mathbf{u})\, \varepsilon_{ij}(\delta u)\, d\Sigma - \omega^2 \rho \int_\Gamma \varphi\, n_i \delta u_i\, d\Gamma = 0$$

for all $\delta u \in \mathcal{U}$,

$$-\omega^2 \int_\Omega \frac{1}{c^2} \varphi\, \delta p\, d\Omega + \frac{1}{\rho} \int_\Omega \frac{1}{c^2} p\, \delta p\, d\Omega = 0$$

for all $\delta p \in \mathcal{P}$ and:

$$-\rho \int_\Omega \frac{\partial \varphi}{\partial x_i} \frac{\partial \delta \varphi}{\partial x_i}\, d\Omega + \int_\Omega \frac{1}{c^2} p\, \delta \varphi\, d\Omega + \rho \int_\Gamma u_i n_i\, \delta \varphi\, d\Gamma = 0$$

for all $\delta \varphi$ in $\mathcal{F}$, with: $\mathcal{U} = \{\mathbf{u} \in H^1(\Sigma), \mathbf{u}|_{\Gamma_o} = \mathbf{0}\}$, $\mathcal{P} = \{p \in L^2(\Omega)\}$ and $\mathcal{F} = \{\varphi \in H^1(\Omega), \varphi|_{\Gamma'} = 0\}$.

The fluid contribution is obtained by multiplying Equation (5.14) by a virtual displacement potential field $\delta \varphi$ and by multiplying Equation (5.15) by a virtual pressure field δp. The first integral is integrated by parts using the boundary conditions (5.16) and (5.17), while the second integral is left unchanged.

The finite element discretisation is performed using the fluid and structure matrices, as well as the fluid–structure coupling matrix. These matrices arise from the various terms involved in the weighted integral formulation, namely:

$$\int_\Omega \frac{\partial \varphi}{\partial x_i} \frac{\partial \delta \varphi}{\partial x_i}\, d\Omega \rightarrow \delta\Phi^\mathsf{T} \mathbf{K}_F \Phi$$

$$\int_\Omega \frac{1}{c^2} p\, \delta\varphi\, d\Omega \rightarrow \delta\Phi^\mathsf{T} \mathbf{M}_F \mathbf{P} \qquad \int_\Omega \frac{1}{c^2} \varphi\, \delta p\, d\Omega \rightarrow \delta\mathbf{P}^\mathsf{T} \mathbf{M}_F \Phi$$

$$\int_\Gamma \varphi\, n_i \delta u_i\, d\Gamma \rightarrow \delta\mathbf{U}^\mathsf{T} \mathbf{R}\Phi \qquad \int_\Gamma u_i n_i\, \delta\varphi\, d\Gamma \rightarrow \delta\Phi^\mathsf{T} \mathbf{R}^\mathsf{T} \mathbf{U}$$

The finite element formulations for the first and second integrals are[1]

$$-\omega^2 \delta\mathbf{U}^\mathsf{T} \mathbf{M}_S \mathbf{U} + \delta\mathbf{U}^\mathsf{T} \mathbf{K}_S \mathbf{U} - \omega^2 \rho \delta\mathbf{U}^\mathsf{T} \mathbf{R}\Phi = 0 \qquad (5.18)$$

$$-\omega^2 \delta\Phi^\mathsf{T} \mathbf{M}_F \Phi + 1/\rho\, \delta\mathbf{P}^\mathsf{T} \mathbf{M}_F \mathbf{P} = 0 \qquad (5.19)$$

The third integral finite element formulation is multiplied by ω^2:

$$\omega^2 \rho \delta\Phi^\mathsf{T} \mathbf{K}_F \Phi - \omega^2 \delta\Phi^\mathsf{T} \mathbf{M}_F \mathbf{P} - \omega^2 \rho \delta\Phi^\mathsf{T} \mathbf{R}^\mathsf{T} \mathbf{U} = 0$$

Since the former relations are valid for any $\delta\mathbf{U}$, $\delta\mathbf{P}$ and $\delta\Phi$, the following matrix system is arrived at:

$$-\omega^2 \begin{bmatrix} \mathbf{M}_S & \mathbf{0} & \rho\mathbf{R} \\ \mathbf{0} & \mathbf{0} & \mathbf{M}_F \\ \rho\mathbf{R}^\mathsf{T} & \mathbf{M}_F & -\rho\mathbf{K}_F \end{bmatrix} \begin{Bmatrix} \mathbf{U} \\ \mathbf{P} \\ \Phi \end{Bmatrix}$$

$$+ \begin{bmatrix} \mathbf{K}_S & \mathbf{0} & \mathbf{0} \\ \mathbf{0} & 1/\rho\mathbf{M}_F & \mathbf{0} \\ \mathbf{0} & \mathbf{0} & \mathbf{0} \end{bmatrix} \begin{Bmatrix} \mathbf{U} \\ \mathbf{P} \\ \Phi \end{Bmatrix} = \begin{Bmatrix} \mathbf{0} \\ \mathbf{0} \\ \mathbf{0} \end{Bmatrix} \qquad (5.20)$$

The $(\mathbf{u}, p, \varphi)$ formulation of the vibro-acoustic coupling leads to symmetric matrices at the cost of doubling the size of the final problem and of arising some extra numerical difficulties because the mass matrix is no more definite positive.

The boundary condition (5.16) can be taken into account either by using a Lagrange multiplier or by eliminating the contribution of the constrained nodes in Equation (5.20). In this case, the problem is well posed for $\omega = 0$ and the stiffness matrix $\mathbf{K}_F$ is non-singular.[2]

---

[1] Equations (5.18) and (5.19) are obtained by using the same finite element for p and φ, which is a simplifying assumption – $\mathcal{F}$ is only included in $\mathcal{P}$, $\mathcal{F}$ and $\mathcal{P}$ being distinct functional spaces. In a general manner, different shape functions may be used for the approximation of the pressure and the displacement potential.

[2] For vibro-acoustic problems without free surface, the $(\mathbf{u}, p, \varphi)$ formulation yields an ill-posed problem for $\omega = 0$. Another definition of the displacement potential is required. In such a case, the pressure is written as $p = \rho\omega^2\varphi + \pi$, where π may be interpreted as the Lagrange multiplier associated with the compatibility condition $\int_\Omega p/\rho c^2 d\Omega + \int_\Gamma u_i n_i d\Gamma = 0$. The $(\mathbf{u}, p, \pi)$ formulation is then shown to be well posed for $\omega = 0$; see Morand and Ohayon (1995) for a deeper insight into the subject.

Using the third row of Equation (5.20) enables the calculation of $\boldsymbol{\Phi}$ according to

$$\boldsymbol{\Phi} = 1/\rho \mathbf{K}_F^{-1}\mathbf{M}_F\mathbf{P} + \mathbf{K}_F^{-1}\mathbf{R}^\mathsf{T}\mathbf{U}$$

Condensing $\boldsymbol{\Phi}$ on $\mathbf{U}$ yields a symmetric formulation in terms of pressure/displacement, which reads as follows:

$$-\omega^2 \begin{bmatrix} \mathbf{M}_S + \rho\mathbf{R}\mathbf{K}_F^{-1}\mathbf{R}^\mathsf{T} & \mathbf{R}\mathbf{K}_F^{-1}\mathbf{M}_F \\ \mathbf{M}_F\mathbf{K}_F^{-1}\mathbf{R}^\mathsf{T} & 1/\rho\mathbf{M}_F\mathbf{K}_F^{-1}\mathbf{M}_F \end{bmatrix} \begin{Bmatrix} \mathbf{U} \\ \mathbf{P} \end{Bmatrix}$$

$$+ \begin{bmatrix} \mathbf{K}_S & \mathbf{0} \\ \mathbf{0} & 1/\rho\mathbf{M}_F \end{bmatrix} \begin{Bmatrix} \mathbf{U} \\ \mathbf{P} \end{Bmatrix} = \begin{Bmatrix} \mathbf{0} \\ \mathbf{0} \end{Bmatrix} \qquad (5.21)$$

This formulation requires the computation of the added mass matrix $\rho\mathbf{R}\mathbf{K}_F^{-1}\mathbf{R}^\mathsf{T}$, and as highlighted in the previous chapter, this numerical operation may become prohibitive for complex geometries and/or for large-size problems. In addition, handling fully populated matrices makes most of the time-integration schemes excessively cumbersome, so that the condensed formulation is usually not used in practice.

Nonetheless, it remains feasible in some applications requiring a limited number of degrees-of-freedom, as in the case of bi-dimensional problems, see for instance Sigrist and Garreau (2007). In such cases not only the size of the problem remains small but also the stiffness matrix associated with both the symmetric and anti-symmetric Fourier components may always be inverted – this will be illustrated by the example presented in Section 5.4.

From the theoretical standpoint, the condensed $(\mathbf{u}, p)$ formulation points out the physical nature of the vibro-acoustic coupling and gives a criterion to decide whether fluid compressibility may be discarded or not in a given FSI problem. According to Equation (5.21), vibro-acoustic coupling is the result of the interaction between the following modes:

- the elastic modes of the structure with the fluid inertial effect solely. Indeed, the first line of Equation (5.21) without the coupling term yields the following eigenvalue problem:

$$(-\omega^2(\mathbf{M}_S + \rho\mathbf{R}\mathbf{K}_F^{-1}\mathbf{R}^\mathsf{T}) + \mathbf{K}_S)\mathbf{U} = \mathbf{0}$$

- the pure acoustic modes of the fluid, which are the solutions of the eigenvalue problem $1/\rho\mathbf{M}_F\mathbf{P} = 1/\rho\mathbf{M}_F\mathbf{K}_F^{-1}\mathbf{M}_F$ (this relation is obtained with the aid of the second line of Equation (5.21) without the coupling term). It is equivalent to the standard pressure formulation of acoustic modes:

$$(-\omega^2\mathbf{M}_F + \mathbf{K}_F)\mathbf{P} = \mathbf{0}$$

Hence, when the fluid acoustic modes and the structure modes with inertial effects lie in dissociated frequency ranges, the compressibility effect in the fluid may be discarded and the FSI is mainly dominated by inertial coupling. An illustration of this criterion is presented in Section 5.4.

Remark 5.2 Alternative symmetric formulations

An alternative symmetric formulation in terms of the $(\mathbf{u}, p, \varphi)$ variables is obtained when the fluid loading on the structure is expressed in terms of pressure instead of displacement potential on the one hand, and when the pressure-displacement potential relation is stated *in terms of gradient, on the other hand, writting*

$$\frac{\partial p}{\partial x_i} = \rho \omega^2 \frac{\partial \varphi}{\partial x_i}$$

The formulation also accounts for gravity waves through the boundary condition:

$$\frac{\partial \varphi}{\partial x_i} n_i = \frac{p}{\rho g} \quad on \ \Gamma'$$

The weighted integral formulation of the problem reads as follows.[3]

Find $\mathbf{u} \in \mathcal{U}$, $p \in \mathcal{P}$, $\varphi \in \mathcal{F}$ and ω such that:

$$-\omega^2 \int_\Sigma \rho u_i \ \delta u_i \ d\Sigma + \int_\Sigma \sigma_{ij}(\mathbf{u}) \ \varepsilon_{ij}(\delta\mathbf{u}) \ d\Sigma - \int_\Gamma p \ n_i \delta u_i \ d\Gamma = 0$$

for all $\delta\mathbf{u} \in \mathcal{U}$,

$$\int_\Omega \frac{\partial \varphi}{\partial x_i} \frac{\partial \delta p}{\partial x_i} \ d\Omega - \frac{1}{\rho} \int_\Omega \frac{1}{c^2} p \ \delta p \ d\Omega - \int_\Gamma u_i n_i \ \delta p \ d\Gamma - \frac{1}{\rho} \int_{\Gamma'} \frac{1}{g} p \ \delta p \ d\Omega = 0$$

for all $\delta p \in \mathcal{P}$ and:

$$-\omega^2 \rho \int_\Omega \frac{\partial \varphi}{\partial x_i} \frac{\partial \delta \varphi}{\partial x_i} \ d\Omega + \int_\Omega \frac{\partial p}{\partial x_i} \frac{\partial \delta \varphi}{\partial x_i} \ d\Omega = 0$$

for all $\delta\varphi$ in $\mathcal{F}$, with: $\mathcal{U} = \{\mathbf{u} \in H^1(\Sigma), \mathbf{u}|_{\Gamma_o} = 0\}$, $\mathcal{P} = \{p \in H^1(\Omega)\}$ and $\mathcal{F} = \{\varphi \in H^1(\Omega)\}$.

The corresponding finite element equation is written as follows:

$$-\omega^2 \begin{bmatrix} \mathbf{M}_S & 0 & 0 \\ 0 & 0 & 0 \\ 0 & 0 & \rho\mathbf{K}_F \end{bmatrix} \begin{Bmatrix} \mathbf{U} \\ \mathbf{P} \\ \mathbf{\Phi} \end{Bmatrix}$$

$$+ \begin{bmatrix} \mathbf{K}_S & -\mathbf{R} & 0 \\ -\mathbf{R}^T & -1/\rho(\mathbf{M}_F + \mathbf{M'}_F) & \mathbf{K}_F \\ 0 & \mathbf{K}_F & 0 \end{bmatrix} \begin{Bmatrix} \mathbf{U} \\ \mathbf{P} \\ \mathbf{\Phi} \end{Bmatrix} = \begin{Bmatrix} 0 \\ 0 \\ 0 \end{Bmatrix} \qquad (5.22)$$

It is similar to Equation (5.20). However, in Equation (5.22), coupling terms are present in the stiffness matrix and the $\mathbf{P}/\mathbf{\Phi}$ relation is expressed with matrix $\mathbf{K}_F$ instead of matrix $\mathbf{M}_F$. This requires the finite element shape functions associated with pressure and displacement potential approximation be at least linear.

---

[3] It is worthy to note that p belongs here to $H^1(\Omega)$, while p belongs to $L^2(\Omega)$ in the standard $(\mathbf{u}, p, \varphi)$ formulation.

Formulation (5.20) is the so-called $(\mathbf{u}, p, \varphi)$ *with mass coupling, while formulation (5.22) is referred to as* $(\mathbf{u}, p, \varphi)$ *with stiffness coupling. Elimination of* $\boldsymbol{\Phi}$ *in Equation (5.22) yields a condensed symmetric formulation (Morand and Ohayon, 1995).*

Using a displacement-based formulation *for both the structure and the fluid yields a symmetric representation of vibro-acoustic coupling. As introduced in Section 3.3.2, the vibration of the fluid is described by*

$$-\omega^2 \rho \xi_i - \frac{\partial^2 \xi_j}{\partial x_i \partial x_j} = 0 \ \ in \ \Omega \tag{5.23}$$

where ξ *is the displacement field. The acoustic-free surface condition is recalled as follows:*

$$\frac{\partial \xi_j}{\partial x_j} = 0 \ \ on \ \Gamma' \tag{5.24}$$

As the pressure in the fluid is $p = -\rho c^2 \dfrac{\partial \xi_j}{\partial x_j}$, *this condition is equivalent to* $p = 0$ *on* Γ'. *The coupling condition with the structure is expressed as follows:*

$$u_j n_j = \xi_j n_j \ \ on \ \Gamma \tag{5.25}$$

while the coupling condition of the structure with the fluid is expressed as follows:

$$\sigma_{ij}(\mathbf{u}) n_j = -\rho c^2 \frac{\partial \xi_j}{\partial x_j} n_i \ \ on \ \Gamma \tag{5.26}$$

The weighted integral formulation of the structure problem reads as follows:

$$-\omega^2 \int_\Sigma \rho u_i \ \delta u_i \ d\Sigma + \int_\Sigma \sigma_{ij}(\mathbf{u}) \epsilon_{ij}(\delta \mathbf{u}) \ d\Sigma = -\int_\Gamma \rho c^2 \frac{\partial \xi_j}{\partial x_j} \delta u_i n_i \ d\Gamma$$

while the weighted integral formulation of the fluid problem reads as follows:

$$-\omega^2 \int_\Omega \rho \xi_i \ \delta \xi_i \ d\Omega + \int_\Omega \rho c^2 \frac{\partial \xi_i}{\partial x_i} \frac{\partial \delta \xi_j}{\partial x_j} \ d\Omega = \int_\Gamma \rho c^2 \frac{\partial \xi_j}{\partial x_j} \delta \xi_i n_i \ d\Gamma$$

The weighted integral formulation of the vibro-acoustic problem is stated as follows.

Find $(\mathbf{u}, \xi) \in \mathcal{U} \mathcal{X}$ *and* ω *such that:*

$$-\omega^2 \int_\Sigma \rho u_i \ \delta u_i \ d\Sigma - \omega^2 \int_\Omega \rho \xi_i \ \delta \xi_i \ d\Omega$$

$$+ \int_\Sigma \sigma_{ij}(\mathbf{u}) \epsilon_{ij}(\delta \mathbf{u}) \ d\Sigma + \int_\Omega \rho c^2 \frac{\partial \xi_i}{\partial x_i} \frac{\partial \delta \xi_j}{\partial x_j} \ d\Omega = 0$$

for all $(\delta \mathbf{u}, \delta \xi) \in \mathcal{U} \mathcal{X}$, *with:*

$$\mathcal{U} \mathcal{X} = \{\mathbf{u} \in H^1(\Omega), \xi \in H^1(\Omega), \ (\mathbf{u} - \xi) \cdot \mathbf{n}|_\Gamma = 0, \ \nabla \times \xi = \mathbf{0}\}.$$

The corresponding finite element equation is

$$\left(-\omega^2 \begin{bmatrix} \mathbf{M}_S & \mathbf{0} \\ \mathbf{0} & \mathbf{M}_F \end{bmatrix} + \begin{bmatrix} \mathbf{K}_S & \mathbf{0} \\ \mathbf{0} & \mathbf{K}_F \end{bmatrix} \right) \begin{Bmatrix} \mathbf{U} \\ \mathbf{X} \end{Bmatrix} = \begin{Bmatrix} \mathbf{0} \\ \mathbf{0} \end{Bmatrix}$$

The mass and stiffness matrices $\mathbf{M}_F$ and $\mathbf{K}_F$ correspond to the displacement-based formulation of the fluid problem, as introduced in Section 3.3.2. With the $(\mathbf{u}, \boldsymbol{\xi})$ formulation, the fluid and the structure are modelled in a similar manner, and the resulting matrices are symmetric. Numerical methods for structural dynamics are directly applicable in the context of FSI, as detailed in Chapter 6. Unfortunately, the displacement/displacement formulation suffers in practice from several limitations, partly because the implementation of the irrotationality condition $\nabla \times \boldsymbol{\xi} = \mathbf{0}$ associated with the fluid displacement turns out to be a difficult numerical problem. It is often omitted in many applications, and as a result, the coupled formulation produces spurious vibro-acoustic eigenmodes in addition to the physically meaningful modes[4]. If not identified as such, their presence may seriously spoil the results derived from modal projection techniques. ∎

5.2.2.2 Hydro-Elastic Sloshing

A symmetric formulation of hydro-elastic sloshing is obtained with a displacement potential/elevation formulation of the fluid problem, as detailed in Remark 3.6. As the fluid flow is potential, φ verifies:

$$\frac{\partial^2 \varphi}{\partial x_i^2} = 0 \text{ in } \Omega \tag{5.27}$$

The Laplace equation is endowed with the following boundary conditions:

- Coupling with the structure:

$$\frac{\partial \varphi}{\partial x_i} n_i = u_i n_i \text{ on } \Gamma \tag{5.28}$$

- Elevation of the fluid at the free surface:

$$\frac{\partial \varphi}{\partial x_i} n_i = \eta \text{ on } \Gamma' \tag{5.29}$$

Equations (5.27) to (5.29) define a problem which is well posed in terms of φ when complemented by the following conditions:

- φ is uniquely determined when complying with an additional constraint $\lambda(\varphi) = 0$. A zero mean value condition for φ on Γ' serves as a typical constraint for hydro-elastic sloshing problems:

$$\int_{\Gamma'} \varphi \, d\Gamma' = 0 \tag{5.30}$$

---

[4] See for instance Bermudez *et al.* (1995).

- The fluid volume conservation is ensured when **u** and η satisfy the compatibility condition:

$$\int_{\Gamma} u_i n_i \, d\Gamma + \int_{\Gamma'} \eta \, d\Gamma' = 0 \tag{5.31}$$

In addition, the elevation is related to the displacement potential according to

$$\rho g \, \eta - \omega^2 \rho \varphi = 0 \quad \text{on } \Gamma' \tag{5.32}$$

The vibrations of the structure are described by Equations (5.1)–(5.3), together with the coupling condition (5.13), and the weighted integral formulation of the problem is as follows.

Find $\mathbf{u} \in \mathcal{V}, \varphi \in \mathcal{F}, \eta \in \mathcal{H}$ and ω such that:

$$-\omega^2 \int_{\Sigma} \rho u_i \, \delta u_i \, d\Sigma + \int_{\Sigma} \sigma_{ij}(\mathbf{u}) \varepsilon_{ij}(\delta \mathbf{u}) \, d\Sigma - \rho \omega^2 \int_{\Gamma} \varphi \, n_i \delta u_i \, d\Gamma = 0$$

for all $\delta \mathbf{u} \in \mathcal{V}$,

$$-\omega^2 \int_{\Gamma'} \rho \varphi \, \delta \eta \, d\Gamma' + \int_{\Gamma'} \rho g \, \eta \delta \eta \, d\Gamma' = 0$$

for all $\delta \eta \in \mathcal{H}$, and:

$$\int_{\Omega} \frac{\partial \varphi}{\partial x_i} \frac{\partial \delta \varphi}{\partial x_i} \, d\Omega - \int_{\Gamma} u_i n_i \, \delta \varphi \, d\Gamma - 1/\rho \int_{\Gamma'} \rho \eta \, \delta \varphi \, d\Gamma' = 0$$

for all $\delta \varphi \in \mathcal{F}$, with:

$$\int_{\Gamma'} \varphi \, d\Gamma' = 0 \quad \text{and} \quad \int_{\Gamma} u_i n_i \, d\Gamma + \int_{\Gamma'} \eta \, d\Gamma' = 0$$

and with: $\mathcal{V} = \{\mathbf{u} \in H^1(\Sigma), \mathbf{u}|_{\Gamma_o} = \mathbf{0}\}, \mathcal{H} = \{\eta \in L^2(\Gamma)\}$ and $\mathcal{F} = \{\varphi \in H^1(\Omega)\}$.

The fluid contribution in the integral formulation is obtained firstly by multiplying Equation (5.32) by a test function $\delta \eta$ and integrating on Γ'; secondly by multiplying Equation (5.27) by a test function $\delta \varphi$ and integrating by parts on Ω, while taking into account the definition of η as per Equation (5.29) and the coupling condition expressed by Equation (5.28).

The corresponding finite element formulation is obtained with matrices $\mathbf{M}'_F$ and $\mathbf{K}'_F$ presented in Remark 3.6, so that:

$$\int_{\Gamma'} \rho g \, \eta \delta \eta \, d\Gamma' \rightarrow \delta \mathbf{H}^{\mathsf{T}} \mathbf{K}'_F \mathbf{H}$$

$$\int_{\Gamma'} \rho \, \varphi \delta \eta \, d\Gamma' \rightarrow \delta \mathbf{H}^{\mathsf{T}} \mathbf{M}'_F \mathbf{\Phi} \qquad \int_{\Gamma'} \rho \, \eta \delta \varphi \, d\Gamma' \rightarrow \delta \mathbf{\Phi}^{\mathsf{T}} \mathbf{M}'^{\mathsf{T}}_F \mathbf{H}$$

The discretisation of Equations (5.30) and (5.31) are $\mathbf{L}_F \mathbf{\Phi} = \mathbf{0}$ and $\mathbf{L}_S \mathbf{U} + \mathbf{L}_F \mathbf{P} = \mathbf{0}$, where $\mathbf{L}_F = \langle \mathbf{1} \rangle \mathbf{M}'_F / \rho$, $\mathbf{L}_S = \langle \mathbf{1} \rangle \mathbf{R}^{\mathsf{T}}$ and $\mathbf{L}'_F = \langle \mathbf{1} \rangle \mathbf{K}'_F / (\rho g)$.

Denoting Λ_φ and $\Lambda_{u\eta}$ the Lagrange multipliers associated with the constraints (5.30) and (5.31), the finite element formulation of the problem is given as follows:

$$-\omega^2 \begin{bmatrix} M_S & 0 & 0 \\ 0 & 0 & 0 \\ 0 & 0 & 0 \end{bmatrix} \begin{Bmatrix} U \\ H \\ \Lambda_{u\eta} \end{Bmatrix} + \begin{bmatrix} K_S & 0 & -L_S^T \\ 0 & K_F' & -L_F'^T \\ -L_S & -L_F' & 0 \end{bmatrix} \begin{Bmatrix} U \\ H \\ \Lambda_{u\eta} \end{Bmatrix}$$

$$= \begin{Bmatrix} \rho\omega^2 R\Phi \\ \omega^2 M_F' \Phi \\ 0 \end{Bmatrix} \tag{5.33}$$

and:

$$\begin{bmatrix} K_F & -L_F^T \\ -L_F & 0 \end{bmatrix} \begin{Bmatrix} \Phi \\ \Lambda_\varphi \end{Bmatrix} = \begin{Bmatrix} R^T U + 1/\rho M_F' H \\ 0 \end{Bmatrix} \tag{5.34}$$

Since the problem formulated in terms of φ is well posed, the inversion of Equation (5.34) is possible:

$$\begin{Bmatrix} \Phi \\ \Lambda_\varphi \end{Bmatrix} = \begin{bmatrix} K_F & -L_F^T \\ -L_F & 0 \end{bmatrix}^{-1} \begin{Bmatrix} R^T U + 1/\rho M_F' H \\ 0 \end{Bmatrix}$$

For the sake of clarity, the latter relation is formulated without reference to the Lagrange multipliers. Thus,

$$\Phi = K_F^{-1} R^T U + 1/\rho K_F^{-1} M_F' H$$

Substituting this relation in Equation (5.33) yields the condensed formulation in terms of structure displacement and fluid elevation:[5]

$$-\omega^2 \begin{bmatrix} M_S + \rho R K_F^{-1} R^T & R K_F^{-1} M_F'^T & 0 \\ M_F' K_F^{-1} R^T & 1/\rho M_F' K_F^{-1} M_F'^T & 0 \\ 0 & 0 & 0 \end{bmatrix} \begin{Bmatrix} U \\ H \\ \Lambda_{u\eta} \end{Bmatrix}$$

$$+ \begin{bmatrix} K_S & 0 & -L_S^T \\ 0 & K_F' & -L_F'^T \\ -L_S & -L_F' & 0 \end{bmatrix} \begin{Bmatrix} U \\ H \\ \Lambda_{u\eta} \end{Bmatrix} = \begin{Bmatrix} 0 \\ 0 \\ 0 \end{Bmatrix}$$

---

[5] This formulation is equivalent to the $(\mathbf{u}, \eta, \pi)$ formulation proposed in Morand and Ohayon (1995), with π interpreted as the Lagrange multiplier associated with the constraint $\int_\Gamma \mathbf{u} \cdot \mathbf{n} \, d\Gamma + \int_{\Gamma'} \eta \, d\Gamma' = 0$.

For the sake of convenience, the preceding relation is also formulated without reference to the Lagrange multipliers:

$$-\omega^2 \begin{bmatrix} \mathbf{M}_S + \rho\mathbf{R}\mathbf{K}_F^{-1}\mathbf{R}^\mathsf{T} & \mathbf{R}\mathbf{K}_F^{-1}\mathbf{M}'^\mathsf{T}_F \\ \mathbf{M}'_F\mathbf{K}_F^{-1}\mathbf{R}^\mathsf{T} & 1/\rho\mathbf{M}'_F\mathbf{K}_F^{-1}\mathbf{M}'^\mathsf{T}_F \end{bmatrix} \begin{Bmatrix} \mathbf{U} \\ \mathbf{H} \end{Bmatrix}$$

$$+ \begin{bmatrix} \mathbf{K}_S & \mathbf{0} \\ \mathbf{0} & \mathbf{K}'_F \end{bmatrix} \begin{Bmatrix} \mathbf{U} \\ \mathbf{H} \end{Bmatrix} = \begin{Bmatrix} \mathbf{0} \\ \mathbf{0} \end{Bmatrix} \tag{5.35}$$

The condensed symmetric $(\mathbf{u}, \eta)$ formulation requires the computation of the added mass matrix $\rho\mathbf{R}\mathbf{K}_F^{-1}\mathbf{R}^\mathsf{T}$, with the practical limitations discussed in Chapter 4.

According to Equation (5.35), hydro-elastic sloshing may be interpreted as the interaction between the following modes:

- the elastic modes of the structure with fluid-added mass, as evidenced by the fact that the first line of Equation (5.35) without coupling simplifies into the following:

$$(-\omega^2(\mathbf{M}_S + \rho\mathbf{R}\mathbf{K}_F^{-1}\mathbf{R}^\mathsf{T}) + \mathbf{K}_S)\mathbf{U} = \mathbf{0}$$

- the pure sloshing modes of the fluid, as evidenced by the fact that the second line of Equation (5.35) without coupling becomes:

$$(-\omega^2/\rho\mathbf{M}'_F\mathbf{K}_F^{-1}\mathbf{M}'^\mathsf{T}_F + \mathbf{K}'_F)\mathbf{H} = \mathbf{0}$$

Hence, when the fluid sloshing modes and the structure modes with inertial effects lie in dissociated frequency ranges, the gravity can be discarded in fluid–structure coupling, while the sloshing modes can be calculated without taking into account the vibration of the structure, which means that sloshing is practically decoupled from the motion of the structure.

5.3 Exterior Problem: Vibro-Acoustic

The finite element modelling of exterior vibro-acoustic problems requires a truncation of the fluid domain at finite distance, as schematically represented in Figure 5.4. The geometry of the enclosing volume can be varied depending on the problem to be tackled – for instance, a sphere (such as illustrated in Figure 3.11) or a half-sphere is typically associated with a spherical wave radiation condition, a cylinder or a half-cylinder (such as illustrated in Figure 2.9) with a cylindrical wave radiation condition.

On the outer fluid boundary Γ_∞, conditions are imposed in order to represent the physics related to an unbounded domain, which turns out to be the absence of any reflected wave from the artificial boundary which has to behave as a perfectly matched asymptotic impedance. It is recalled that this condition can be approximated at finite distance, such as discussed in Chapter 3. In the following, the so-called BGT condition of zero order is considered: this condition corresponds to a direct application of the Sommerfeld condition on boundary Γ_∞. Using the pressure field, it is written as follows:

$$\frac{\partial p}{\partial x_j}n_j = -\frac{i\omega}{c}p \quad \text{on } \Gamma_\infty \tag{5.36}$$

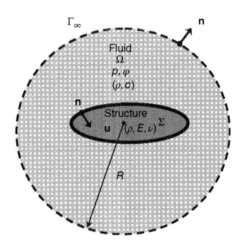

Figure 5.4   Truncation of the fluid domain for exterior vibro-acoustic problems

The formulation of the vibro-acoustic problem depicted in Figure 5.4 is given here in terms of displacement and pressure. It is obtained with the aid of Equations (5.1) and (5.5) together with the coupling conditions (5.4), (5.7) and the boundary condition (5.36). The weighted integral formulation of the problem is as follows.

Find $\mathbf{u} \in \mathcal{U}$, $p \in \mathcal{P}$ and ω such that:

$$-\omega^2 \int_{\Sigma} \rho u_i \, \delta u_i \, d\Sigma + \int_{\Sigma} \sigma_{ij}(\mathbf{u}) \, \varepsilon_{ij}(\delta \mathbf{u}) \, d\Omega = \int_{\Gamma} p n_i \, \delta u_i \, d\Gamma$$

for all $\delta \mathbf{u} \in \mathcal{U} = 0$ and:

$$-\omega^2 \int_{\Omega} \frac{p \, \delta p}{c^2} \, d\Omega + i\omega \int_{\Gamma_{\infty}} \frac{p \, \delta p}{c} \, d\Gamma_{\infty} + \int_{\Omega} \frac{\partial p}{\partial x_i} \frac{\partial \delta p}{\partial x_i} \, d\Omega = \rho \omega^2 \int_{\Gamma} u_i n_i \, \delta p \, d\Gamma$$

for all $\delta p, \in \mathcal{P}$, with $\mathcal{U} = \{\mathbf{u} \in H^1(\Sigma)\}$ and $\mathcal{P} = \{p \in H^1(\Omega)\}$.

The problem is well posed for $\omega = 0$ when eliminating the structure rigid body modes $\mathbf{u}^o$. They are otherwise contained in the preceding formulation, so that $(\mathbf{u}^o, p = 0, \omega = 0)$ is a trivial solution of the problem. The corresponding finite element equation is as follows:

$$\left(-\omega^2 \begin{bmatrix} \mathbf{M}_S & 0 \\ \rho \mathbf{R}^{\mathsf{T}} & \mathbf{M}_F \end{bmatrix} + i\omega \begin{bmatrix} 0 & 0 \\ 0 & \mathbf{C}_F \end{bmatrix} + \begin{bmatrix} \mathbf{K}_S & -\mathbf{R} \\ 0 & \mathbf{K}_F \end{bmatrix} \right) \begin{Bmatrix} \mathbf{U} \\ \mathbf{P} \end{Bmatrix} = \begin{Bmatrix} 0 \\ 0 \end{Bmatrix} \tag{5.37}$$

where matrix $\mathbf{C}_F$ accounts for acoustic damping. It is obtained from the finite element discretisation of the radiation condition on Γ_{∞}:

$$\int_{\Gamma_{\infty}} \frac{p \, \delta p}{c} \, d\Gamma_{\infty} \rightarrow \mathbf{C}_F$$

Using a $(\mathbf{u}, p)$-based formulation results in a non-symmetric problem. As for interior vibro-acoustic problems or vibro-acoustic problems, symmetry may be retrieved by using a mixed (p, φ) formulation of the acoustic part of the problem. Accordingly, the fluid is described using Equations (5.14) and (5.15), together with a coupling condition with the structure, Equation (5.13), while the radiation condition is expressed in terms of displacement potential, hence for the BGT-0 condition:

$$\frac{\partial \varphi}{\partial x_j} n_j = -\frac{i\omega}{c} \varphi \text{ on } \Gamma_\infty \tag{5.38}$$

In the weighted integral formulation of the $(\mathbf{u}, p, \varphi)$ problem, the integral on the displacement potential is

$$-\rho \int_\Omega \frac{\partial \varphi}{\partial x_i} \frac{\partial \delta \varphi}{\partial x_i} d\Omega + \int_\Omega \frac{p \, \delta \varphi}{c^2} d\Omega - i\rho\omega \int_{\Gamma_\infty} \frac{\varphi \, \delta \varphi}{c} d\Gamma_\infty + \rho \int_\Gamma u_i n_i \, \delta \varphi \, d\Gamma = 0$$

for all $\delta \varphi \in \mathcal{F}$ with $\mathcal{F} = \{\varphi \in H^1(\Omega)\}$.

The finite element discretisation of this integral produces the following equation:

$$\rho \, \delta \mathbf{\Phi}^T \mathbf{K}_F \mathbf{\Phi} - \delta \mathbf{\Phi}^T \mathbf{M}_F \mathbf{P} + i\omega\rho \mathbf{\Phi}^T \mathbf{C}_F \mathbf{\Phi} - \rho \delta \mathbf{\Phi}^T \mathbf{R}^T \mathbf{U} = 0$$

The above equation may be recast into its final form by multiplying by ω^2 and making use of Equations (5.18) and (5.19), giving thus the symmetric matrix equation:

$$\left(-\omega^2 \begin{bmatrix} \mathbf{M}_S & \mathbf{0} & \rho\mathbf{R} \\ \mathbf{0} & \mathbf{0} & \mathbf{M}_F \\ \rho\mathbf{R}^T & \mathbf{M}_F & -\rho\mathbf{K}_F \end{bmatrix} + i\omega^3 \begin{bmatrix} \mathbf{0} & \mathbf{0} & \mathbf{0} \\ \mathbf{0} & \mathbf{0} & \mathbf{0} \\ \mathbf{0} & \mathbf{0} & \rho\mathbf{C}_F \end{bmatrix} \right.$$

$$\left. + \begin{bmatrix} \mathbf{K}_S & \mathbf{0} & \mathbf{0} \\ \mathbf{0} & 1/\rho\mathbf{M}_F & \mathbf{0} \\ \mathbf{0} & \mathbf{0} & \mathbf{0} \end{bmatrix} \right) \begin{Bmatrix} \mathbf{U} \\ \mathbf{P} \\ \mathbf{\Phi} \end{Bmatrix} = \begin{Bmatrix} \mathbf{0} \\ \mathbf{0} \\ \mathbf{0} \end{Bmatrix} \tag{5.39}$$

Equation (5.37) involves non-symmetric matrices in an eigenvalue problem formulated as a quadratic function of the pulsation, while Equation (5.39) involves symmetric matrices in an eigenvalue problem formulated as a cubic function of the pulsation. From the practical point of view, the extraction of the eigenvalues and eigenvectors from Equation (5.39) turns out to be a burdensome task. Therefore, in practice, to carry out a modal analysis, the use of the non-symmetric Equation (5.37) is preferable to that of Equation (5.39), while the latter may be advantageously used to obtain the forced response of the fluid–structure system as computed directly in the frequency domain.

A symmetric formulation may also be obtained if acoustic waves are described in terms of the *velocity potential*, denoted here χ^6. It is related to pressure according to $p = -i\omega\chi$ and to

---

[6] The $(\mathbf{u}, \chi)$ formulation derived in what follows may also serve as an alternative for modelling vibro-acoustic for interior problem with symmetric matrices.

displacement potential according to $\chi = i\omega\varphi$ (Everstine, 1981). In such a case, the coupling conditions with the structure are expressed as follows:

$$\sigma_{ij}(\mathbf{u})n_j = -i\omega\rho\chi n_i \quad \text{and} \quad \frac{\partial\chi}{\partial x_i}n_i = i\omega u_i n_i \quad \text{on } \Gamma$$

while the impedance condition is stated as follows:

$$\frac{\partial\chi}{\partial x_i}n_i = -\frac{i\omega}{c}\chi \quad \text{on } \Gamma_\infty$$

The weighted integral formulation of the problem is then formulated as follows.

Find $\mathbf{u} \in \mathcal{U}$, $\chi \in \mathcal{Y}$ and ω such that:

$$-\omega^2\int_\Sigma \rho u_i\ \delta u_i\ d\Sigma + i\omega\rho\int_\Gamma \chi n_i\ \delta u_i\ d\Gamma + \int_\Sigma \sigma_{ij}(\mathbf{u})\ \varepsilon_{ij}(\delta\mathbf{u})\ d\Sigma = 0$$

for all $\delta\mathbf{u} \in \mathcal{U}$ and:

$$\omega^2\rho\int_\Omega \frac{\chi\ \delta\chi}{c^2}\ d\Omega - i\omega\rho\int_{\Gamma_\infty} \frac{\chi\ \delta\chi}{c}\ d\Gamma_\infty$$

$$+i\omega\rho\int_\Gamma u_i n_i\ \delta\chi\ d\Gamma + \rho\int_\Omega \frac{\partial\chi}{\partial x_i}\frac{\partial\delta\chi}{\partial x_i}\ d\Omega = 0$$

for all $\delta\chi \in \mathcal{Y}$, with: $\mathcal{U} = \{\mathbf{u} \in H^1(\Sigma)\}$ and $\mathcal{Y} = \{\chi \in H^1(\Omega)\}$.

The corresponding finite element formulation yields a quadratic eigenvalue problem with symmetric matrices:

$$\left(-\omega^2\begin{bmatrix} \mathbf{M}_S & \mathbf{0} \\ \mathbf{0} & -\rho\mathbf{M}_F \end{bmatrix} + i\omega\begin{bmatrix} \mathbf{0} & \rho\mathbf{R} \\ \rho\mathbf{R}^\mathsf{T} & -\rho\mathbf{C}_F \end{bmatrix}\right.$$

$$\left.+\begin{bmatrix} \mathbf{K}_S & \mathbf{0} \\ \mathbf{0} & -\rho\mathbf{K}_F \end{bmatrix}\right)\begin{Bmatrix} \mathbf{U} \\ \chi \end{Bmatrix} = \begin{Bmatrix} \mathbf{0} \\ \mathbf{0} \end{Bmatrix} \tag{5.40}$$

Remark 5.3 FEM and BEM coupling *As highlighted in Section 3.4, an integral formulation of the compressibility wave equation has the major advantage of automatically complying with the Sommerfeld condition of non-reflection at infinity, while requiring the sole discretisation of the physical boundary limiting the fluid. Hence, combining FEM and BEM offers an appealing numerical technique to deal with vibro-acoustic coupling, especially for exterior problems.*

Coupling FEM with BEM is briefly described here in the context of the direct method and is applied in Section 5.5 to treat the bi-dimensional problem of a circular ring vibrating in-plane while immersed in an unbounded compressible fluid.

In the weighted integral formulation of structural dynamics, the fluid loading on the structure is accounted for with the term

$$\int_\Gamma p \, \delta\mathbf{u} \cdot \mathbf{n} \, d\Gamma$$

With the use of a BEM approach, the fluid–structure interface Γ is divided in E faces Γ_e, so that:

$$\int_\Gamma p \, \delta\mathbf{u} \cdot \mathbf{n} \, d\Gamma = \sum_{e=1}^{e=E} \int_{\Gamma_e} p \, \delta\mathbf{u} \cdot \mathbf{n} \, d\Gamma_e$$

On Γ_e, the pressure is approximated as $p|_{\Gamma_e} \simeq \mathbf{N}_\mu^e \mathbf{P}_e$, as detailed in Section 3.4.3, and the displacement is calculated to be $\mathbf{u}|_{\Gamma_e} \simeq \mathbf{N}_e \mathbf{U}_e$ so that:

$$\int_{\Gamma_e} p \, \delta\mathbf{u} \cdot \mathbf{n} \, d\Gamma_e = \delta\mathbf{U}^\mathsf{T} \mathbf{\Lambda}_e^\mathsf{T} \int_{\Gamma_e} \mathbf{N}_e^\mathsf{T} \mathbf{n} \mathbf{N}_\mu^e \, d\Gamma_e \mathbf{\Lambda}_\mu^e$$

where $\mathbf{\Lambda}_e$ and $\mathbf{\Lambda}_\mu^e$ are the localisation matrices associated with the structure and the fluid for element Γ_e. Hence,

$$\int_\Gamma p \, \delta\mathbf{u} \cdot \mathbf{n} \, d\Gamma = \delta\mathbf{U}^\mathsf{T} \mathbf{R} \mathbf{P}$$

where the fluid–structure coupling matrix is

$$\mathbf{R} = \sum_{e=1}^{e=E} \mathbf{\Lambda}_e^\mathsf{T} \int_{\Gamma_e} \mathbf{N}_e^\mathsf{T} \mathbf{n} \mathbf{N}_\mu^e \, d\Gamma_e \mathbf{\Lambda}_\mu^e$$

The pressure is calculated according to Equation (3.5.3), which gives

$$\mathbf{P} = \mathbf{D}^{-1}\mathbf{S}\frac{\partial\mathbf{P}}{\partial n}$$

The pressure gradient is derived from Equation (5.7), and to formulate the FEM/BEM coupling, it is expressed as follows:

$$\frac{\partial\mathbf{P}}{\partial n} = \rho\omega^2 \mathbf{R}^\mathsf{T} \mathbf{U}$$

with

$$\mathbf{R} = \sum_{e=1}^{e=E} \mathbf{\Lambda}_\mu^{e\,\mathsf{T}} \int_{\Gamma_e} \mathbf{N}_\sigma^{e\,\mathsf{T}} \mathbf{n}^\mathsf{T} \mathbf{N}_e \, d\Gamma_e \mathbf{\Lambda}_e$$

Hence,

$$\int_\Gamma p \, \delta\mathbf{u} \cdot \mathbf{n} \, d\Gamma = \rho\omega^2 \delta\mathbf{U}^\mathsf{T} \mathbf{M}_A(\omega)\mathbf{U}$$

with

$$\mathbf{M}_A(\omega) = \mathbf{R}\mathbf{D}^{-1}\mathbf{S}\mathbf{R}^\mathsf{T}$$

Therefore, the vibro-acoustic coupling is described by the matrix equation:

$$-\omega^2(\mathbf{M}_S + \mathbf{M}_A(\omega))\mathbf{U} + \mathbf{K}_S\mathbf{U} = 0 \qquad (5.41)$$

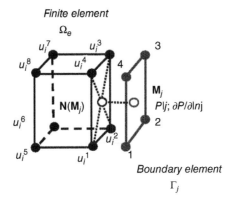

Finite element
Ω_e

Boundary element

Γ_j

Figure 5.5   FEM/BEM coupling with a direct zero-order BEM scheme. *The fluid–structure inter-face Γ is divided in J faces Γ_j. With the direct BEM, the collocation of each integral on Γ_j centred on a single point M_j gives $\int_{\Gamma_j} p\, \delta\mathbf{u}\cdot\mathbf{n}\, d\Gamma = \delta\mathbf{U}_j^T\mathbf{N}_j^T\mathbf{n}_j|\Gamma_j|P_j$, with $\mathbf{N}_j = \mathbf{N}(M_j)$. $\mathbf{N}$ is the shape function matrix for the finite element which is coupled with Γ_j; $|\Gamma_j|$ stands for the area of surface element Γ_j. P_j is the pressure value on M_j, calculated from Equation (3.5.3) according to $P_j = \sum_{i=1}^{i=J}(\mathbf{D}^{-1}\mathbf{S})_{ji}\left.\dfrac{\partial P}{\partial n}\right|_i$. The nor-mal pressure gradient is approximated as $\left.\dfrac{\partial P}{\partial n}\right|_i = \rho\omega^2\mathbf{n}_i^T\mathbf{N}_i\mathbf{U}_i$. Hence, $\int_\Gamma p\, \delta\mathbf{u}\cdot\mathbf{n}\, d\Gamma = \rho\omega^2\delta\mathbf{U}^T\mathbf{M}_A(\omega)\mathbf{U}$ making use of the vibro-acoustic matrix defined as $\mathbf{M}_A = \sum_{j=1}^{j=J}\sum_{i=1}^{i=J}\delta\mathbf{U}_j^T\mathbf{N}_j^T\mathbf{n}_j|\Gamma_j|(\mathbf{D}^{-1}\mathbf{S})_{ji}\mathbf{n}_i^T\mathbf{N}_i\mathbf{U}_i$*

where $\mathbf{M}_A(\omega)$ is the vibro-acoustic matrix, *which is frequency dependent:*

$$\mathbf{M}_A(\omega) = \mathbf{R}^T\,\mathbf{\Gamma}\,\mathbf{D}^{-1}\mathbf{S}\,\mathbf{R} \tag{5.42}$$

When coupling FEM with indirect BEM, the vibro-acoustic matrix is shown to be

$$\mathbf{M}_A(\omega) = \rho\mathbf{R}\mathbf{L}\mathbf{R}^T$$

with the coupling matrix defined as

$$\mathbf{R} = \int_\Gamma \mathbf{N}^T\mathbf{n}\mathbf{N}_\mu\, d\Gamma \qquad \mathbf{R}^T = \int_\Gamma \mathbf{N}_\sigma{}^T\mathbf{n}^T\mathbf{N}\, d\Gamma$$

$\mathbf{N}$ *is the structure element shape function and* $\mathbf{N}_\mu, \mathbf{N}_\sigma$ *are the fluid shape functions for the double/single layer potentials* μ, σ, *while* $\mathbf{L}$ *is expressed as follows:*

$$\mathbf{L} = \mathbf{\Lambda}_\mu - \mathbf{\Sigma}^T\mathbf{\Lambda}_\sigma^{-1}\mathbf{\Sigma}$$

where matrices $\mathbf{\Lambda}_\sigma, \mathbf{\Lambda}_\mu$ *and* $\mathbf{\Sigma}$ *have already been introduced in Section 3.4.3. An illustration of FEM/BEM coupling is proposed in Figure 5.5 in the context of zero-order BEM schemes.*

The vibro-acoustic matrix is frequency dependent, fully populated and non-symmetric. It accounts for inertial effect on the vibrating structure, acoustic wave radiation from the fluid–structure vibrating interface and radiative damping describing the absence of wave reflection at infinity.

Formulation (5.41) involves less degrees-of-freedom than its FEM/FEM symmetric or non-symmetric counterparts, appearing in Equations (5.37) and (5.39). However, the *manipulations involved in the symmetric formulation to assemble and store* $\mathbf{M}_A(\omega)$ *and then to solve Equation (5.41) or to solve* $(-\omega^2(\mathbf{M}_S + \mathbf{M}_A(\omega)) + \mathbf{K}_S)\mathbf{U}(\omega) = \mathbf{F}(\omega)$ *at each frequency[7] often seems more computationally expensive than making use of the non-symmetric formulation.*

Once the pressure and pressure gradient are known at the boundary, the evaluation of the pressure in the fluid domain ensues from Equation (3.54). Using the normal acceleration of the structure at the interface, this equation is stated as follows:

$$p(M_o) = \Pi(M_o)\mathbf{P} + \rho\omega^2\Gamma(M_o)\mathbf{R}^{\mathsf{T}}\mathbf{U} \tag{5.43}$$

A similar expression is obtained from Equation (3.58) if the FEM/BEM indirect coupling formulation is used.

■

5.4   Example: Vibro-Acoustic Coupling and Hydro-Elastic Sloshing

The problem under concern here is depicted in Figure 5.6: a beam of circular cross section is immersed in a compressible liquid up to height H less than the tube length L and gravity is not discarded at the free level. The bending modes of the tube are described using the Euler–Bernoulli model with the distributed pressure force on each wetted section, that is, for $z \in [0, H]$.

As both compressibility and gravity effects are accounted for, the pressure field is governed by the sound wave equation in the fluid domain ($[R, R'] \times [0, 2\pi] \times [0, H]$), while its value at the free level ($[R, R'] \times [0, 2\pi] \times \{z = H\}$) must comply with the gravity condition.

The illustrative case presented here is carried out with the following geometrical data: $R = 0.1$ m, $R' = 0.2$ m, $L = 1$ m, $h = 0.05$ m, $H = 0.75$ m and $g = 9.81$ m/s^2; the material properties are that of steel for the structure ($\rho = 7,800$ kg/m^3, $E = 2.1 \times 10^{11}$ Pa) and water for the fluid ($\rho = 1,000$ kg/m^3 and $c = 1,500$ m/s).

The discrete model is built by coupling beam finite elements and axisymmetric fluid finite elements, according to the technique depicted in Figure 5.7.

As a preliminary to understand the behaviour of the coupled system, it is found useful to first perform its modal analysis, discarding fluid compressibility and gravity. The mode shapes of the first four bending modes with fluid-added mass are displayed in Figures 5.8, while the first mode shape of the pure sloshing and acoustic modes are represented in terms of elevation and pressure, respectively, in Figure 5.9.

In the present case, the sloshing modes occur in a frequency range much lower than the first natural modes of vibration of the structure with fluid inertia. As a consequence, they are practically decoupled which means that sloshing occurs with negligible motion of the structure, while the natural vibration of the structure occurs with negligible vertical motion of the fluid-free level. On the other hand, the motion of the liquid within the annular space surrounding the tube accounts for the added mass effect.

In contrast with the decoupling of the fluid sloshing and structure bending modes found in the preliminary study, when compressibility of water is taken into account, it is found that the first acoustic and structural modes (with fluid inertia) lie in the same frequency range, indicating strong coupling.

A symmetric formulation of the coupled problem is used here to include hydro-elastic sloshing and vibro-acoustic coupling in a single model. Such an approach proves to be of interest

---

[7] As exposed in the next chapter, this comes into play when computing the vibro-acoustic frequency response of coupled systems.

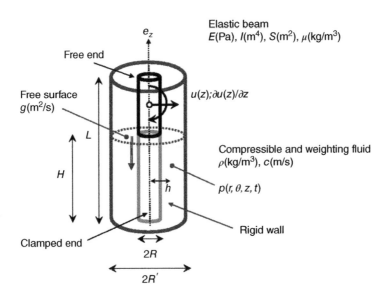

Figure 5.6 Elastic beam of circular cross section coupled to a compressible and weighting fluid filling a coaxial cylindrical tank

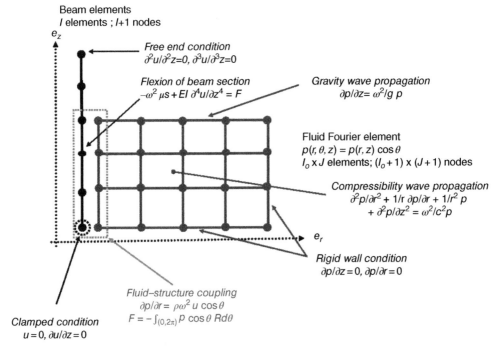

Figure 5.7 1D beam element and 2D Fourier fluid element coupling (displacement/pressure formulation)

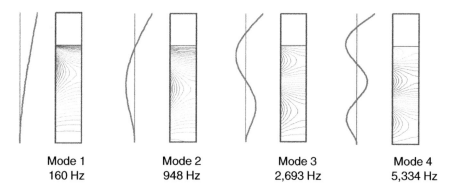

Mode 1
160 Hz

Mode 2
948 Hz

Mode 3
2,693 Hz

Mode 4
5,334 Hz

Figure 5.8   Structure modes with fluid inertial effect

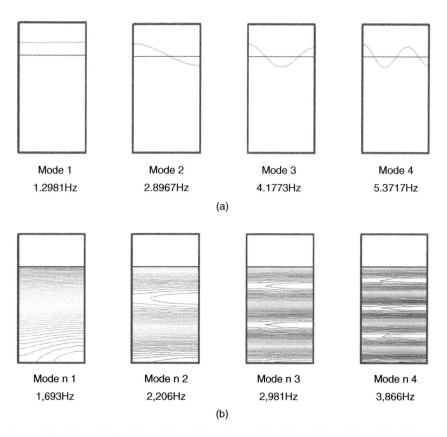

Mode 1
1.2981Hz

Mode 2
2.8967Hz

Mode 3
4.1773Hz

Mode 4
5.3717Hz

(a)

Mode n 1
1,693Hz

Mode n 2
2,206Hz

Mode n 3
2,981Hz

Mode n 4
3,866Hz

(b)

Figure 5.9   Fluid (a) sloshing (elevation formulation) and (b) acoustic (pressure formulation) modes

when the dynamics of the system is investigated along an extended frequency range. The condensed $(\mathbf{u}, \eta, p)$ finite element equation of the problem is written as follows:

$$
-\omega^2
\begin{bmatrix}
\mathbf{M}_S + \rho\mathbf{R}\mathbf{K}_F^{-1}\mathbf{R}^\mathsf{T} & \mathbf{R}\mathbf{K}_F^{-1}\mathbf{M}'^\mathsf{T}_F & \mathbf{R}\mathbf{K}_F^{-1}\mathbf{M}_F \\
\mathbf{M}'_F\mathbf{K}_F^{-1}\mathbf{R}^\mathsf{T} & 1/\rho\mathbf{M}'_F\mathbf{K}_F^{-1}\mathbf{M}'^\mathsf{T}_F & 0 \\
\mathbf{M}_F\mathbf{K}_F^{-1}\mathbf{R}^\mathsf{T} & 0 & 1/\rho\mathbf{M}_F\mathbf{K}_F^{-1}\mathbf{M}_F
\end{bmatrix}
\begin{Bmatrix} \mathbf{U} \\ \mathbf{H} \\ \mathbf{P} \end{Bmatrix}
$$

$$
+
\begin{bmatrix}
\mathbf{K}_S & 0 & 0 \\
0 & \mathbf{K}'_F & 0 \\
0 & 0 & 1/\rho\mathbf{M}_F
\end{bmatrix}
\begin{Bmatrix} \mathbf{U} \\ \mathbf{H} \\ \mathbf{P} \end{Bmatrix}
=
\begin{Bmatrix} 0 \\ 0 \\ 0 \end{Bmatrix}
\tag{5.44}
$$

In the present case, making use of Equation (5.44) proves particularly adapted.

- Adopting a bi-dimensional and axisymmetric model for the problem alleviates the computational cost substantially. Hence, a direct computation of the added mass is possible. Figure 5.10 shows how the mass and stiffness matrices derived from Equation (5.44) together with the element connectivity presented in Tables 2.6 and 3.9 are populated, by mapping their non-zero coefficients. It is verified that the stiffness matrix remains sparse with a limited bandwidth, while the mass matrix is densely populated.
- As the compatibility conditions are automatically verified by any Fourier component (symmetric or antisymmetric) of the solution, there is no need to complement Equation (5.44) with Equations (5.9) and (5.31) – that is, no Lagrange multiplier is involved in the problem.

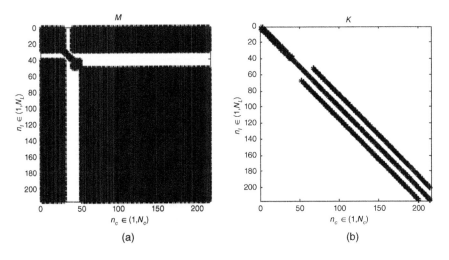

Figure 5.10 Structure of the (a) mass and (b) stiffness matrices with the $(\mathbf{u}, \eta, p)$ formulation

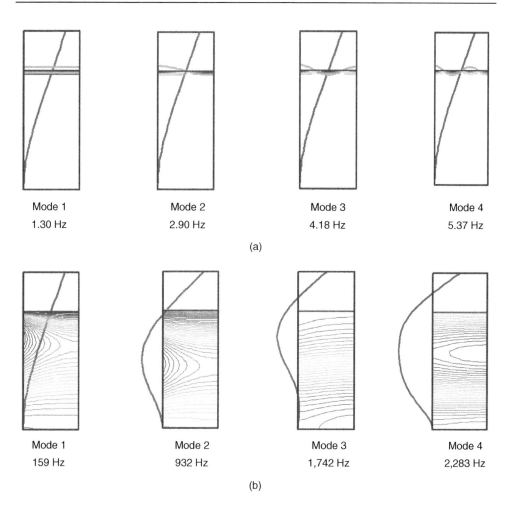

Mode 1   Mode 2   Mode 3   Mode 4
1.30 Hz   2.90 Hz   4.18 Hz   5.37 Hz

(a)

Mode 1   Mode 2   Mode 3   Mode 4
159 Hz   932 Hz   1,742 Hz   2,283 Hz

(b)

Figure 5.11 Fluid–structure coupling in the $(\mathbf{u}, \eta, p)$ formulation: (a) 'low' (sloshing) and (b) 'high' (vibro-acoustic coupling) frequency modes

Figure 5.11 displays the first four modes in the 'low-frequency' range (under 10 Hz) and in the 'high-frequency' range (above 100 Hz) separately.

As expected, the first modes are entirely dominated by gravity effects in the fluid. Accordingly, their shapes and frequencies are almost identical to those of the pure sloshing modes (see Figure 5.11 (a)). In contrast, the first and second bending modes of the structure are significantly modified by coupling with the liquid. The first mode excepted, where coupling remains essentially inertial in nature, all the other modes at higher frequencies are vibro-acoustic in nature, that is, affected by both fluid inertia and compressibility.

The frequency of the first mode is unchanged, whereas the frequency of the second mode is decreased due to fluid compressibility. The first and second acoustic modes of the fluid are affected by the elasticity of the wall: their frequencies are raised on account of additional potential energy brought in the system by structural deformations.

The computation of the effective masses also provides a good illustration of the properties of vibro-acoustic coupling. Adopting the condensed $(\mathbf{u}, p)$ formulation, as per Equation (5.21), the effective mass μ_n of mode $\mathbf{X}_n$ is calculated for a given direction Δ. In the present case, the direction of concern is parallel to the beam bending axis, as explained by Figure 5.12.

As highlighted in Remark 4.2, in the assembled added mass matrix all coupling terms on fluid boundaries are included, so that the total fluid mass may be retrieved as the sum of all components of $\mathbf{M}_A$ in the considered direction.

The summation of the effective mass reads as follows:

$$\sum_n \mu_n = \langle \Delta^\mathsf{T} 0 \rangle \begin{bmatrix} \mathbf{M}_S + \rho \mathbf{R} \mathbf{K}_F^{-1} \mathbf{R}^\mathsf{T} & \mathbf{R} \mathbf{K}_F^{-1} \mathbf{M}_F \\ \mathbf{M}_F \mathbf{K}_F^{-1} \mathbf{R}^\mathsf{T} & 1/\rho \mathbf{M}_F \mathbf{K}_F^{-1} \mathbf{M}_F \end{bmatrix} \begin{Bmatrix} \Delta \\ 0 \end{Bmatrix}$$

$$- \Delta^\mathsf{T} (\mathbf{M}_S + \rho \mathbf{R} \mathbf{K}_F^{-1} \mathbf{R}^\mathsf{T}) \Delta = m_S + m_F = m$$

which stands for the extension of Equation (2.28) for a coupled fluid–structure system.

In the present numerical application, only the vibro-acoustic modes are considered and the tank is assumed to be full of fluid and closed at the top (with $H = L = 1$ m). Analytical and numerical values of the structure and fluid masses are presented in Table 5.1 and in Figure 5.13 are plotted the accumulated effective masses over the whole frequency range swept in the computation.

The first three bending modes of the structure in vacuo account for more than 85% of the structure mass, as already outlined in Figure 2.22 and in Table 2.5, while only a few acoustic modes account for the fluid mass: they are the modes with vertical isobars and correspond to the first and the eleventh cavity modes with frequencies $1,617$ Hz and $7,850$ Hz, as displayed in Figure 5.14: their effective mass accounts for some 90% of the fluid mass and all other modes have a null effective mass.

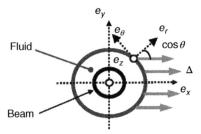

Figure 5.12   Symmetric loading and axisymmetric system. A seismic loading is supposed to be applied on the beam/cylindrical tank system, which has already been shown in Figure 5.6. Δ stands for the direction of the prescribed acceleration on the system. As the loading is symmetric with respect to the $(0, x)$ axis, the symmetric Fourier components are of interest, as explained in Figure 2.26. The expansion of Δ as a Fourier series involves the component $m = 1$ solely. Evaluating the effective mass allows to identify the modes of the system which are prone to respond to the imposed acceleration. For the Fourier component equal to one, it is precisely verified that $\sum_n \mu_n$ yields the mass of the system, as per Equation (2.28)

Table 5.1 Calculation of the structure and fluid mass with matrices $\mathbf{M}_S$ and $\rho\mathbf{R}\mathbf{K}_F^{-1}\mathbf{R}^\mathsf{T}$

Mass	Analytical solution (kg)	Numerical computation (kg)
Structure	$m_S = 46.56$	$\mathbf{\Delta}^\mathsf{T}\mathbf{M}_S\mathbf{\Delta} = 45.83$
Fluid	$m_F = 94.25$	$\mathbf{\Delta}^\mathsf{T}\rho\mathbf{R}\mathbf{K}_F^{-1}\mathbf{R}^\mathsf{T}\mathbf{\Delta} = 93.75$

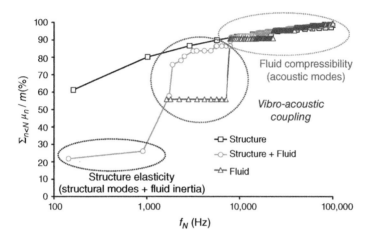

Figure 5.13 Vibro-acoustic coupling: effective mass versus frequency (the accumulated effective masses of the structure, fluid and fluid–structure modes are plotted versus the frequency as an illustration of vibro-acoustic coupling for the problem depicted in Figure 5.6)

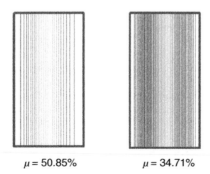

$\mu = 50.85\%$ $\mu = 34.71\%$

Figure 5.14 First and eleventh acoustic modes with non-null effective mass at frequencies $1{,}617$ and $7{,}850\,\mathrm{Hz}$

Fluid–structure modes are of different nature. On the one hand, the first two modes, which come at the lower end of the frequency spectrum, correspond to bending modes of the structure with fluid inertia, and in contrast with the corresponding modes in vacuo, their effective mass is found to be substantially lowered on account of the inertial force associated with the bulk motion of the accelerated fluid – the corresponding mass stands for some 25% of the total mass (a simple model which accounts for this observation is exposed in Remark 6.1). On the other hand, the acoustic modes, which come at the upper bound of the frequency spectrum, are barely affected by structural elasticity and account for some 15% of the total mass, as for the pure acoustic modes, which are found in the same frequency range and which account for 15% of the fluid mass within this interval. Vibro-acoustic modes, which combine structural elasticity and fluid elasticity properties, are lying in the intermediate frequency range and account roughly for 60% of the total mass, with the first two coupled modes, respectively, at frequencies 1,742 Hz and 2,283 Hz, as displayed in Figure 5.11, standing for 50% of the total mass.

5.5 Example: Acoustic Damping

5.5.1 Analytical Modelling

The vibro-acoustic behaviour of a circular elastic ring immersed in an unbounded acoustic fluid is considered: the problem is treated according to the bi-dimensional configuration depicted in Figure 5.15.

The problem is formulated in terms of displacement and pressure, by coupling together the equation of motion of the ring, as presented in Section 2.4, and the sound wave equation, as presented in Section 3.8. The external load applied to the ring is written as a right-hand-side term in the radial equation of motion:

- the pointwise forcing term on the structure is conveniently written as $\varphi(\theta, \omega) = \phi(\omega)\delta(\theta)$ using the singular Dirac distribution $\delta(\theta)$;
- the fluid pressure force per unit area is $-p|_{r=R}$.

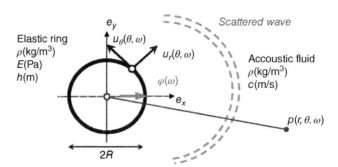

Figure 5.15 Elastic ring immersed in a compressible fluid. The elastic ring is modelled using a bi-dimensional shell of radius R and thickness h, with Young's modulus E and density ρ, as detailed in Figure 4.28. It is immersed in a compressible fluid of density ρ and speed of sound c. A pointwise radial force $\varphi(\theta, \omega)$ is applied on the ring at $\theta = 0$, with an amplitude which may depend on the pulsation ω. $u_r(\theta, \omega)$ and $u_\theta(\theta, \omega)$ denote the radial and ortho-radial displacements of the shell, while $p(r, \theta, \omega)$ denotes the pressure in the fluid at pulsation ω

The imposed acceleration on the fluid is $\omega^2 u_r(\omega)$, so that the coupling condition for the pressure field is expressed as follows:

$$\frac{\partial p}{\partial r}\bigg|_{r=R} = \rho \omega^2 u_r(\omega)$$

By expanding $u_r(\theta)$, $u_\theta(\theta)$ and $p(r, \theta)$ as Fourier series, as detailed in Sections 2.4.2 and 3.8.1, the frequency response of the system is obtained as a linear superposition of the partial response of any Fourier component m; the latter is obtained by solving the following system:

$$(-\omega^2 \mathbf{M}_m(\omega) + i\omega \mathbf{C}_m(\omega) + \mathbf{K}_m)\mathbf{U}_m = \mathbf{\Phi}_m(\omega),$$

$$\text{with } \mathbf{U}_m = \begin{Bmatrix} u_r^m \\ u_\theta^m \end{Bmatrix} \text{ and } \mathbf{\Phi}_m(\omega) = \begin{Bmatrix} \phi(\omega) \\ 0 \end{Bmatrix}.$$

- $\mathbf{M}_m(\omega)$ is the mass matrix and stands for the structure inertia and the fluid inertial effect, which is represented by $\Re(\mu_m(\omega R/c))$, where $\mu_m(\omega R/c)$ is, defined in Equation (3.67):

$$\mathbf{M}_m(\omega) = \begin{bmatrix} \mu h + \rho R \Re(\mu_m(\omega R/c)) & 0 \\ 0 & \mu h \end{bmatrix}$$

- $\mathbf{C}_m(\omega)$ is the damping matrix which accounts for the dissipation by acoustic radiation; its non-zero term is $-\omega R/c \Im(\mu_m^c(\omega R/c))$, while the matrix itself reads as follows:

$$\mathbf{C}_m(\omega) = \begin{bmatrix} -\rho c \omega R/c \Im(\mu_m(\omega R/c)) & 0 \\ 0 & 0 \end{bmatrix}$$

- $\mathbf{K}_m$ is the stiffness matrix which couples the bending and membrane motion of the ring, as exposed in Section 2.4:

$$\mathbf{K}_m = \frac{Eh}{R^2} \begin{bmatrix} \left(1 + m^4 \frac{h^2}{12R^2}\right) & m\left(1 + m^2 \frac{h^2}{12R^2}\right) \\ m\left(1 + m^2 \frac{h^2}{12R^2}\right) & m^2\left(1 + \frac{h^2}{12R^2}\right) \end{bmatrix}$$

The real and imaginary parts of $\mu_m(\omega R/c)$ are plotted for the Fourier components $m = 1, 2, 3$ in Figure 5.16. Such plots are well suited to highlight the property of FSI: the inertial effect is obviously dominant in the 'low-frequency' range $\omega R/c < 0.1$, while the acoustic damping is predominant in the 'high-frequency' range. The transition between the two regimes occurs for $\omega R/c \sim 1$. Furthermore, it is worthy to note that

- $\mu_m(\omega R/c) \simeq 1/m$ for $\omega R/c \ll 1$, where $1/m$ corresponds to the added mass coefficient for pure inertial effects;
- $\mu_m(\omega R/c) \simeq -i\omega \rho c \, u_r^m$ for $\omega R/c \gg 1$. In the 'high-frequency' range, the acoustic damping is of plane wave type for any Fourier component: it is therefore expected that a BGT-0 type condition will be suited to model radiation damping in this problem;
- at intermediate frequencies, that is, in the present case within the interval $10^{-1} \leq \omega R/c \leq 10^{+1}$, the inertial effect of the compressible fluid overrates that is observed in the low-frequency range – or that is evaluated with an incompressible fluid model;

similarly, in this frequency range the acoustic damping is also more pronounced than in the high-frequency range.

The frequency response of a structure made of steel ($R = 0.5$m, $h = 0.05$ m, $\rho = 7,800$ kg/m³, $E = 2.1 \times 10^{11}$ Pa), immersed in water ($\rho = 1,000$ kg/m³ and $c = 1,500$ m/s) and subjected to a unit force ($\phi(\omega) = 1$, $\forall \omega$) is computed in the frequency range $0 - 5,000$ Hz. Figure 5.17 plots the response function in terms of radial displacement at $\theta = 0$ for three distinct cases, namely, the ring in vacuo, the ring immersed in an hypothetically incompressible fluid, and finally the more realistic case of immersion in a compressible fluid.

The relative importance of inertial and acoustic damping effects in FSI are highlighted by looking at the distinct graphs displayed in Figure 5.17. In the present case, the validity of the incompressible fluid model is limited up to $1,500$ Hz (which corresponds to $\omega R/c \approx 5$), the acoustic damping being important in terms of both maximum amplitude and width at the resonances peaks.

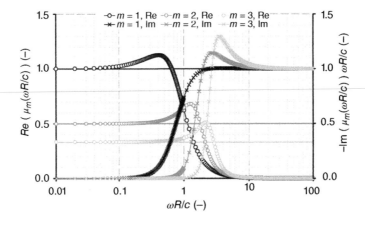

Figure 5.16   Real and imaginary parts of the added mass coefficient $\mu_m(\omega R/c)$

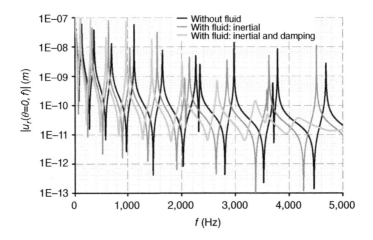

Figure 5.17   Frequency response of the structure with and without fluid coupling

Figure 5.18 Directivity diagram $\theta \mapsto p(r = 2R, \theta, \omega = 2\pi f)$ at frequency $f = 575$ Hz and $f = 1{,}950$ Hz

It is also of interest to calculate the fluid response, for instance, in terms of pressure.[8] In the present case, the pressure is a linear combination $\sum_m p_m(r, \omega) \cos(m\theta)$ of the Fourier components, which are derived from the structure normal acceleration $\omega^2 u_r^m(\omega)$ according to the following formula:

$$p_m(r, \omega) = \frac{\rho c}{\omega} \frac{J_m(\omega r/c) - iY_m(\omega r/c)}{J'_m(\omega R/c) - iY'_m(\omega R/c)} \omega^2 u_r^m(\omega)$$

Figure 5.18 gives the directivity diagram of the radiated pressure field at distance $2R$ away from the structure and for frequencies $f = 575$ Hz and $f = 1{,}950$ Hz.

5.5.2 Numerical Computation

The numerical computation of the frequency response and the directivity diagram is carried out using a finite element model, by coupling 0D structure and 1D fluid axisymmetric elements, according to Figure 4.30. $I = 30$ finite elements are used in the fluid domain, which is limited to $R' = 3R$. Zero- and first-order boundary conditions (plane or cylindrical wave radiation conditions) are imposed on node $I + 1$ at $r = R'$.

The frequency response is computed from the individual response of each Fourier component m by solving the following equation:

$$\left(-\omega^2 \begin{bmatrix} M_S^m & 0 \\ \rho R^T & M_F^m \end{bmatrix} + i\omega \begin{bmatrix} 0 & 0 \\ 0 & C_F^m \end{bmatrix} + \begin{bmatrix} K_S^m & -R \\ 0 & K_F^m \end{bmatrix} \right) \begin{Bmatrix} U(\omega) \\ P(\omega) \end{Bmatrix} = \begin{Bmatrix} \Phi^m(\omega) \\ 0 \end{Bmatrix}$$

---

[8] Other quantities might be of interest for industrial applications, such as the fluid velocity $\mathbf{v}$ or the *acoustic intensity* $\mathbf{I}$. The velocity is related to the pressure according to

$$\rho i\omega \mathbf{v}(r, \theta, \omega) = -\nabla p(r, \theta, \omega)$$

$\mathbf{I}$ is the mechanical power associated with the pressure force and it is calculated as follows:

$$\mathbf{I}(r, \theta, \omega) = p(r, \theta, \omega)\mathbf{v}(r, \theta, \omega)$$

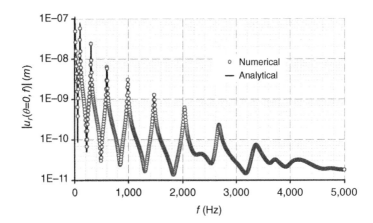

Figure 5.19 Vibro-acoustics of the submerged ring: analytical solution and numerical computation of the frequency response

In order to take into account the contribution of all modes in the frequency range of interest, $M = 15$ Fourier components are needed at least.

Figure 5.19 evidences a remarkable agreement between the analytical solution and the numerical computation of the frequency response. The zero- and first-order radiation conditions produce similar results: as expected, the plane wave radiation condition for each Fourier component of pressure suffices to accurately describe the acoustic damping. In the present case, the numerical computation is slightly dependent on the fluid domain truncation and on its discretisation: with $R' = 1.25R$ and $I = 5$, the finite element computation endowed with the zero-order radiation condition still compares favourably with the analytical solution.

The accuracy of the numerical simulation may be assessed in a more quantitative manner using various criteria, whether defined locally or globally. The following may, for instance, be suited to the frequency response at a given point $\mathbf{x}$ and over a frequency range of interest $[\omega_-, \omega_+]$.

- The *relative error* $\varepsilon(\omega, \mathbf{x})$ gives a measure of the accuracy of the numerical computation versus the analytical solution:

$$\varepsilon(\omega, \mathbf{x}) = \frac{|H_n(\omega, \mathbf{x}) - H_a(\omega, \mathbf{x})|}{|H_a(\omega, \mathbf{x})|} \qquad \forall \omega \in [\omega_-, \omega_+]$$

$H_a(\omega, \mathbf{x})$ and $H_n(\omega, \mathbf{x})$ refer, respectively, to the analytical solution and to the numerical computation of the frequency response. Figure 5.20 plots the relative error over the frequency range of interest: it suggests that the overall accuracy of the numerical computation can be bettered with a finer finite element mesh combined with a larger fluid domain.
- The so-called *frequency response assurance criteria* (FRAC) gives an indication of the correlation between the analytical and the numerical frequency responses. According to

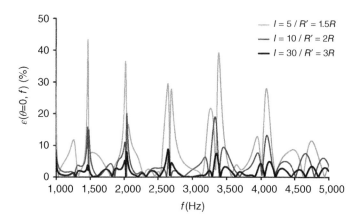

Figure 5.20 Relative error on the frequency response at $\theta = 0$ for various discretisations

Table 5.2 Frequency Response Assurance Criteria at $\theta = 0$ for various discretisations:

Finite element mesh	$I = 5/R' = 1.5R$	$I = 10/R' = 2R$	$I = 30/R' = 3R$
$\psi(\theta = 0)$	96.54%	98.81%	99.92%

Allemang (2003), it may be defined as follows:

$$\psi(\mathbf{x}) = \frac{\left|\sum_\omega \bar{H}_a(\omega, \mathbf{x})H_n(\omega, \mathbf{x})\right|^2}{\left(\sum_\omega \bar{H}_a H_a(\omega, \mathbf{x})\right)\left(\sum_\omega \bar{H}_n H_n(\omega, \mathbf{x})\right)}$$

for a given point $\mathbf{x}$. Table 5.2 gives the FRAC calculated at point $\theta = 0$ and evidences that, as expected, the correlation between the analytical and numerical frequency responses also increases with a finer mesh and a larger fluid domains.

Remark 5.4 Conditioning and accuracy *Symmetric* $(\mathbf{u}, p, \varphi)$ *and* $(\mathbf{u}, \chi)$ *formulations are equivalent from the physical standpoint, but exhibit rather different behaviour when it comes to numerical accuracy. On the one hand, the matrix produced by the* $(\mathbf{u}, p, \varphi)$ *formulation is* $\mathbf{A}(\omega) = i\omega^3\mathbf{C} - \omega^2\mathbf{M} + \mathbf{K}$*; it tends to be rather badly scaled with respect to matrix inversion, all the more so as ω is increased. On the other hand, the* $(\mathbf{u}, \chi)$ *formulation produces the matrix* $\mathbf{A}(\omega) = -\omega^2\mathbf{M} + i\omega\mathbf{C} + \mathbf{K}$*, which is better conditioned.*

It is recalled that the conditioning of a problem quantifies the sensitivity of the solution to various internal or external parameters: when it is poor, the validity of the results is often questionable. As for the inversion of a linear system $\mathbf{A}\mathbf{x} = \mathbf{b}$*, with* $\mathbf{A}$ *a square invertible matrix,* *the conditioning of the problem is evaluated with the* conditioning number $\kappa(A)$ *defined as follows:*

$$\kappa(A) = |||A^{-1}|||\; |||A|||$$

where $||| \bullet |||$ is a matrix norm.[9]

When matrix A *varies by* δA, *the solution of the linear system varies by* δx *and it can be stated that*

$$\frac{||\delta x||}{||x + \delta x||} \leq \kappa(A)\frac{||\delta A||}{||A||}$$

Hence, the smaller $\kappa(A)$, the less influence may the relative error on matrix A have on the solution.

In general, matrices involved in a finite element vibro-acoustic model exhibit large variations of their components, and as a result, the conditioning of the coupled formulations is likely to be degraded.

As an illustration, a computation of the frequency response is performed with both the (u, p, φ) and the (u, χ) formulations by solving the linear system $Ax = b$ with an algorithm which is very sensitive to the conditioning of matrix A. Figure 5.21 shows that the loss of accuracy observed in the example with the (u, p, φ) formulation is rather spectacular.

There are several ways to remedy the problem, and in most finite element codes, either one or both options presented below may be used.

Scaling matrices yields a non-dimensional linear system $A^*(\omega^*)x^*(\omega^*) = b^*(\omega^*)$ with a lower conditioning number. The non-dimensional equations are written using reference physical quantities which scale the variables of the problem. The dimensional form for the vibro-acoustic equation is as follows:

$$-\frac{\omega^2}{c^2}p + \frac{\partial^2 p}{\partial x_i^2} = 0 \qquad \frac{\partial p}{\partial x_i}n_i = \rho\omega^2 u_i n_i$$

$$-\rho\omega^2 u_i - \frac{\partial \sigma_{ij}(u)}{\partial x_j} = 0 \qquad \sigma_{ij}(u) = p n_i$$

---

[9] By definition, the matrix norm $||| \bullet |||$ associated with the scalar product $\langle \bullet | \bullet \rangle$ is

$$|||A||| = \max_{x \neq 0} \frac{\langle Ax|Ax \rangle}{\langle x|x \rangle}$$

With classical linear algebra results, it can be shown that the conditioning number of A evaluated with the scalar product $\langle \bullet | \bullet \rangle = \sum_{i,j} \bar{\bullet}_{i,j} \bullet_{i,j}$ takes the explicit form:

$$\kappa(A) = \frac{\max_{n \in [1,N]}(\lambda_n)}{\min_{n \in [1,N]}(\lambda_n)}$$

where $(\lambda_n)_{n \in [1,N]}$ are the eigenvalues of $\bar{A}^T A$ – with $\bar{\bullet}$ and $\bullet^T$ the complex conjugate and the transpose of $\bullet$. Over an interval within which the pulsation ω varies, the eigenvalues of the matrix $\bar{A}^T A(\omega)$ undergo variations of possibly wide amplitude so that the conditioning number is likely to increase significantly. The inversion of $A(\omega)x(\omega) = b(\omega)$ progressively loses accuracy as ω is increased, as illustrated in Figure 5.21. In the present example, the conditioning of matrix A ranges from 10^{+15} for $f < 1,000$ Hz to 10^{+18} for $f > 4,000$ Hz; with matrix scaling or conditioning, as discussed in what follows, it reduces to 10^{+7} to 10^{+11} for the same frequency range, resulting in a better accuracy of the frequency response computation, as highlighted in Figure 5.22.

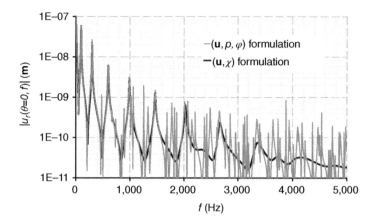

Figure 5.21 *Finite element computation of a frequency response with symmetric* $(\mathbf{u}, p, \varphi)$ *and* $(\mathbf{u}, \chi)$
formulations

c_o, ρ_o and x_o denoting a characteristic velocity, density and length of the problem, the scaled
variables are expressed as follows:

$$p^* = \frac{p}{\rho_o c_o^2} \quad \sigma_{ij}^*(\mathbf{u}^*) = \frac{\sigma_{ij}(\mathbf{u})}{\rho_o c_o^2} \quad u_i^* = \frac{u_i}{x_o} \quad x_i^* = \frac{x_i}{x_o} \quad \rho^* = \frac{\rho}{\rho_o}$$

The non-dimensional vibro-acoustic equations are found to be:

$$-\omega^{*2} p^* + \frac{\partial^2 p^*}{\partial x_i^{*2}} = 0 \quad \frac{\partial p^*}{\partial x_i^*} n_i = \rho^* \omega^{*2} u_i n_i$$

$$-\rho^* \omega^{*2} u_i^* - \frac{\partial \sigma_{ij}^*(\mathbf{u}^*)}{\partial x_j^*} = 0 \quad \sigma_{ij}^*(\mathbf{u}^*) = p^* n_i$$

The accuracy of the frequency response computed with the $(\mathbf{u}, p, \varphi)$ *formulation is improved
here when scaling variables with* $\rho_o = 1{,}000$ *kg/m³,* $c_o = 1{,}500$ *m/s and* $x_o = 0.5$ *m, as
evidenced in Figure 5.22.*

Conditioning matrices *produces a modified linear system* $\bar{\mathbf{A}}(\omega)\bar{\mathbf{x}}(\omega) = \bar{\mathbf{b}}(\omega)$ *where* $\bar{\mathbf{A}}(\omega)$ *is a
preconditioned matrix with* $\kappa(\bar{\mathbf{A}}) \leq \kappa(\mathbf{A})$. *The preconditioning is represented with matrix*
$\mathbf{C}^{-1}(\omega)$ *such that*

$$\bar{\mathbf{A}}(\omega) = \mathbf{C}^{-1}(\omega)\mathbf{A}(\omega)\mathbf{C}(\omega) \qquad \bar{\mathbf{x}}(\omega) = \mathbf{C}^{-1}(\omega)\mathbf{x}(\omega) \qquad \bar{\mathbf{b}}(\omega) = \mathbf{C}^{-1}(\omega)\mathbf{b}(\omega)$$

*In the present example, it is found that computing the frequency response with a precon-
ditioning based on the diagonal values of matrix* $\mathbf{A}(\omega)$, *that is, with* $\mathbf{C}_{ij}(\omega) = \sqrt{\mathbf{A}_{ij}(\omega)} \delta_{ij}$,
betters the accuracy obtained with the $(\mathbf{u}, p, \varphi)$ *formulation, as evidenced in Figure 5.22.* ∎

Figure 5.23 compares the analytical solution with the numerical computation of the directivity diagram at distance $2R$ and with truncation of the fluid domain at $R' = 3R$, using $I = 10$
or $I = 30$ fluid elements. The computation of the pressure is more sensitive to discretisation

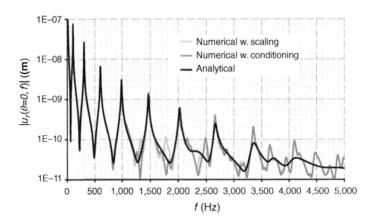

Figure 5.22   Scaling and preconditioning. *Computation of the frequency response with the symmetric* $(\mathbf{u}, p, \varphi)$ *formulation is displayed in the graph. Scaling or preconditioning matrices significantly improves the accuracy of the numerical results – scaling may be interpreted as a physical pre-conditioning*

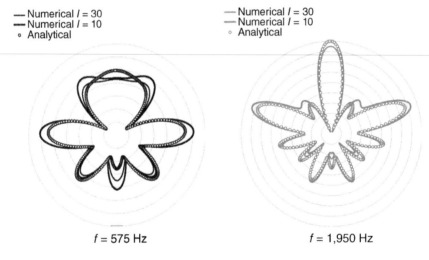

Figure 5.23   Vibro-acoustics of the submerged ring: analytical solution and numerical computation of the directivity diagram

than the computation of the displacement is. A similar trend is observed as far as the influence of the truncation and the discretisation of the fluid domain are concerned, which is found to be more important in the computation of pressure than in the computation of displacement.

The numerical results obtained with several models, involving a larger fluid domain and a finer element mesh, are compared in Figure 5.24: it is noted that, as expected, convergence is obtained at the expense of an increased computational cost.

An integral representation of the Helmholtz equation overcomes the question of domain truncation: as depicted in Figure 5.25, FEM/BEM coupling may be used in the present case as an alternative to the technique of coupling FEM/IEM for the computation of the vibro-acoustic response.

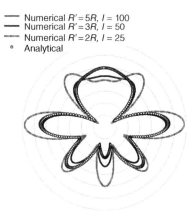

Figure 5.24   Influence of fluid domain truncation and discretisation

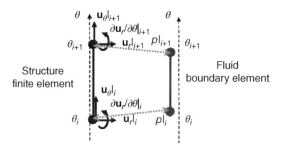

Figure 5.25   Coupling structure finite elements and fluid boundary elements

The numerical representation of the problem is derived from 1D straight finite elements – with bending/membrane degrees-of-freedom, as discussed in Remark 2.7 – being coupled to 1D straight boundary element, as discussed in Section 3.8.3.

The real and imaginary parts of the vibro-acoustic coefficient $\mu_m(\omega R/c)$ are plotted in Figure 5.26 versus the wave number $\omega R/c$. The analytical values are compared with the numerical values obtained by using Equation (3.68) using a FEM/BEM coupling based on a discretisation with $I = 180$ elements over the circumference of the ring.

Numerical computations are performed with and without CHIEF points in the integral formulation of the acoustic problem: as evidenced by the plots, the use of CHIEF points improves the quality of the computation, particularly when it comes to accurately account for acoustic damping, which is represented here by the imaginary part of μ_m for $\omega R/c \geq 1$.

The vibro-acoustic modes may be computed from Equation (5.41): as the vibro-acoustic matrix is frequency dependent, it is necessary to use an iterative procedure, detailed here by Algorithm 1, to solve the non-linear eigenvalue problem.

The numerical computations of the first eigenfrequencies with various meshes are presented in Table 5.3, which indicates that a remarkable accuracy is obtained with a FEM/BEM mesh composed of $I = 90$ circumferential elements.

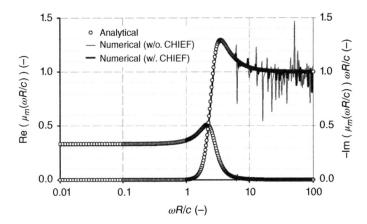

Figure 5.26 BEM computation of the vibro-acoustic coefficient $\mu_m(\omega R/c)$ for Fourier component $m = 3$

Algorithm 1 Solving eigenvalue problem $(-\omega^2\mathbf{M}(\omega) + \mathbf{K})\mathbf{X} = \mathbf{0}$ with an iterative method

Require: N, ε and Q.

Assemble matrix $\mathbf{K}$.
Assemble matrix $\mathbf{M}(0)$ from FEM/BEM coupling.
Solve eigenvalue problem:
$$(-\omega^2\mathbf{M}(0) + \mathbf{K})\mathbf{X} = \mathbf{0}$$
Store the N first eigenvectors $(\mathbf{X}_n^o)_{n\in[1,N]}$ and eigenpulsations $(\omega_n^o)_{n\in[1,N]}$.
for $n \in [1,N]$ **do**
    Set initial values $q = 0$, $\varepsilon_n^q = 1$, $\omega_n^q = \omega_n^o$ and $\mathbf{X}_n^q = \mathbf{X}_n^o$.
    while $\varepsilon_n^q < \varepsilon$ and $q < Q$ **do**
        Assemble matrix $\mathbf{M}(\omega_n^q)$ from FEM/BEM coupling.
        Compute the nth eigenpulsation ω_n^{q+1} and eigenvector $\mathbf{X}_n^{q+1}$ of the eigenvalue problem in $\mathbb{C}$:
$$(-\omega^2\mathbf{M}(\mathfrak{R}(\omega_n^q)) + \mathbf{K})\mathbf{X} = \mathbf{0}$$
        Evaluate the convergence criterion $\varepsilon_n^q = \dfrac{|\omega_n^{q+1} - \omega_n^q|}{|\omega_n^q|}$.
        Increment loop $q = q + 1$.
    end while
    Store the nth mode $\mathbf{X}_n = \mathbf{X}_n^{q+1}$ and eigenpulsation $\omega_n = \mathfrak{R}(\omega_n^{q+1})$.
end for

To conclude with the comparison, Figure 5.27 displays the pressure directivity diagram at distance $2R$ for $f = 575\,\text{Hz}$, the pressure being computed according to Equation (5.43), and thereby indicates that the numerical results are in overall good agreement with the analytical solution.

FEM/BEM and FEM/IEM coupling techniques often compete for solving vibro-acoustic problems and for the mechanical engineer, the choice of one method or the other proceeds

Table 5.3 Vibro-acoustic eigenfrequencies: analytical solution and numerical computation with FEM/BEM coupling

Frequency	Analytical (Hz)	Numerical (Hz)		
		$I = 90$	$I = 180$	$I = 360$
f_0	0	0	0	0
f_1	103.71	103.89	103.76	103.73
f_2	307.20	306.29	305.80	305.67
f_3	607.42	603.72	602.81	602.75
f_4	990.35	993.69	991.27	990.62

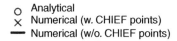

Figure 5.27 Directivity diagram $\theta \mapsto p(r = 2R, \theta, f)$ at frequency $f = 575\,\text{Hz}$ with a FEM/BEM computation

less from theoretical recommendations than from practical considerations – among which the experience gained in handling FEM, BEM or IEM, the availability of numerical tools, or the existence of computational resources and so on.

References

Allemang RJ 2003 The modal assurance criterion – twenty years of use and abuse. *Sound and Vibration Magazine*, **37**, 14–20.

Axisa F 2007 *Modelling of Mechanical Systems – Fluid-Structure Interaction*. Elsevier.

Béchade E, Chaigne A, Derveau G, and Joly P 2005 Numerical simulation of a guitar. *Computers and Structures*, **83**, 107–126.

Bermudez A, Duran R, Muschietti MA, Rodriguez R, and Solomin J 1995 Finite element vibration analysis of fluid-solid systems without spurious modes. *Journal of Numerical Analysis*, **32**, 1280–1295.

Boujot J 1987 Mathematical formulation of fluid-structure interaction problems. *Mathematical Modeling and Numerical Analysis*, **21**, 239–260.

Everstine GC 1981 A symmetric potential formulation for fluid-structure interaction. *Journal of Sound and Vibration*, **79**, 157–160.

Morand H JP and Ohayon R 1995 *Fluid-Structure Interaction*. John Wiley & Sons, Inc.

Ohayon R 2001 Reduced symmetric models for modal analysis of internal structural-acoustic and hydroelastic-sloshing systems. *Computer Methods in Applied Mechanics and Engineering*, **190**, 3009–3019.

Rajakumar C and Rogers CG 1991 The Lanczos algorithm applied to unsymmetric generalized eigenvalue problem. *International Journal of Numerical Methods in Engineering*, **32**, 1009–1026.

Sigrist JF and Garreau S 2007 Dynamic analysis of fluid-structure interaction problems with modal methods using pressure-based fluid finite elements. *Finite Element in Analysis and Design*, **43**, 287–300.

Stammberger M and Voss H 2010 On an unsymmetric eigenvalue problem governing free vibrations of fluid-solid structures. *Electronic Transaction on Numerical Analysis*, **39**, 113–125.

6

Structural Dynamics with Fluid–Structure Interaction

Numerical methods for structural dynamics are revisited in this chapter in the context of vibro-acoustics, either for time- or frequency-domain analysis. The computation of the response of the system under dynamic loading can be carried out with direct methods, that is, by solving a finite element equation, which accounts for the dynamic behaviour of the system. An inversion of large-size matrix systems is in this case necessary: for industrial applications, a limit is often reached in terms of numerical efficiency. With modal approaches, a reduced-order model (ROM) is obtained from the projection of the original equations onto a few appropriate modes. Involving less degrees-of-freedom, the ROM is more versatile and allows various scenarios to be studied at a minimum computational cost. Both methods are presented and illustrated in this chapter on simple cases of application.

Figure 6.1 Structural dynamics with fluid–structure interaction. The dynamic response of a structure subjected to dynamic loading and fluid coupling can be strongly influenced by FSI. In some situations, such as a seismic event, FSI manifestations might appear counter-intuitive. In contrast with the structural dynamics without fluid coupling, compelling modifications of the seismic response can be observed when the fluid is present, as explained later in Section 6.4. *Source*: ©Jean-François Sɪɢʀɪsᴛ

Fluid–Structure Interaction: An Introduction to Finite Element Coupling, First Edition. Jean-François Sɪɢʀɪsᴛ.
© 2015 John Wiley & Sons, Ltd. Published 2015 by John Wiley & Sons, Ltd.
Companion Website: www.wiley.com/go/sigrist

6.1   Introduction

Various numerical methods to compute the vibro-acoustic response of structures to external dynamical loads, described either in the time or frequency domain, are described in this chapter. First, it is of interest to distinguish between the so-called *direct methods* and the *ROM methods*, because of the large difference in the size of the problem to be solved in each case.

According to the direct methods, the finite element model is directly used to compute the solution of the forced problem at all the physical nodes and at each time- or frequency-step of interest. In contrast, ROM methods consist of using the finite element model to build a set of a rather few number of functional vectors serving as a projection basis for computing an approximated solution to the forced problem in the functional space instead of the physical space, thus drastically reducing the size of the matrix equations to be solved. In this book, ROM methods are essentially discussed in the more restricted context of the so-called *modal expansion method*, where the natural modes of vibration of the mechanical system, assumed to be conservative, are selected as the base vectors. In order to put the presentation in a wider perspective, some other methods to derive ROM are also exposed and illustrated on a few specific examples.

In many industrial applications, direct methods are of common use despite of the heavy computational cost usually involved in the many time repetitive manipulations of large-sized matrices. Using a direct method avoids the difficult problem of establishing a ROM, validity of which may depend largely on the educated guesses worked out by the analyser to select the appropriate set of vectors for projection[1]. Actually, this advantage is even more indisputable when the forced problem involves various non-linearities such as large and/or plastic deformations, which are more or less uniformly distributed over the structure, or difficult to locate a priori.

By contrast, an ROM method stands as a rather smart and elegant technique, which allows not only drastic saving in computational cost but may also serve to produce enlightening information about the dynamical behaviour of the mechanical system at hand as well as useful clues to improve its performance. In particular, modal expansion methods are especially fruitful when one has to deal with large and complex mechanical systems whose dynamics is in fact governed by a restricted and well-defined number of natural modes of vibration. The method is not restricted to the linear domain of responses, provided the places where non-linearities may occur can be foreseen at the stage of building the finite element model and the modal expansion basis – see for instance Moumni and Axisa (2004).

In the field of computational mechanics, a particularly abundant literature is devoted to the various direct and reduced-order methods made available to compute the response of complex systems to external dynamic loading, either in the time- or frequency-domain. The purpose of this chapter is to provide an overview of some methods which may be used by the engineering community for dealing with forced dynamical problems in mechanical systems involving fluid–structure interaction (FSI).

However, it is worth emphasising already here that the presentation will be restricted to the methods adapted to the so-called *low-frequency regime*, with the meaning that in this range,

---

[1] This is the case at least from the theoretical standpoint when, in the process of building an appropriate projection basis, the ROM is not built according to algorithms which are able to handle estimates of errors adapted to the problem at hand.

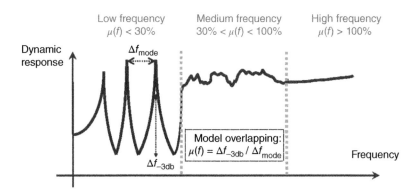

Figure 6.2   Definition of the low-frequency regime. The dynamic response of a mechanical system in the frequency-domain typically falls into three distinct ranges. In the so-called *low-frequency regime*, the response is dominated by the contribution of few dominating natural modes of vibration, whereas in the *medium-frequency* or *high-frequency* regimes, the influence of each mode in the global response tends to subside. Let Δf_{mode} be the distance between two consecutive modal frequencies and let Δf_{-3dB} be the frequency bandwidth for the modal response. The modal density, that is, the number of modes per frequency, is $n(f) = 1/\Delta f_{mode}$ and the mode overlapping is $\mu(f) = \Delta f_{-3dB}/\Delta f_{mode}$. According to Lion and Dejong (1995), who investigate the Statistical Energy Analysis as a method to tackle the high-frequency range, the low-frequency regime extends for frequencies f such as $\mu(f) < 30\%$, whereas the modal density and mode overlapping increase in the medium-frequency and high-frequency regimes. The medium-frequency range mainly concerns complex mechanical systems (i.e. structures with complex geometry and boundary conditions, which may also be composed of various elements and materials) and does not exist for simple structures, such as the ones studied in this book (e.g. beams and shells). More on the medium-frequency range in the context of numerical methods for vibro-acoustics may be found, for instance, in Soize and Ohayon (1998) and Soize and Ohayon (2014)

the dynamical response of the mechanical system is in fact controlled by a few number of modes at sufficiently separated frequencies to produce distinct resonance peaks in the spectral response, as represented in Figure 6.2.

This chapter focusses in addition on the dynamic behaviour of coupled systems in the context of *deterministic* modelling: the system geometrical and material properties, as well as the loads (force or acceleration) to which it responds, are supposed to be well characterised. It is however an ideal representation of actual configurations since differences arise between the designed and the real mechanical system, on account of the manufacturing process. In addition, the mathematical modelling process induces errors in the system representation, so that uncertainty and variability, as defined in Figure 6.3, may be out of grasp of deterministic modelling.

The *probabilistic* approach offers a general framework which is suited to tackle the uncertainty in the computational model and the variability of the real system. Uncertainty quantification with stochastic modelling is discussed in the context of FSI in Ohayon and Soize (2012) and references herein, using a reduced-order modelling of vibro-acoustic systems with dissipative effects.

A presentation of time- or frequency-domain analysis for deterministic/low-frequency problems remains a pertinent introduction, prior to probabilistic modelling. Some direct and

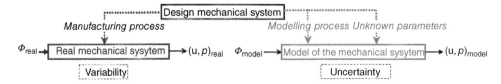

Figure 6.3 Variability and uncertainty. According to Ohayon and Soize (2012), variability is observed between the designed mechanical systems and its many real instances, on account of the manufacturing process. Uncertainties in the numerical representation arise both from the errors which are inevitably induced by the modelling process and from the intrinsical lack of knowledge in various parameters (such as geometrical and/or material properties) and configurations (such as boundary and/or operating conditions)

modal-based techniques are discussed hereafter; they are presented using modal equations and further illustrated on simple problems.

6.2 Time-Domain Analysis

6.2.1 Direct Methods

6.2.1.1 Approximation Schemes for Differential Equations

Space discretisation of the local equations governing the dynamical behaviour of any continuous mechanical system produces a matrix differential equation of second order in time, which describes the dynamical equilibrium – see for instance Belytschko et $al.$ (2000). Denoting $\mathbf{X}(t)$ the degrees-of-freedom of the system, the balance between the inertial force $\mathbf{M}\ddot{\mathbf{X}}(t)$ and a resisting force $-\kappa(t, \mathbf{X}(t))$, a viscous-type dissipative force $-\mathbf{C}\dot{\mathbf{X}}(t)$ and an external load $\mathbf{\Phi}(t)^2$ is stated as follows:

$$\mathbf{M}\ddot{\mathbf{X}}(t) + \mathbf{C}\dot{\mathbf{X}}(t) + \kappa(t, \mathbf{X}(t)) = \mathbf{\Phi}(t) \qquad (6.1)$$

The internal force $\kappa(t, \mathbf{X}(t))$ corresponds to the element stresses in the current configuration at time t; for linear problems, it reduces to

$$\kappa(t, \mathbf{X}(t)) = \mathbf{K}\mathbf{X}(t)$$

Equation (6.1) is complemented with initial conditions expressed as prescribed vectors of displacement and velocity:

$$\mathbf{X}(t = 0) = \mathbf{X}_o \qquad \dot{\mathbf{X}}(t = 0) = \dot{\mathbf{X}}_o \qquad (6.2)$$

$Direct$ $methods$ consist of integrating Equation (6.1) using a step-by-step procedure, which means that the equilibrium equation is satisfied at discrete time steps $(t_n)_{n \in 1, N}$, with $\delta t =$

---

[2] Equation (6.1) is stated without the presence of $follower$ $forces$, that is, external loads which depend on the system displacement for their definition. Such forces come into play for instance for buckling of thin elastic structures subjected to compressive loads, see, for instance, Hibbit (1979) for a short introduction on the topic.

$t_{n+1} - t_n$ the time increment. Within the time interval $[t_n, t_{n+1}]$, the variations of displacements, velocities and accelerations are written using an approximation scheme, which gives rise to various integration algorithms. To be workable, the proposed numerical scheme should possess the following mathematical properties.

Consistency   A numerical scheme is *consistent* if it reduces to the original differential equation as the increment vanishes to zero.

Let $\Pi u = 0$ be a differential equation on a time interval $[0, T]$, with initial conditions at time $t = 0$. A numerical approximation of the equation is denoted $\Pi_{\delta t}$; it is consistent with Π if for any smooth function ψ, $\Pi_{\delta t}\psi - \Pi\psi \rightarrow 0$ when $\delta t \rightarrow 0$. $\Pi_{\delta t}\psi - \Pi\psi$ is the *truncation error*: the numerical scheme is said to be *of order p* if the truncation error is $\mathcal{O}(\delta t^p)$ – meaning that the approximation is all the more accurate for higher p. As an illustration, consider the partial differential equation:

$$\Pi u = \dot{u} - \lambda u \tag{6.3}$$

for $\lambda > 0$, with the initial condition $u(t = 0) = u_o$. The so-called forward Euler scheme $\Pi_{\delta t}\bullet = \dfrac{\bullet(t + \delta) - \bullet(t)}{\delta t} - \lambda \bullet (t)$ is a first-order consistent scheme with Π. Indeed, for any smoothed function ψ, a second-order Taylor's development reads as follows:

$$\psi(t + \delta t) = \psi(t) + \delta t \dot{\psi}(t) + \mathcal{O}(\delta t^2)$$

Hence,

$$\Pi_{\delta t}\psi - \Pi\psi = \frac{\psi(t + \delta) - \psi(t)}{\delta t} - \dot{\psi}(t) = \mathcal{O}(\delta t)$$

Stability   A numerical scheme is *stable* if the error caused by a small perturbation in the numerical solution at any time t remains bounded for all $t' > t$.

As for consistency, stability is established with a dedicated analysis. For a linear differential equation $\Pi u = 0$, a step-by-step approximation scheme leads to a recursive matrix relation of the kind $\mathbf{u}_{n+1} = \pi_{\delta t}\mathbf{u}_n$, where $\mathbf{u}_n$ gathers the values of u (and its time derivative if needed by the construction of the approximation scheme) at time step t_n. $\pi_{\delta t}$ is an approximation of $\Pi_{\delta t}$ and stands for the so-called *amplification matrix* of the time discretised equations. It follows that $\mathbf{u}_n = \pi_{\delta t}^n \mathbf{u}_0$, with $\mathbf{u}_0$ standing for the initial condition: the stability of the algorithm depends on the eigenvalues λ of the amplification matrix. Denoting $\rho = \max(|\lambda|)$, the stability is in particular ensured when $\rho \leq 1$, which gives a condition formulated in terms of δt.

Turning to Equation (6.3) as an illustrative case, the forward Euler scheme gives the following relation:

$$\frac{u_{n+1} - u_n}{\delta t} - \lambda u_n = 0 \tag{6.4}$$

Thus, $u_n = u_o(1 - \lambda \delta t)^n$ for all n: stability is granted when $|1 - \lambda \delta t| < 1$. The condition on the time step is then expressed as:

$$\delta t < \frac{2}{\lambda}$$

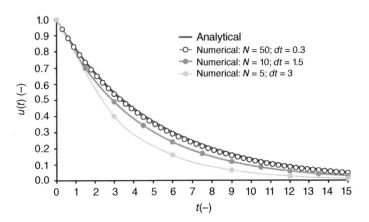

Figure 6.4   Convergence of a numerical scheme. The first-order forward Euler scheme, as defined by Equation (6.4), is used here to solve the first-order differential equation $\dot{u} - \lambda u = 0$ on $[0, T]$ with $u(0) = u_o$. Computations are performed on the time interval of interest with a N–points scheme; the time step is constant $\delta t = T/N$. The forward Euler scheme is consistent and the stability condition is fulfilled with $\delta t < 2/\lambda$: convergence is therefore granted. This point is evidenced by increasing the number of time steps N, thereby decreasing the time step δt (in the present example, $\lambda = 1/5$, $u_o = 1$ and $T = 15$)

Convergence   A numerical scheme is *convergent* when the solution it produces approaches the solution to the partial differential equation as the increment vanishes to zero.

For linear numerical schemes applied to well-posed linear differential equations, it is demonstrated that a consistent and stable scheme is convergent. However, for non-linear problems, general results remain very scarce, somehow putting convergence issues at the kernel of concern in numerical scheme analysis. The rate of convergence of the forward Euler scheme for Equation (6.3) is illustrated in Figure 6.4.

6.2.1.2   Explicit and Implicit Schemes

Direct time integration methods can be further subdivided into two main categories, the choice of which mainly depends on the type of loading and the frequency content of the loading and of the loaded mechanical system.

Explicit methods In explicit methods (Dokainish and Subbaraj, 1989), the solution at time $t + \delta t$ is obtained by considering the force balancing at previous discrete times $t' < t$. Accordingly, an explicit equation is to be solved in order to calculate by extrapolating the solution at the next computed time $\mathbf{X}(t + \delta t)$:

$$\mathbf{X}(t + \delta t) = \psi(\mathbf{X}(t))$$

Furthermore, the computational effort may be significantly reduced provided the mass matrix is recast in a diagonal form in such a way that the integration scheme does not require any matrix inversion.

Nevertheless, a consequence of paramount importance of proceeding by extrapolation is that the time step used in the integration procedures must be less than a critical value to produce meaningful results. A linear analysis of the problem highlights that explicit time integration schemes are necessarily *conditionally stable*. The maximum time step is shown to be inversely proportional to the highest frequency present in the finite element model[3] – in explicit methods, the choice of time step is therefore governed by *stability* considerations. As soon as the critical value is exceeded, the computed solution blows up exponentially with time. There is a lack of general criteria concerning the behaviour of the algorithms – either explicit or implicit – in the non-linear domain of regimes. However, the criterion seems not to be less restrictive in the non-linear domain than in the linear one.

Despite conditional stability, the explicit algorithms are often preferred to the implicit algorithms because the programming task is generally easier and the efficiency is usually better for dealing with problems which involve short-lived transients. Impact problems are typical example of application: they may involve a single and violent transient excitation, as the response of marine platform to underwater explosions or hydrodynamics impact (as illustrated in the latter case in Figures 1.3 and 3.8) or the more or less steady non-linear vibrations of a structure impacting repeatedly against neighbouring structure, as the rattling of heat exchanger tubes hitting loose supports (the case being particular concern because of the fretting-wear so induced (Pettigrew, 2003a, 2003b)).

Implicit methods Contrasting with the explicit methods, the implicit schemes (Subbaraj and Dokainish, 1989) make use of the force balance verified at the same time step at which the displacement is computed, avoiding the risk of unrealistic prediction inherent to any extrapolation procedure. The solution of the equation of motion at time $t + \delta t$ is expressed in terms of the variation of displacements, velocities and accelerations at the same time step, and, accordingly, $\mathbf{X}(t + \delta t)$ is found to be the solution of an implicit equation:

$$\psi(\mathbf{X}(t + \delta t)) = 0$$

The implicit procedure allows to build *unconditionally stable* algorithms – so that in implicit methods, the choice of time step is usually governed by *accuracy* considerations. However, the computational cost to pay this improvement is the necessity to carry out an inversion of a matrix at each time step.

Despite this drawback, implicit algorithms are preferentially selected in many linear and non-linear applications for dealing with complex or refined finite element models so that the maximum time step required by an explicit algorithm would reveal much shorter than the physical timescale of interest in the problem. A typical example of application of implicit time integration procedures involving FSI is the seismic analysis of mechanical systems such as fluid-filled reservoirs (see, for instance, Rebouillat and Liksonov (2010)) or pressure vessels (as exposed, for instance, in Section 4.3.3 in the case of a tube bundle).

6.2.1.3 Example of Implicit Non-Linear Schemes

The Newmark-type schemes are, among many implicit methods, some of the most popular in computational structural dynamics (CSD) (Newmark, 1959). Able to tackle both linear and non-linear problems, the common starting point consists of the following approximation of acceleration and velocity.

---

[3] The time step is therefore conditioned by the size of the smallest element in the mesh. When large deformations of the mesh are to be handled, for instance, in the crashworthiness assessment of vehicles, the time step should be updated at each time iteration so as to be consistent with the stability condition.

- On the one hand, the displacement field $\mathbf{X}(t)$ is expanded as a third-order Taylor's series for $t = t_{n+1}$ using a free parameter denoted β:

$$\mathbf{X}(t_{n+1}) = \mathbf{X}(t_n) + \delta t \dot{\mathbf{X}}(t_n) + \delta t^2 \left[\beta \ddot{\mathbf{X}}(t_{n+1}) + \left(\frac{1}{2} - \beta \right) \ddot{\mathbf{X}}(t_n) \right] + \mathcal{O}(\delta t^3)$$

A first-order approximation of the acceleration at time step t_{n+1} is thus deduced:

$$\ddot{\mathbf{X}}(t_{n+1}) = \frac{\mathbf{X}(t_{n+1}) - \mathbf{X}(t_n)}{\beta \delta t^2} - \frac{\dot{\mathbf{X}}(t_n)}{\beta \delta t} - \left(\frac{1}{2\beta} - 1 \right) \ddot{\mathbf{X}}(t_n) + \mathcal{O}(\delta t) \tag{6.5}$$

- On the other hand, the velocity field $\dot{\mathbf{X}}(t)$ is expanded as a first-order Taylor's series for $t = t_{n+1}$, using another free parameter denoted γ:

$$\dot{\mathbf{X}}(t_{n+1}) = \dot{\mathbf{X}}(t_n) + \delta t [\gamma \ddot{\mathbf{X}}(t_{n+1}) + (1 - \gamma) \ddot{\mathbf{X}}(t_n)] + \mathcal{O}(\delta t) \tag{6.6}$$

The values selected for parameters β and γ control the stability and the accuracy of the algorithm and serve to specify the variant of the Newmark algorithm, as illustrated in Table 6.1. Equation (6.1) is stated at time step t_{n+1}:

$$\mathbf{M}\ddot{\mathbf{X}}_{n+1} + \mathbf{C}\dot{\mathbf{X}}_{n+1} + \kappa(\mathbf{X}_{n+1}) = \mathbf{\Phi}_{n+1}$$

where $\ddot{\mathbf{X}}_{n+1} = \ddot{\mathbf{X}}(t_{n+1})$, $\dot{\mathbf{X}}_{n+1} = \dot{\mathbf{X}}(t_{n+1})$, $\mathbf{X}_{n+1} = \mathbf{X}(t_{n+1})$ and $\mathbf{\Phi}_{n+1} = \mathbf{\Phi}(t_{n+1})$. Substituting the former approximations of the acceleration and velocity in the preceding equation yields the following relation:

$$\left(\frac{\mathbf{M}}{\beta \delta t^2} + \frac{\mathbf{C}\gamma}{\beta \delta t} \right) \mathbf{X}_{n+1} + \kappa(\mathbf{X}_{n+1}) = \mathbf{\Phi}_{n+1} + \left(\frac{\mathbf{M}}{\beta \delta t^2} + \frac{\mathbf{C}\gamma}{\beta \delta t} \right) \mathbf{X}_n$$
$$+ \left(\frac{\mathbf{M}}{\beta \delta t} - \mathbf{C}\left(1 - \frac{\gamma}{\beta} \right) \right) \dot{\mathbf{X}}_n + \left(\mathbf{M}\left(\frac{1}{2\beta} - 1 \right) + \mathbf{C}\delta t \left(\frac{\gamma}{2\beta} - 1 \right) \right) \ddot{\mathbf{X}}_n \tag{6.7}$$

This relation takes the form of an implicit non-linear equation of the unknown $\mathbf{X}_{n+1}$: the right-hand side of the equation gathers quantities which are evaluated at the current time step (as for the external load) or which have been calculated at the previous time step (such as the acceleration, velocity and displacement).

Equation (6.7) may also be recast as $\psi_n(\mathbf{X}_{n+1}) = \mathbf{0}$, where ψ_n is a non-linear function. Solving the latter equation may be achieved with various methods; among many, the

Table 6.1 Example of Newmark time integration schemes. Adjusting parameters β and γ in the Newmark-type schemes produces various time integration algorithms. The combination proposed here is among the most commonly used in structural dynamics. They all achieve a second-order convergence, but they are not equivalent when it comes to numerical damping or stability (Hughes and Belytschko, 1983)

γ	β	Method	Type	Damping	Accuracy	Stability
$1/2$	0	Centred differences	Explicit	No	$\mathcal{O}(\delta t^2)$	$\omega \delta t < 2$
$1/2$	$1/6$	Linear acceleration	Implicit	Yes	$\mathcal{O}(\delta t^2)$	$\omega \delta t < 2\sqrt{3}$
$1/2 - \alpha$	$(1 - \alpha)^2/4$	Average acceleration	Implicit	Yes	$\mathcal{O}(\delta t^2)$	$\forall \delta t$

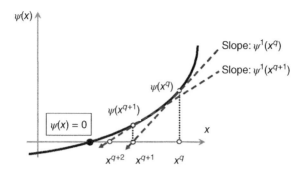

Figure 6.5 Solving $\psi(x) = 0$ with the Newton–Raphson algorithm. The iterative procedure $x^{q+1} = x^q - \psi'(x^q)^{-1}\psi(x^q)$ generates a sequence $(x^q)_{q>0}$ which converges towards the solution of the equation $\psi(x) = 0$. The convergence of the procedure depends on various conditions, which are discussed in many textbooks on numerical analysis: suffice it here to recall that the rate of convergence is at least quadratic near the solution. As powerful as a numerical technique, the Newton–Raphson procedure is also limited by some shortcomings: the calculation of the function derivative ψ' might not be analytically possible in all instances, while the method might also fail to converge in certain cases – for example, because of the existence of a stationary point x_o for ψ (as ψ' is close to zero in the vicinity of x_o, $1/\psi'$ becomes nearly singular), or because of the sensitivity of the algorithm to the starting point x^0, and so on

Newton–Raphson algorithm is of broad use in CSD (Wriggers, 2008). As depicted in Figure 6.5, finding roots of scalar function ψ is performed with the following iterative scheme:

$$x^{q+1} = x^q - \psi'(x^q)^{-1}\psi(x^q)$$

starting with a given x^0.

Applying the Newton–Raphson procedure to Equation (6.7) yields the following iterative scheme:

$$\left(\frac{\mathbf{M}}{\beta \delta t^2} + \frac{\mathbf{C}\gamma}{\beta \delta t} + \left. \frac{\partial \boldsymbol{\kappa}}{\partial \mathbf{X}} \right|_{n+1,q} \right) \mathbf{X}_{n+1}^{q+1} = \boldsymbol{\Phi}_{n+1} - \boldsymbol{\kappa}(\mathbf{X}_{n+1}^q)$$

$$+ \left. \frac{\partial \boldsymbol{\kappa}}{\partial \mathbf{X}} \right|_{n+1,q} \mathbf{X}_{n+1}^q + \left(\frac{\mathbf{M}}{\beta \delta t^2} + \frac{\mathbf{C}\gamma}{\beta \delta t} \right) \mathbf{X}_n$$

$$+ \left(\frac{\mathbf{M}}{\beta \delta t} - \mathbf{C}\left(1 - \frac{\gamma}{\beta}\right) \right) \dot{\mathbf{X}}_n + \left(\mathbf{M}\left(\frac{1}{2\beta} - 1 \right) + \mathbf{C}\delta t \left(\frac{\gamma}{2\beta} - 1 \right) \right) \ddot{\mathbf{X}}_n \qquad (6.8)$$

The displacement at the previous time step can be used as an initial value for the iterations, that is, $\mathbf{X}_{n+1}^0 = \mathbf{X}_n$. For each time step, the procedure is repeated until a convergence criterion is verified or stopped when a maximum number of iterations Q is reached; see Algorithm 2. A convergence criterion may be formulated in various ways, among which is the following (Bathe, 1982).

- A convergence criterion on displacement may be written as follows:

$$\frac{\|\mathbf{X}_{n+1}^{q+1} - \mathbf{X}_{n+1}^q\|}{\|\mathbf{X}_{n+1}^q\|} \leq \varepsilon$$

where $\| \bullet \|$ is a vectorial norm.

Algorithm 2 Implicit time integration scheme for non-linear problems (Newmark scheme)

Require: Time interval $[0, T]$, scheme parameters β and γ and time increment δt.

Set initial conditions $\mathbf{X}_o$, $\dot{\mathbf{X}}_o$ and $\ddot{\mathbf{X}}_o = \mathbf{\Phi}(0) - \mathbf{M}^{-1}(\mathbf{KX}_o + \mathbf{C}\dot{\mathbf{X}}_o)$.

Set number of time iterations N from $T = N\delta t$ and of maximum non-linear iterations Q within each time step.

Assemble matrices $\mathbf{M}$, $\mathbf{C}$ and $\mathbf{K}$ from elementary matrices $\mathbf{m}^e$, $\mathbf{c}^e$ and $\mathbf{k}^e$.

Calculate pseudo-mass, damping and stiffness matrices:

$$\hat{\mathbf{M}} = \mathbf{M}\left(\frac{1}{2\beta} - 1\right) + \mathbf{C}\delta t\left(\frac{\gamma}{2\beta} - 1\right)$$

$$\hat{\mathbf{C}} = \frac{\mathbf{M}}{\beta\delta t} - \mathbf{C}(1 - \frac{\gamma}{\beta})$$

$$\hat{\mathbf{K}} = \frac{\mathbf{M}}{\beta\delta t^2} + \frac{\mathbf{C}\gamma}{\beta\delta t}$$

for $n \in [1, N]$ **do**

  Set initial values $q = 0$, $\varepsilon_n^q = 1$, $\mathbf{X}_{n+1}^q = \mathbf{X}_n$.

  while $\varepsilon_n^q < \varepsilon$ and $q < Q$ **do**

    Compute $\kappa(\mathbf{X}_{n+1}^q)$.

    Compute $\dfrac{\partial \kappa}{\partial \mathbf{X}}(\mathbf{X}_{n+1}^q)$.

    Compute the pseudo-load $\hat{\mathbf{\Phi}}_{n+1}^q$:

$$\hat{\mathbf{\Phi}}_{n+1}^q = \mathbf{\Phi}_{n+1} - \kappa(\mathbf{X}_{n+1}^q) + \frac{\partial \kappa}{\partial \mathbf{X}}(\mathbf{X}_{n+1}^q) + \hat{\mathbf{M}}\ddot{\mathbf{X}}_n + \hat{\mathbf{C}}\dot{\mathbf{X}}_n + \hat{\mathbf{K}}\mathbf{X}_n$$

    Compute the pseudo-stiffness $\hat{\mathbf{K}}_{n+1}^q$:

$$\hat{\mathbf{K}}_{n+1}^q = \frac{\mathbf{M}}{\beta\delta t^2} + \frac{\mathbf{C}\gamma}{\beta\delta t} + \left.\frac{\partial \kappa}{\partial \mathbf{X}}\right|_{n+1,q}$$

    Solve the pseudo-static problem:

$$\hat{\mathbf{K}}_{n+1}^q \mathbf{X}_{n+1}^{q+1} = \hat{\mathbf{\Phi}}_{n+1}^q$$

    Evaluate the convergence criterion $\varepsilon_n^q = \dfrac{\|\mathbf{X}_{n+1}^{q+1} - \mathbf{X}_{n+1}^q\|}{\|\mathbf{X}_{n+1}^q\|}$.

    Increment loop $q = q + 1$.

  end while

  Update displacement at time step t_{n+1} after Q non-linear iterations: $\mathbf{X}_{n+1} = \mathbf{X}_{n+1}^Q$.

  Compute velocity and acceleration at time t_{n+1} with current displacement $\mathbf{X}_{n+1}$ and previous acceleration, velocity and displacement $\ddot{\mathbf{X}}_n$, $\dot{\mathbf{X}}_n$, $\mathbf{X}_n$:

$$\dot{\mathbf{X}}_{n+1} = \left(1 - \frac{\gamma}{2\beta}\right)\delta t\ddot{\mathbf{X}}_n + \left(1 - \frac{\gamma}{\beta}\right)\dot{\mathbf{X}}_n + \frac{\gamma}{\beta\delta t}(\mathbf{X}_{n+1} - \mathbf{X}_n)$$

$$\ddot{\mathbf{X}}_{n+1} = \left(1 - \frac{1}{2\beta}\right)\ddot{\mathbf{X}}_n - \frac{1}{\beta\delta t}\dot{\mathbf{X}}_n + \frac{1}{\beta\delta t^2}(\mathbf{X}_{n+1} - \mathbf{X}_n)$$

end for

- Other convergence criteria which are of practical interest in non-linear structural dynamics may be formulated in terms of force as follows:

$$\frac{\|\mathbf{\Phi}_{n+1} - \kappa(\mathbf{X}_{n+1}^{q+1}) - \mathbf{M}\ddot{\mathbf{X}}_{n+1}^{q+1} - \mathbf{C}\dot{\mathbf{X}}_{n+1}^{q+1}\|}{\|\mathbf{\Phi}_{n+1}\|} \leq \varepsilon$$

or in terms of energy as follows:

$$\frac{(\mathbf{X}_{n+1}^{q+1} - \mathbf{X}_{n+1}^q)^\mathsf{T}(\mathbf{\Phi}_{n+1} - \kappa(\mathbf{X}_{n+1}^{q+1}) - \mathbf{M}\ddot{\mathbf{X}}_{n+1}^{q+1} - \mathbf{C}\dot{\mathbf{X}}_{n+1}^{q+1})}{(\mathbf{X}_{n+1}^1 - \mathbf{X}_n)^\mathsf{T}(\mathbf{\Phi}_{n+1} - \kappa(\mathbf{X}_n) - \mathbf{M}\ddot{\mathbf{X}}_n - \mathbf{C}\dot{\mathbf{X}}_n)} \leq \varepsilon$$

The iterative procedure is computationally demanding because it requires the evaluation of the non-linear stiffness $\kappa(\mathbf{X})$ and the *tangent stiffness matrix* $\frac{\partial \kappa}{\partial \mathbf{X}}(\mathbf{X})$. In addition, a resolution of a *pseudo-static problem* $\hat{\mathbf{K}}_{n+1}^q \mathbf{X}_{n+1}^{q+1} = \hat{\mathbf{\Phi}}_{n+1}^q$ is also required for each iteration q, $\hat{\mathbf{K}}_{n+1}^q$ stands here for the so-called *pseudo-stiffness matrix* and it is defined as follows:

$$\hat{\mathbf{K}}_{n+1}^q = \frac{\mathbf{M}}{\beta \delta t^2} + \frac{\mathbf{C}\gamma}{\beta \delta t} + \frac{\partial \kappa}{\partial \mathbf{X}}\Big|_{n+1,q}$$

Solving a linear system $\mathbf{A}\mathbf{X} = \mathbf{b}$ is generally performed using an appropriate decomposition of matrix $\mathbf{A}$, judiciously chosen according to a few basic properties of the matrix.

- The LU decomposition applies to square matrices and reads as follows:

$$\mathbf{A} = \mathbf{L}\mathbf{U}$$

where $\mathbf{L}$ and $\mathbf{U}$ are lower and upper triangular matrices, respectively.
- The LDL decomposition applies for symmetric matrices:

$$\mathbf{A} = \mathbf{L}\mathbf{D}\bar{\mathbf{L}}^\mathsf{T}$$

with $\mathbf{L}$ a lower triangular matrix and $\mathbf{D}$ a diagonal matrix; $\bar{\bullet}^\mathsf{T}$ denotes the conjugate transpose of $\bullet$.
- The LL decomposition applies for positive symmetric matrices:

$$\mathbf{A} = \mathbf{L}\bar{\mathbf{L}}^\mathsf{T}$$

It stands as a particular LDL decomposition where the diagonal matrix reduces to the identity matrix.

Adopting, for instance, the LDL decomposition of $\mathbf{A}$, the solution of the linear system $\mathbf{A}\mathbf{X} = \mathbf{b}$ is obtained as follows:

$$\mathbf{L}\mathbf{D}\bar{\mathbf{L}}^\mathsf{T}\mathbf{X} = \mathbf{b} \iff \mathbf{D}\mathbf{Y} = \mathbf{c} \qquad \mathbf{Y} = \bar{\mathbf{L}}^\mathsf{T}\mathbf{X} \quad \mathbf{c} = \mathbf{L}^{-1}\mathbf{b}$$

where solving $\mathbf{D}\mathbf{Y} = \mathbf{c}$ is immediate since $\mathbf{D}$ is a diagonal matrix.

For structural dynamics without FSI, the matrices are usually symmetric and definite positive[4], whereas in the context of FSI modelling, the coupled formulations yield matrices of various nature (either symmetric or non-symmetric, either non-negative or positive definite). Algorithms to solve linear systems, as well as eigenvalue problems, are in general less efficient with non-symmetric matrices than with symmetric matrices of the same size: CSD with FSI remains in general rather computationally demanding.

6.2.1.4 Example of Implicit Linear Scheme

For linear problems, Equation (6.7) reduces to

$$\left(\frac{\mathbf{M}}{\beta \delta t^2} + \frac{\mathbf{C}\gamma}{\beta \delta t} + \mathbf{K} \right) \mathbf{X}_{n+1} =$$

$$\left[\mathbf{\Phi}_{n+1} + \left(\frac{\mathbf{M}}{\beta \delta t^2} + \frac{\mathbf{C}\gamma}{\beta \delta t} \right) \mathbf{X}_n + \left(\frac{\mathbf{M}}{\beta \delta t} - \mathbf{C}\left(1 - \frac{\gamma}{\beta} \right) \right) \dot{\mathbf{X}}_n \right.$$

$$\left. + \left(\mathbf{M} \left(\frac{1}{2\beta} - 1 \right) + \mathbf{C}\delta t \left(\frac{\gamma}{2\beta} - 1 \right) \right) \ddot{\mathbf{X}}_n \right]$$

so that $\mathbf{X}_{n+1}$ is derived from the displacement, velocity and acceleration fields at the previous time step t_n.

The stability of the Newmark-type schemes may be carried out by considering a harmonic vibration of pulsation ω – it is reminded that the algorithm is unstable if the computed amplitude of the vibration grows exponentially with time, and it is stable otherwise – and shows that the stability condition of the Newmark-type schemes is

$$\left(\gamma + \frac{1}{4} \right)^2 - 4\beta \leq \frac{4}{(\omega \delta t)^2} \tag{6.9}$$

In the particularly interesting case of *critical stability* (which corresponds to $\beta = 1/4$ and $\gamma = 1/2$, see Figure 6.6), the algorithm does not introduce any numerical damping, so that the amplitude of free oscillations remains constant.

As other implicit algorithms, some of which being proposed in Table 6.1, the Newmark scheme introduces numerical dissipation – except for the *mean acceleration* variant, derived with parameters $\beta = 1/4$ and $\gamma = 1/2$ – which may be used in practice to damp out the unwanted contribution of high-frequency modes, as suggested by Hughes and Belytschko (1983)[5].

The Hilber–Hughes–Taylor scheme (Hilber *et al.*, 1977) is an improvement of the Newmark scheme, which is aimed at controlling the dissipative properties of the integration procedure. It produces an approximation of $\mathbf{M}\ddot{\mathbf{X}}(t) + \mathbf{C}\dot{\mathbf{X}}(t) + \mathbf{K}\mathbf{X}(t) = \mathbf{\Phi}(t)$ according to:

$$\mathbf{M}\ddot{\mathbf{X}}_{n+1} + (1 + \alpha)\mathbf{C}\dot{\mathbf{X}}_{n+1} - \alpha\mathbf{C}\dot{\mathbf{X}}_n + (1 + \alpha)\mathbf{K}\mathbf{X}_{n+1} - \alpha\mathbf{K}\mathbf{X}_n = \mathbf{\Phi}_{n+1}$$

where $\ddot{\mathbf{X}}_{n+1}$ and $\dot{\mathbf{X}}_{n+1}$ are given by Equations (6.5) and (6.6).

---

[4] This holds for *conservative* problems; for non-conservative problems, stemming, for instance, from the existence of follower forces, matrices involved in the finite element formulation are not symmetric.
[5] The numerical approximation also introduces a phase lag in the system response, which has to be limited in order not to degrade the accuracy of the computed response.

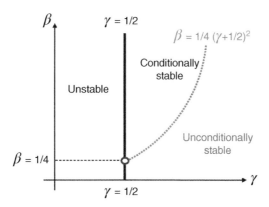

Figure 6.6 Stability diagram. The stability analysis of linear time integration schemes is developed by Bathe and Wilson (1973). It is shown that (i) the Newmark scheme is unstable for $\gamma < 1/2$; (ii) it is conditionally stable, Equation (6.9), for $\gamma > 1/2$, $\beta > 1/4(\gamma + 1/2)^2$; (iii) it is unconditionally stable for $\gamma > 1/2$, $\beta < 1/4(\gamma + 1/2)^2$ – the particular combination $\beta = 1/4$ and $\gamma = 1/2$ corresponds to the so-called *critical stability*

The time iterations of the Hilber–Hughes-Taylor scheme are presented in Algorithm 3 in the linear case, with α, β and γ the three parameters of the algorithm. α and γ govern the numerical dissipation, which does not affect the low-frequency modes too significantly and can be lowered to zero if needed.

Linear implicit algorithms require the inversion of the pseudo-stiffness matrix $\hat{\mathbf{K}}$; for the Newmark-type schemes, it is calculated as follows:

$$\hat{\mathbf{K}} = \mathbf{M} + (1 + \alpha)\gamma \delta t \mathbf{C} + (1 + \alpha)\beta \delta t^2 \mathbf{K}$$

In linear implicit schemes, the decomposition of matrix $\hat{\mathbf{K}}$ is performed once for all before time iterations, as evidenced in Algorithm 3, whereas non-linear implicit schemes require the resolution of a linear system at each non-linear iteration.

With the continuous development of efficient numerical procedures and the ever-increasing storage capability of computational machines, direct time integration methods are commonly used in mechanical engineering. The Newmark and Hilber–Hughes–Taylor implicit schemes therefore appear as standard time integration procedures in most general-purpose finite element codes.

Nevertheless, solving large-scale tri-dimensional problems – as highlighted, for instance, in Section 4.3 in the context of FSI modelling in tube bundles – still requires some computational resources and may remain beyond the capabilities of many computation centres.

6.2.2 Modal Methods

Model reduction techniques often provide an alternative to the simulations carried out using direct methods, which are often very demanding in terms of computational cost. As they imply a significant reduction in the number of degrees-of-freedom in comparison with the full finite

---

Algorithm 3 Implicit time integration scheme for linear problems (Hilber–Hughes–Taylor scheme)

Require: Time interval $[0, T]$, time increment δt, scheme parameters α, β and γ with:

$$\beta = (1 - \alpha)^2/4 \qquad \gamma = 1/2 - \alpha$$

Set initial conditions $\mathbf{X}_o$, $\dot{\mathbf{X}}_o$ and $\ddot{\mathbf{X}}_o = \boldsymbol{\Phi}(0) - \mathbf{M}^{-1}(\mathbf{K}\mathbf{X}_o + \mathbf{C}\dot{\mathbf{X}}_o)$.

Set number of time iterations N from $T = N\delta t$.

Assemble matrices $\mathbf{M}$, $\mathbf{C}$ and $\mathbf{K}$ from elementary matrices $\mathbf{m}^e$, $\mathbf{c}^e$ and $\mathbf{k}^e$.

Calculate the pseudo-stiffness matrix $\hat{\mathbf{K}} = \mathbf{M} + (1 + \alpha)\gamma\,\delta t\mathbf{C} + (1 + \alpha)\beta\delta t^2\mathbf{K}$.

Perform a decomposition of matrix $\hat{\mathbf{K}}$; for instance using the LDL decomposition:

$$\hat{\mathbf{K}} = \mathbf{LDL}^\mathsf{T}$$

with symmetric definite positive matrices.

for $n \in [1, N]$ **do**

    Calculate the pseudo-load at time t_{n+1}:

$$\hat{\boldsymbol{\Phi}}_{n+1} = \boldsymbol{\Phi}_{n+1} - (1 + \alpha)\mathbf{C}(\dot{\mathbf{X}}_n + (1 - \gamma)\delta t\ddot{\mathbf{X}}_n) + \alpha\mathbf{C}\dot{\mathbf{X}}_n$$

$$-(1 + \alpha)\mathbf{K}(\mathbf{X}_n + \delta t\dot{\mathbf{X}}_n + \left(\frac{1}{2} - \beta\right)\delta t^2\ddot{\mathbf{X}}_n) + \alpha\mathbf{K}\mathbf{X}_n$$

    Solve $\hat{\mathbf{K}}\ddot{\mathbf{X}}_{n+1} = \hat{\boldsymbol{\Phi}}_{n+1}$ to evaluate the acceleration at time t_{n+1}; for instance using again the LDL decomposition:

$$\mathbf{LDL}^\mathsf{T}\ddot{\mathbf{X}}_{n+1} = \hat{\boldsymbol{\Phi}}_{n+1}$$

    Compute velocity and displacement at time t_{n+1} with current acceleration $\ddot{\mathbf{X}}_{n+1}$ and previous acceleration, velocity and displacement $\ddot{\mathbf{X}}_n$, $\dot{\mathbf{X}}_n$, $\mathbf{X}_n$:

$$\dot{\mathbf{X}}_{n+1} = \dot{\mathbf{X}}_n + (1 - \gamma)\delta t\ddot{\mathbf{X}}_n + \gamma\delta t\ddot{\mathbf{X}}_{n+1}$$

$$\mathbf{X}_{n+1} = \mathbf{X}_n + \delta t\dot{\mathbf{X}}_n + \left(\frac{1}{2} - \beta\right)\delta t^2\ddot{\mathbf{X}}_n + \beta\delta t^2\ddot{\mathbf{X}}_{n+1}$$

end for

---

element model, they are especially advisable to perform parametric studies aimed at improving the design of a mechanical component.

Restricting to the linear case, in a first step at least, the projection of the system's dynamic equations, as written in the time domain, onto an eigenmode basis produces a set of decoupled equations, which can be solved in a straightforward manner. Without loss of generality regarding the principles of modal methods, the presentation is given here in the context of the dynamic response of a mechanical system to an imposed acceleration. From the engineering standpoint, such problems come upon for instance in seismic analysis.

In this situation, the equation of motion is conveniently written in the *relative frame*, that is, the frame attached to the prescribed motion. Assuming a standard linear dynamical model involving the internal forces of inertia (accounted for with the mass matrix $\mathbf{M}$), of elasticity (accounted for with the stiffness matrix $\mathbf{K}$) and of viscous damping (accounted for with the damping matrix $\mathbf{C}$), the forced motion is governed by the following matrix equation:

$$\mathbf{M\ddot{X}}(t) + \mathbf{C\dot{X}}(t) + \mathbf{KX}(t) = -\mathbf{M}\boldsymbol{\Delta}\gamma(t) \tag{6.10}$$

$\mathbf{X}$ is the system's degrees-of-freedom in the relative frame, $\gamma(t)$ is the acceleration imposed in direction $\boldsymbol{\Delta}$. Initial conditions in the relative frame are expressed by Equation (6.2).

6.2.2.1   Modal Equations

For mechanical systems with symmetric matrices, the eigenmode vectors are real and form an orthogonal basis to expand the solution of Equation (6.10) as a modal series, as already emphasised in Chapter 2, Subsection 2.2.6:

$$\mathbf{X}(\mathbf{x}, t) = \sum_{n \geq 1} \xi_n(t)\mathbf{X}_n(\mathbf{x}) \tag{6.11}$$

The projection of the initial conditions onto the basis $(\mathbf{X}_n)_{n \geq 1}$ is

$$\mathbf{X}_o = \sum_{n \geq 1} \xi_n^o \mathbf{X}_n \qquad \dot{\mathbf{X}}_o = \sum_{n \geq 1} \dot{\xi}_n^o \mathbf{X}_n$$

The projection of Equation (6.10) onto any eigenvector $\mathbf{X}_{n'}$ using a modal expansion according to Equation (6.11) is

$$\mathbf{X}_{n'}^{\mathsf{T}} \sum_{n \geq 1} (\ddot{\xi}_n \mathbf{M}\mathbf{X}_n + \dot{\xi}_n \mathbf{C}\mathbf{X}_n + \xi_n \mathbf{K}\mathbf{X}_n) = -\mathbf{X}_{n'}^{\mathsf{T}} \mathbf{M}\boldsymbol{\Delta}\gamma \tag{6.12}$$

In order to fully take advantage of the orthogonality conditions, as per Equation (2.24), it is often assumed that the damping matrix is a linear combination of the mass and stiffness matrices[6]:

$$\mathbf{C} = \mu\mathbf{M} + \kappa\mathbf{K} \tag{6.13}$$

The above formula, broadly known as the proportional damping matrix[7], is very popular for its mathematical convenience and serves to model lightly damped systems, for instance, mechanical components composed of metallic materials, without having to describe the mechanisms actually responsible for dissipation. Proportional damping may however produce misleading results in the case of heavily damped systems, such as mechanical components embedding visco-elastic materials. This may also be the case in fluid–structure coupled systems where gravity and/or compressibility waves excited by the radiating structure may

---

[6] Coefficients μ and κ apply for the global system; it is also possible to use similar relations for I subsystems (or even at the element scale):

$$\mathbf{C} = \sum_{i \in [1, I]} \mu_i \mathbf{M}_i + \kappa_i \mathbf{K}_i \qquad (\mathbf{c}_e = \mu_e \mathbf{m}_e + \kappa_e \mathbf{k}_e)$$

[7] See for instance Géradin and Rixen (1994).

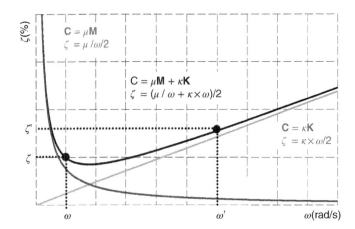

Figure 6.7 Proportional damping. The damping coefficient ζ is plotted as a function of the pulsation with: (i) the damping matrix proportional to the mass matrix; (ii) the damping matrix proportional to the stiffness matrices; (iii) the damping matrix proportional to the mass and stiffness matrices, as per Equation (6.13). In the last case, the damping coefficient can be adjusted for two particular pulsations ω and ω', so that it remains fairly constant within the interval $[\omega, \omega']$. However, for pulsations outside the interval $[\omega, \omega']$, ζ might vary to unrealistic values, evidencing another limit of validity of the proportional damping model

convey a relatively significant amount of mechanical energy. In such situations, it is advisable to replace Equation (6.13) by a more refined model closer to reality than the proportional damping model – see for instance Section 6.5.

Using Equation (6.13) together with Equation (6.12) yields a set of decoupled ordinary differential equations in terms of ξ_n:

$$\ddot{\xi}_n(t) + 2\zeta_n\omega_n \, \dot{\xi}_n(t) + \omega_n^2 \, \xi_n(t) = -q_n\gamma(t) \tag{6.14}$$

ζ_n is the so-called *modal damping coefficient*[8]; it depends on the mode pulsation ω_n:

$$\zeta_n = \frac{1}{2}\left(\frac{\mu}{\omega_n} + \kappa\omega_n\right)$$

As indicated by the heavy dots in Figure 6.7, μ and κ are obtained by setting two values of the damping coefficients ζ and ζ' for two pulsations ω and ω':

$$\mu = \frac{2\omega'\omega}{\omega'^2 - \omega^2}(\omega'\zeta - \omega\zeta') \qquad \kappa = \frac{2\omega'\omega}{\omega'^2 - \omega^2}\left(\frac{\zeta'}{\omega} - \frac{\zeta}{\omega'}\right)$$

The initial conditions for each modal equation are

$$\xi_n(0) = \xi_n^o \qquad \dot{\xi}_n(0) = \dot{\xi}_n^o$$

---

[8] ζ may represent damping arising from various effects. In the context of FSI, it can be used to account for dissipation observed in the vibration of structures coupled to viscous flows, such as tubes in heat exchangers (Pettigrew *et al.*, 2011).

The time integration can be performed numerically with direct methods, or even analytically according to

$$\xi_n(t) = \exp\left(-\zeta_n \omega_n t\right)\phi_n(t) + \frac{1}{\omega_n'} \int_0^t -q_n \gamma(t-\tau) \exp\left(-\zeta_n \omega_n \tau\right) \sin(\omega_n' \tau)\, d\tau \qquad (6.15)$$

where $\omega_n' = \omega_n \sqrt{1 - \zeta_n^2}$.

The second term of Equation (6.15) is the convolution product of the modal excitation by the Green function of the damped harmonic oscillator, which is endowed with the properties of the natural modes of vibration considered, identified by the index n. The first term takes into account the influence of non-zero initial conditions, with

$$\phi_n(t) = \frac{1}{\omega_n}[(\dot{\xi}_n^o + \zeta_n \omega_n \xi_n^o)\sin(\omega_n' t) + \xi_n^o \cos(\omega_n' t)]$$

As the system's response combines linearly the response according to each natural mode of vibration, it is worthy of interest to investigate the behaviour of single degree-of-freedom system subjected to vibrations or shocks.

6.2.2.2   Single Degree-of-Freedom (SDOF) Systems: Response Spectrum

Prior to the development of numerical methods which made CSD affordable for tri-dimensional systems, the linear analysis of seismic-type problems could be performed using the response spectrum method (RSM).[9] The RSM is based on the responses of single degree-of-freedom systems subjected to a prescribed acceleration signal, as depicted in Figure 6.8.

The equation of motion of the single degree-of-freedom (SDOF) system is stated in the relative frame:

$$\ddot{\xi}(t) + 2\zeta\omega\,\dot{\xi}(t) + \omega^2\,\xi(t) = -\gamma(t)$$

with $\omega = \sqrt{\dfrac{k}{m}}$ and $\zeta = \dfrac{c}{2\sqrt{km}}$.

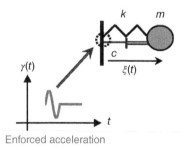

Enforced acceleration

Figure 6.8   Acceleration enforced on a single degree-of-freedom system subjected

---

[9] More on the use of the RSM in seismic or shock engineering is proposed for instance by Gupta (1990) and Harris (1998).

For design purposes, computing the maximum values over a sufficiently large extent of time is of interest. To that end, the *response spectrum* plots the maximum response of SDOF system as a function of its pulsation ω and for a given damping coefficient ζ (Biot, 1941).

There are as many representations of a response spectrum as quantities under concern (displacement, velocity, acceleration and force). In the following, a description in terms of *pseudo-acceleration* is considered, so that the response spectrum is referred to as the mapping:

$$\omega \mapsto \Gamma(\omega, \zeta) = \omega^2 \max_{t>0} |\xi(t, \omega, \zeta, \gamma)| \qquad (6.16)$$

The representation of a dynamic loading through its response spectrum enables a simple comparison of various loads in the frequency domain: $\Gamma(\omega, \zeta, \gamma) > \Gamma(\omega, \zeta, \gamma')$ indicates that the load γ is more severe than the load γ' for a SDOF system behaving as a natural mode of vibration with pulsation ω and damping ζ.

An example of response spectrum is provided on Figure 6.9 for a sine wave acceleration; the typical features of the system's response are noticeable on the plot.

- The *quasi-inertial* response is observed when kinetic energy dominates, that is, for 'low' frequencies. The system's response is ruled by $\ddot{\xi}(t) = -\gamma(t)$ $\xi(t) = -d(t)$ so that the pseudo-acceleration is

$$\Gamma(\omega) = \omega^2 \max_{t>0} |d(t)|$$

where $d(t)$ is the imposed displacement.
- The *quasi-static* behaviour is evidenced when potential energy dominates, that is, for 'high' frequencies. The system's response is described by $\omega^2 \xi(t) = -\gamma(t)$; hence, the pseudo-acceleration is

$$\Gamma(\omega) = \max_{t>0} |\gamma(t)|$$

- An amplification is observed at intermediate frequencies, when resonant effects are predominant. In such cases, the pseudo-acceleration can largely exceed the maximum value of the acceleration imposed on the system:

$$\Gamma(\omega) > \max_{t>0} |\gamma(t)|$$

By definition, a response spectrum gives an indicative measurement of the maximum amplitude of the motion of an SDOF system, in terms of displacement, velocity and acceleration. In a more complicated case of a multi-degrees-of-freedom (MDOF) system, the method can be extended by considering the contribution of each individual mode of vibration to the response of the system. The validity of this *spectral approach* essentially lies in the linearity of the system and the ability to clearly identify each mode individually, which requires the modal density to be small – meaning that the number of modes per unit frequency be sufficiently small to make the individual peaks of resonance discernible, as represented in Figure 6.2. Provided that the two conditions outlined above are fulfilled, the maximum value of each modal coordinate may be deduced from the response spectrum of γ according to

$$\max_{t>0} |\xi_n(t)| = |q_n| \frac{\Gamma(\omega_n, \zeta_n)}{\omega_n^2}$$

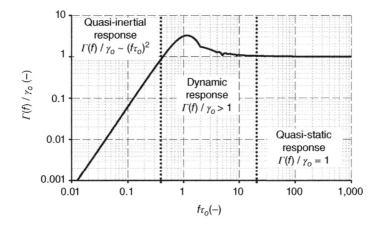

Figure 6.9   Response spectrum. The response spectrum of a sine wave acceleration of amplitude γ_o and duration τ_o is represented in terms of 'pseudo-acceleration', as defined by Equation (6.16). The sine wave acceleration corresponds to $\gamma(t) = \gamma_o \sin 2\pi t/\tau_o$ for $t \in [0, \tau_o]$; see also Figure 6.15, so that $1/\tau_o$ is the characteristic frequency of the excitation. Depending on its frequency f, the response of a SDOF to the excitation is as follows: (i) for 'high-frequency' modes $f \gg 1/\tau_o$, the response is quasi-static, the pseudo-acceleration is independent of the frequency: $\Gamma(f)/\gamma_o \sim 1$; (ii) for 'low-frequency modes' $f \ll 1/\tau_o$, the response is quasi-inertial, the pseudo-acceleration evolves with the square of the frequency: $\Gamma(f)/\gamma_o \sim (f\tau_o)^2$; (iii) for 'medium-frequency modes' $f \sim 1/\tau_o$, the response is dynamic: the pseudo-acceleration overrates the imposed acceleration $\Gamma(f)/\gamma_o > 1$. The 'cut-off frequency' $\bar{f}$ may be defined as the lower bound of the pseudo-static regime: for the sine wave acceleration, a reasonable evaluation of $\bar{f}$ is $\bar{f}\tau_o \sim 10$.

Thus, the maximum value of the MDOF system's displacement at a given location may be estimated with a combination of each eigenmode contribution, stated in a general form as follows:

$$\max_{t>0} \|\mathbf{X}(\mathbf{x},t)\| \simeq C\left(\|\mathbf{X}_n(\mathbf{x})\| \, |q_n| \frac{\Gamma(\omega_n, \zeta_n)}{\omega_n^2} \right)_{n \in [1,N]}$$

The combination denoted here $C(\bullet)$ applies on any quantity ψ (displacement, acceleration, force, stress, strain, etc.) by using the corresponding modal quantities ψ_n. A comparative analysis concerning the domain of validity and the relative accuracy of the various combination rules proposed is the object of an abundant literature in seismic engineering. Without entering into details of these very specified topic, the definition of three combination methods is given here for illustration. The so-called ABS, SRSS and CQC are incorporated in many finite element codes as standard combination procedures for spectral analysis.

- The *Absolute* (ABS) combination rule assumes that the maximum modal values occur at the same time for all modes. Thus,

$$\psi \simeq \sum_n |\psi_n|$$

ABS may produce a so conservative estimate of ψ, so that in practice, other combination rules are often preferred. SRSS and CQC have received a wide acceptance among the seismic engineering community. Based on the random vibration theory, SRSS and CQC combinations allow for a more realistic evaluation of $\max_{t>0}\|\psi(\mathbf{x},t)\|$, for any quantity of interest.

- The *square root of the sum of the square* (SRSS) combination rule, by contrast with the ABS rule, assumes that the maximum modal values are statistically independent:

$$\psi \simeq \sqrt{\sum_n |\psi_n|^2}$$

It turns out that the accuracy of the SRSS rule is satisfactory when modal frequencies are well separated, that is, when

$$r = \frac{\omega_n}{\omega_{n'}} \gg 1$$

for any (n, n').

- The *Complete quadratic combination* (CQC) improves the SRSS combination for systems with closely separated modal frequencies (Wilson *et al.*, 1981):

$$\psi \simeq \sqrt{\sum_{n,n'} \rho_{n,n'} |\psi_n\| \psi_{n'}|}$$

$\rho_{n,n'}$ is the so-called *cross-modal coefficient*; it is evaluated from the modal damping coefficients according to

$$\rho_{n,n'} = \frac{8\sqrt{\zeta_n \zeta_{n'}}(\zeta_n + r\zeta_{n'})r^{3/2}}{(1 - r^2)^2 + 4\zeta_n \zeta_{n'} r(1 + r) + (\zeta_n^2 + \zeta_{n'}^2)r^2}$$

Among modal quantities, the resultant modal force at the system anchor is of practical interest for the design of interfaces. The modal force is $\mathbf{F}_n(t) = \xi_n(t)\mathbf{KX}_n$ and its projection in the direction of the imposed loading is $\phi_n(t) = \boldsymbol{\Delta}^{\mathsf{T}}\mathbf{F}_n(t) = \omega_n^2 \xi_n(t)\boldsymbol{\Delta}^{\mathsf{T}}\mathbf{MX}_n$. The maximum value of the projected modal force contribution is then

$$\max_{t>0} \omega_n^2 |\xi_n(t)| \; \|\boldsymbol{\Delta}^{\mathsf{T}}\mathbf{MX}_n\|$$

Taking into account the definitions of the pseudo-acceleration $\Gamma(\omega_n, \zeta_n)$ on the one hand, and of the participation factor q_n, on the other hand, yields

$$\max_{t>0} |\phi_n(t)| = \mu_n \Gamma(\omega_n, \zeta_n)$$

which gives a physical interpretation of the effective mass μ_n.

6.2.2.3 Static Correction

As a basic rule, modes with a tendency to respond and/or to amplify the excitation – that is, modes with dominant participation factor q_n and/or pseudo-acceleration $\Gamma(\omega_n, \zeta_n)$ in the context of seismic analysis – have to be retained in order to ensure the accuracy of the approximate solution with modal projection.

The contribution of the modes which are rejected from the reduced model, and which are often in large number, may still be non-negligible. As these modes are in the high-frequency range, that is, higher than the upper frequency of the excitation signal, their response is quasi-static in nature and can be accounted for with an intermediate computation of a static problem performed with the finite element model. In the case of a seismic loading, the modal coordinates of the quasi-static modes are

$$\xi_n(t) = -\frac{q_n}{\omega_n^2}\gamma(t)$$

Considering the contribution of all modes, the system's response may be calculated as the following series:

$$\mathbf{X}(\mathbf{x}, t) = \sum_{n=1}^{n=N} \xi_n(t)\mathbf{X}_n(\mathbf{x}) + \gamma(t) \sum_{n \geq N+1} -\frac{q_n}{\omega_n^2}\mathbf{X}_n(\mathbf{x})$$

The first term in this expression corresponds to the contribution of the N modes with quasi-inertial or resonant response, while the second term corresponds to the contribution of all the remaining modes present in the finite element model and exhibiting a quasi-static response to the excitation.

The latter may be derived from the static problem written as $\mathbf{KX}_o = -\mathbf{M\Delta} \iff \mathbf{X}_o = -\mathbf{K}^{-1}\mathbf{M\Delta}$. Furthermore, the static solution $\mathbf{X}_o$ may be expanded as a modal series:

$$\mathbf{X}_o(\mathbf{x}) = -\mathbf{K}^{-1}\mathbf{M\Delta}(\mathbf{x}) = \sum_{n \geq 1} -\frac{q_n}{\omega_n^2}\mathbf{X}_n(\mathbf{x})$$

In addition, there is no need to compute the high-frequency modes, as their global quasi-static contribution to the solution may be expressed in the following convenient form:

$$\sum_{n \geq N+1} -\frac{q_n}{\omega_n^2}\mathbf{X}_n(\mathbf{x}) = -\mathbf{K}^{-1}\mathbf{M\Delta} + \sum_{n=1}^{n=N} \frac{q_n}{\omega_n^2}\mathbf{X}_n(\mathbf{x})$$

Thus,

$$\mathbf{X}(\mathbf{x}, t) = \sum_{n=1}^{n=N} \xi_n(t)\mathbf{X}_n(\mathbf{x}) + \gamma(t) \left(-\mathbf{K}^{-1}\mathbf{M\Delta}(\mathbf{x}) + \sum_{n=1}^{n=N} \frac{q_n}{\omega_n^2}\mathbf{X}_n(\mathbf{x}) \right)$$

The vector $-\mathbf{K}^{-1}\mathbf{M\Delta}(\mathbf{x}) + \sum_{n=1}^{n=N} \frac{q_n}{\omega_n^2}\mathbf{X}_n(\mathbf{x})$ is the so-called *pseudo-mode*, whose contribution is required for improving the accuracy of the modal response, especially in the computation of quantities derived from the displacement field, for example, strain, stress, acceleration – see the example in Section 6.4.

Remark 6.1 Participation factor and inertial effect *According to Equation (6.14), the modal response is solely dependent on the pulsation, damping ratio and participation factor q_n of the vibration mode. As shown in the example depicted in Figure 6.10 these modal characteristics may be significantly affected by FSI – mainly on account of inertial effect.*

Let $u(t)$ and $\xi(t)$ be the tube displacement in the fixed and moving frames, so that

$$\ddot{u} = \ddot{\xi} + \gamma \tag{6.17}$$

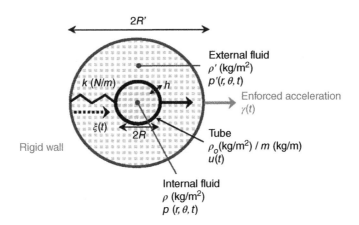

Figure 6.10 *Seismic-type excitation on a single degree-of-freedom system with fluid–structure interaction*

Without FSI, *the equation of motion in the fixed frame is* $m\ddot{u} = -k\xi$, *where the right-hand side of the equation is the spring restoring force. Using Equation (6.17), the equation of motion in the relative frame is*

$$m\ddot{\xi} + k\xi = -m\gamma \tag{6.18}$$

with $m = \rho_o \pi (R^2 - (R - h)^2)$. *The SDOF pulsation, participation factor and effective mass are:*

$$\omega_o = \sqrt{\frac{k}{m}} \qquad q_o = 1 \qquad \mu_o = m$$

With FSI, *the equation of motion in the fixed frame is* $m\ddot{u} = -k\xi + \phi + \phi'$, *where* ϕ *and* ϕ' *are the pressure forces exerted by the inner and outer fluids on the SDOF system:*

$$\phi = +\int_0^{2\pi} p(R - h, \theta)(R - h)\cos\theta \; d\theta \qquad \phi' = -\int_0^{2\pi} p'(R, \theta)R\cos\theta \; d\theta$$

The pressure induced in the inner fluid by the motion of the solid satisfies the equation $\Delta p = 0$, *for* $(r, \theta) \in [0, R - h] \times [0, 2\pi]$, *with the coupling condition:*

$$\left.\frac{\partial p}{\partial r}\right|_{r=R-h} = -\rho\ddot{u}\cos\theta$$

It is calculated as $p(r, \theta) = -\rho r\ddot{u}\cos\theta$, *so that* ϕ *reads as follows:*

$$\phi = -m_f\ddot{u}$$

with $m_f = \rho\pi(R - h)^2$ *the mass of fluid contained in the tube.*
The pressure induced in the outer fluid by the motion of the solid is governed by the equation $\Delta p' = 0$ *for* $(r, \theta) \in [R, R'] \times [0, 2\pi]$, *with coupling conditions on the moving walls:*

$$\left.\frac{\partial p'}{\partial r}\right|_{r=R} = -\rho'\ddot{u}\cos\theta \qquad \left.\frac{\partial p'}{\partial r}\right|_{r=R'} = -\rho'\gamma\cos\theta$$

It is calculated as

$$p'(r, \theta) = -\rho \left(\frac{\gamma R'^2 - \ddot{u} R^2}{R'^2 - R^2} r + \frac{(\gamma - \ddot{u}) R'^2 R^2}{R'^2 - R^2} \frac{1}{r} \right) \cos \theta,$$

ϕ' *is readily:*

$$\phi' = -m_a \ddot{u} + (m_a + m_d) \gamma$$

where $m_a = \rho' \pi R^2 \frac{R'^2 + R^2}{R'^2 - R^2}$ *is the fluid-added mass and* $m_d = \rho' \pi R^2$ *is the fluid-displaced mass, as discussed in Section 4.2.3.*
Writing the former expressions in terms of ξ *and using Equation (6.17) yields the equation of motion in the relative frame:*

$$(m + m_f + m_a)\ddot{\xi} + k\xi = -(m + m_f - m_d)\gamma \tag{6.19}$$

With FSI, the SDOF pulsation, participation factor and effective mass are as follows:

$$\omega = \sqrt{\frac{k}{m + m_f + m_a}} \qquad q = \frac{m + m_f - m_d}{m + m_f + m_a} < q_o$$

and:

$$\mu = \frac{(m + m_f - m_d)^2}{m + m_f + m_a}$$

The importance of taking FSI into account may be highlighted by comparing Equations (6.18) and (6.19).

- *The inner fluid contributes to an additional mass, precisely equal to the physical mass* $+m_f$, *which is present in both sides of Equation (6.19): it is associated with an external inertial force of the responding system on the left-hand side of the force balance and associated with an external exciting force induced by the prescribed acceleration in the right-hand side of the force balance.*
- *The outer fluid conveys an added inertia* $(+m_a$ *in the left-hand side of Equation (6.19)) and also alleviates the external force induced by the prescribed acceleration* $(-m_d$ *in the right-hand side of Equation (6.19)), as a result of the inertial force stated as* $-m_d \gamma$ *and associated with the bulk motion of the fluid in the moving frame.*

The pulsation, participation factor and effective mass of the SDOF system are lowered on account of FSI. The importance of this effect may be usefully illustrated by considering a numerical example, which is representative of a heat exchanger tube. Let $R = 5$ *mm,* $R' = 10$ *mm,* $h = 1$ *mm,* $\rho_o = 8,000$ *kg/m³,* $\rho = 600$ *kg/m³,* $\rho' = 900$ *kg/m³, then:* $m = 0.23$ *kg/m,* $m_f = 0.05$ *kg/m,* $m_d = 0.07$ *kg/m and* $m_a = 0.12$ *kg/m. The tube effective mass with and without FSI are, respectively,* $\mu_o = 0.23$ *kg/m and* $\mu = 0.10$ *kg/m. In some instances, the influence of inertial coupling is more significant, as evidenced, for instance, in Figure 5.13 and Table 6.2 for another simple example.* ∎

6.3 Frequency-Domain Analysis

Analyses of forced motion as carried out in the frequency domain – also called the *spectral domain* – are illustrated here in the context of the linear vibrations of a structure excited by an external load and possibly coupled to a fluid at rest (i.e. in the absence of steady flow). The frequency response of the system to the excitation is supposed to be represented by the following equation:

$$(-\omega^2 \mathbf{M} + \mathbf{K}(\omega))\mathbf{X}(\omega) = \mathbf{\Phi}(\omega) \tag{6.20}$$

The former expression constitutes a modal equation in which $\mathbf{K}(\omega)$ is a complex-valued matrix which may account for dissipation observed in systems comprising frequency-dependent materials, as evoked in Remarks 2.1 and 2.2, and/or for damping represented with a viscous-type force. The latter case is encountered in the context of FSI when radiation damping is present; in this occurrence, the dynamic stiffness matrix may be stated as follows:

$$\mathbf{K}(\omega) = \mathbf{K} + i\omega\mathbf{C}$$

where $\mathbf{K}$ and $\mathbf{C}$ are real-valued matrices.

6.3.1 Direct and Modal Methods

The *direct* resolution to Equation (6.20) requires the inversion of a linear system at each pulsation ω:

$$\mathbf{X}(\omega) = (-\omega^2 \mathbf{M} + \mathbf{K}(\omega))^{-1} \mathbf{\Phi}(\omega) \tag{6.21}$$

It takes into account the dependency of the stiffness matrix on the frequency, which is straightforward but also computationally demanding. When the number of pulsations of interest is large and/or when the size of the problem is important, the direct resolution might not be affordable for engineering applications.

As already illustrated in the context of time-domain analysis, a ROM may be obtained by expanding the system's response $\mathbf{X}(\omega)$ onto a suitable basis,[10] according to:

$$\mathbf{X}(\omega) = \mathbf{X}_N \boldsymbol{\xi}_N(\omega)$$

where $\boldsymbol{\xi}_N(\omega) = \{\xi_n(\omega)\}_{n \in [1,N]}$ is the vector of modal coordinates and $\mathbf{X}_N = \langle \mathbf{X}_n \rangle_{n \in [1,N]}$ is the matrix composed of the vector basis.

The projection of Equation (6.20) onto $\mathbf{X}_N$ reads

$$(-\omega^2 \mathbf{m}_N + \mathbf{k}_N(\omega))\boldsymbol{\xi}_N(\omega) = \boldsymbol{\varphi}_N(\omega) \tag{6.22}$$

where $\mathbf{m}_N$, $\mathbf{k}_N(\omega)$ and $\boldsymbol{\varphi}_N(\omega)$ are matrices of size $N \times N$. They are calculated as

$$\mathbf{m}_N = \mathbf{X}_N^\mathsf{T}\mathbf{M}\mathbf{X}_N \qquad \mathbf{k}_N(\omega) = \mathbf{X}_N^\mathsf{T}\mathbf{K}(\omega)\mathbf{X}_N \qquad \boldsymbol{\varphi}_N(\omega) = \mathbf{X}_N^\mathsf{T}\mathbf{\Phi}(\omega)$$

---

[10] It is worthy to underline here that the projection basis may not necessarily be that of the system's eigenvectors: this choice is, for instance, the most appropriate one when the mode shapes form an orthogonal and *complete* vector basis to expand the solution of the problem under concern, which is the case, for instance, for problems of the type $(-\omega^2\mathbf{M} + \mathbf{K})\mathbf{X}(\omega) = \mathbf{\Phi}(\omega)$. A projection basis which is suited to the problem under concern may be derived for more complex cases with specific algorithms, as illustrated for instance in the conclusive remark of this chapter.

The reduced-order Equation (6.22) still involves a non-linear frequency-dependent operator. For each frequency of interest, a linear system of the type $\mathbf{A}(\omega)\mathbf{x}(\omega) = \mathbf{b}(\omega)$ has to be solved, but this remains computationally manageable with the low-sized matrices produced by the ROM.

For slightly damped systems, it is often assumed that

$$\mathbf{k}_N(\omega) \approx \mathbf{\Omega}_N^2 \mathbf{m}_N \tag{6.23}$$

where $\mathbf{\Omega}_N^2$ is a diagonal matrix:

$$\mathbf{\Omega}_N^2 = \text{diag}(\omega_n^2(1 + i\eta_n)^2)_{n\in[1,N]}$$

ω_n and η_n are, respectively, the pulsation and the *loss factor* of mode $\mathbf{X}_n$. Equation (6.23) may hold when the eigenmodes are loosely dependent on the damping, the latter being represented for each mode with a simple coefficient, namely, η. In this context, the system response is readily

$$X(\mathbf{x}, \omega) = \sum_{n=1}^{n=N} \frac{\mathbf{X}_n^\mathsf{T}\mathbf{\Phi}(\omega)}{\mathbf{X}_n^\mathsf{T}\mathbf{M}\mathbf{X}_n} \frac{\mathbf{X}_n(\mathbf{x})}{-\omega^2 + \omega_n^2(1 + i\eta_n)^2} = \sum_{n=1}^{n=N} \frac{\mathbf{X}_n^\mathsf{T}\mathbf{\Phi}(\omega)}{\mathbf{X}_n^\mathsf{T}\mathbf{M}\mathbf{X}_n} H_n(\omega)\mathbf{X}_n(\mathbf{x}) \tag{6.24}$$

which may be viewed as a rough simplification of Equation (6.22).

If needed, a static correction is performed to achieve better accuracy in the low-frequency range, in a similar manner to that done to account for the high-frequency mode contribution in time-domain analysis. For instance, the following correction

$$\mathbf{K}(\omega_o)^{-1}\mathbf{\Phi}(\omega_o) - \sum_{n=1}^{n=N} \frac{\mathbf{X}_n^\mathsf{T}\mathbf{\Phi}(\omega_o)}{\mathbf{X}_n^\mathsf{T}\mathbf{M}\mathbf{X}_n} \frac{\mathbf{X}_n(\mathbf{x})}{-\omega_o^2 + \omega_n^2(1 + i\eta_n)^2}$$

is exact for $\omega = \omega_o$.

The contribution of a single mode of pulsation ω_o and loss factor η_o is described by the transfer function plotted in Figure 6.11 for illustration. The transfer function is readily

$$H(\omega/\omega_o) = -\frac{1}{\omega_o^2} \frac{1}{(\omega/\omega_o)^2 - (1 + i\eta_o)^2} \tag{6.25}$$

and may also be formulated as

$$H(\omega/\omega_o) = |H(\omega/\omega_o)| \exp\left(i\phi(\omega/\omega_o)\right)$$

- $|H(\omega/\omega_o)|$ denotes the amplitude:

$$|H(\omega/\omega_o)| = \frac{1}{\omega_o^2} \frac{1}{\sqrt{((\omega/\omega_o)^2 + \eta_o^2 - 1)^2 + 4\eta_o^2}}$$

- $\phi(\omega/\omega_o)$ denotes the phase:

$$\phi(\omega/\omega_o) = \arctan\left(\frac{2\eta_o}{(\omega/\omega_o)^2 + \eta_o^2 - 1}\right)$$

When damping of the system is important however, the approximation introduced by Equation (6.24) is baseless.

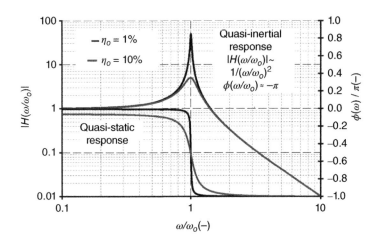

Figure 6.11 Amplitude and phase of the SDOF transfer function. The amplitude $|H(\omega/\omega_o)|$ and the phase $\phi(\omega/\omega_o)$ of the response function $H(\omega/\omega_o)$ are plotted for η_o set to 1% and 10%. The asymptotic regimes of the response are evidenced: (i) the quasi-static regime corresponds to the low-frequency response: when $\omega/\omega_o \to 0$, the asymptotic values of the amplitude and the phase are, respectively, $|H(\omega/\omega_o)| \to 1/\sqrt{(\eta_o^2 - 1)^2 + 4\eta_o/\omega_o} > 1$ and $\phi(\omega/\omega_o) = \arctan(2\eta_o/(\eta_o^2 - 1)) < 0$; (ii) the quasi-inertial regime corresponds to the high-frequency response: when $\omega/\omega_o \to \infty$, the asymptotic values of the amplitude and the phase are, respectively, $1/(\omega/\omega_o)^2$ and $-\pi$. In comparison, the transfer function of a damped SDOF oscillator with energy dissipation modelled as a damping coefficient ζ_o is $H(\omega/\omega_o) = 1/(\omega_o^2 - \omega^2 + 2i\zeta_o\omega_o\omega)$. The asymptotic values of the phase and amplitude in the quasi-inertial regimes remain unchanged. They are 1 for the amplitude and 0 for the phase in the quasi-static regime. For slightly damped systems, the response function can be evaluated using Equation (6.24) with a modal expansion on the damped eigenmodes, in lieu of the direct computation as per Equation (6.21). This approximation assumes that damping may be accounted for by a constant loss factor η for each mode, which is often a restrictive assumption

6.3.2 Computation of the Projection Basis

Many methods have been proposed to derive a suitable projection basis for Equation (6.20). The projection basis is usually obtained using iterative algorithms which solve non-linear eigenvalue problems (Duigou *et al.*, 2003). As an overview of the various approaches would require an entire volume in itself, the following presentation is restricted here to outline a few of the most common methods.

Modal energy methods – such as the modal strain energy (MSE) – consist of solving an eigenvalue problem with the stiffness matrix evaluated for a given pulsation ω_o. $\mathbf{X}_N$ is composed of N eigenvectors, which are the solutions of

$$(-\omega^2\mathbf{M} + \mathfrak{R}(\mathbf{K}(\omega_o)))\mathbf{X} = \mathbf{0}$$

The method holds only to low-damped systems (Moreira and Rodrigues, 2006); it is generally limited to a frequency interval centred around ω_o. It may be extended in order to produce a projection basis which covers a wider frequency range. To that end, several sets of vectors $\mathbf{X}_N^p$ with $1 \leq p \leq P$ are calculated as solution of the eigenvalue problems:

$$(-\omega^2 \mathbf{M} + \mathfrak{R}(\mathbf{K}(\omega_p)))\mathbf{X} = \mathbf{0}$$

$(\mathbf{X}_N^p)_{p \in [1,P]}$ are assembled with an ortho-normalisation procedure to produce a global projection basis $\mathbf{X}_{NP}$, which represents the physics on a large spectrum. This approach is, for instance, successfully evaluated for highly damped structures by Balmès (1996).

Iterative modal energy methods consist of solving a non-linear eigenvalue problem $(-\omega^2 \mathbf{M} + \mathfrak{R}(\mathbf{K}(\omega)))\mathbf{X}(\omega) = \mathbf{0}$ with an iterative procedure, delivering a set of real vectors $(\mathbf{X}_n)_{n \in [1,N]}$. The iterative modal strain energy (IMSE) and modified iterative modal strain energy (MIMSE) belong to this category; they are detailed in Algorithm 4. IMSE and MIMSE allow for a better representation of damping over an extended frequency range; see, for instance, Trindade *et al.* (2000) and Merlette *et al.* (2012).

Iterative modal energy methods are however not suited when damping is predominantly represented by the imaginary part of $\mathbf{K}(\omega)$, which is the case in the context of FSI when radiative damping has to be accounted for, and they are also rather computationally expensive.[11]

In order to improve the accuracy of the solution when damping is important, the projection basis can be completed with a set of suitable vectors. Following, for instance, Plouin and Balmès (1998), an enriched basis $\mathbf{X}_N^*$ takes the contribution of the residual force $\mathbf{T}_n$ generated by the pseudo-mode $\mathbf{X}_n$:

$$\mathbf{X}_N^* = \langle \mathbf{X}_N, (\mathbf{T}_n)_{n \in [1,N]} \rangle$$

with:

$$\mathbf{T}_n = \mathbf{K}(0)^{-1} \mathfrak{I}(\mathbf{K}(\omega_n))\mathbf{X}_n \qquad \forall n \in [1,N]$$

Complex iterative methods – such as the iterative complex eigensolution (ICE) – are based on an iterative resolution of $(-\omega^2 \mathbf{M} + \mathbf{K}(\omega))\mathbf{X}(\omega) = \mathbf{0}$, which gives a set of complex vectors $(\mathbf{X}_n)_{n \in [1,N]}$ (Vasques *et al.*, 2010), as detailed by Algorithm 5. When damping is important, or when damping is exclusively accounted for with an imaginary matrix, complex iterative methods offer a better accuracy than iterative modal energy methods offer. They are however more numerically expensive, since algorithms handling complex matrices prove in general less efficient than algorithms designed for real-valued matrices.

In order to enhance the accuracy of the ROM, the projection basis can be enriched: taking into account the contribution of the residual displacements for each mode is one possible option; in such a case, the enriched basis is

$$\mathbf{X}_N^* = \langle \mathbf{X}_N, (\mathbf{R}_n)_{n \in [1,N]} \rangle$$

where $\mathbf{R}_n$ is

$$\mathbf{R}_n = [(-\omega_n^2 \mathbf{M} + \mathbf{K}(\omega_n))^{-1} - \mathbf{X}_N(-\omega_n^2 \mathbf{m}_N + \mathbf{k}_N(\omega_n))^{-1}\mathbf{X}_N^{\mathsf{T}}]\Phi(\omega_n)$$

The preceding expression indicates that $\mathbf{R}_n$ quantifies the error between $\mathbf{X}_{\text{direct}}(\omega_n)$ and $\mathbf{X}_{\text{modal}}(\omega_n)$.

---

[11] Note that N modes are obtained with the iterative energy methods by solving at least $3N$ eigenvalue problems.

Algorithm 4 Computation of frequency response – Iterative modal energy methods

---

Require: N, **x**, ε and Q

  Solve eigenvalue problem:

  $$(-\omega^2 \mathbf{M} + \mathbf{K}(0))\mathbf{X} = \mathbf{0}$$

  Store the N first eigenvectors $(\mathbf{X}_n^o)_{n\in[1,N]}$ and eigenpulsations $(\omega_n^o)_{n\in[1,N]}$.
  Set the projection basis $\mathbf{X}_0 = \emptyset$.
  for $n \in [1, N]$ **do**
    Set initial values $q = 0$, $\varepsilon_n^q = 1$, $\omega_n^q = \omega_n^o$ and $\mathbf{X}_n^q = \mathbf{X}_n^o$.
    while $\varepsilon_n^q < \varepsilon$ and $q < Q$ **do**
      Compute the n th eigenpulsation ω_n^{q+1} and eigenvector $\mathbf{X}_n^{q+1}$ of the eigenvalue
      problem in $\mathbb{R}$:

      $$(-\omega^2 \mathbf{M} + \mathfrak{R}(\mathbf{K}(\omega_n^q)))\mathbf{X} = \mathbf{0}$$

    for the modal method, or:

    $$(-\omega^2 \mathbf{M} + \mathfrak{R}(\mathbf{K}(\omega_n^q)) + \beta(\omega_n^q)\mathfrak{I}(\mathbf{K}(\omega_n^q)))\mathbf{X} = \mathbf{0}$$

    with:

    $$\beta(\omega) = \frac{\mathrm{Tr}(\mathfrak{R}(\mathbf{K}(\omega))}{\mathrm{Tr}(\mathfrak{I}(\mathbf{K}(\omega))}$$

    for the modified modal method (with $\mathrm{Tr}(\bullet) = \sum_{i,j} \bullet_{i,j}$).
    Evaluate the convergence criterion $\varepsilon_n^q = \dfrac{|\omega_n^{q+1} - \omega_n^q|}{|\omega_n^q|}$.
    Increment loop $q = q + 1$.
    end while
    Update the projection basis $\mathbf{X}_n = \langle \mathbf{X}_{n-1}, \mathbf{X}_n^{q+1} \rangle$.
  end for
  Store the projection basis $\mathbf{X}_N$.
  Calculate the frequency response on the ROM:

  $$(-\omega^2 \mathbf{m}_N + \mathbf{k}_N(\omega))\boldsymbol{\xi}_N(\omega) = \boldsymbol{\varphi}_N(\omega)$$

  Calculate the frequency response at location **x** for pulsation ω:

  $$\mathbf{X}(\omega, \mathbf{x}) = \mathbf{X}_N(\mathbf{x})\boldsymbol{\xi}_N(\omega)$$

Algorithm 5 Computation of frequency response – Complex iterative method

---

Require: N, $\mathbf{x}$, ε and Q

Solve eigenvalue problem:
$$(-\omega^2 \mathbf{M} + \mathbf{K}(0))\mathbf{X} = \mathbf{0}$$

Store the N first eigenvectors $(\mathbf{X}_n^o)_{n\in[1,N]}$ and eigenpulsations $(\omega_n^o)_{n\in[1,N]}$.
Set the projection basis $\mathbf{X}_0 = \emptyset$.

for $n \in [1,N]$ **do**
  Set initial values $q = 0$, $\varepsilon_n^q = 1$, $\omega_n^q = \omega_n^o$ and $\mathbf{X}_n^q = \mathbf{X}_n^o$.
  while $\varepsilon_n^q < \varepsilon$ and $q < Q$ **do**
  Compute the nth eigenpulsation ω_n^{q+1} and eigenvector $\mathbf{X}_n^{q+1}$ of the eigenvalue problem in $\mathbb{C}$:

$$(-\omega^2 \mathbf{M} + \mathbf{K}(\omega_n^q))\mathbf{X} = \mathbf{0}$$

  Evaluate the convergence criterion $\varepsilon_n^q = \dfrac{|\omega_n^{q+1} - \omega_n^q|}{|\omega_n^q|}$.
  Increment loop $q = q + 1$.
  end while
  Update the projection basis $\mathbf{X}_n = \langle \mathbf{X}_{n-1}, \mathbf{X}_n^{q+1}\rangle$.
end for
Store the projection basis $\mathbf{X}_N$.
Calculate the frequency response on the ROM:

$$(-\omega^2 \mathbf{m}_N + \mathbf{k}_N(\omega))\boldsymbol{\xi}_N(\omega) = \boldsymbol{\varphi}_N(\omega)$$

Calculate the frequency response at location $\mathbf{x}$ for pulsation ω:

$$\mathbf{X}(\omega, \mathbf{x}) = \mathbf{X}_N(\mathbf{x})\boldsymbol{\xi}_N(\omega)$$

- $\mathbf{X}_{\text{direct}}(\omega_n)$ is the direct computation of the response at pulsation ω_n:

$$\mathbf{X}_{\text{direct}}(\omega_n) = (-\omega^2 \mathbf{M} + \mathbf{K}(\omega_n))^{-1}\boldsymbol{\Phi}(\omega_n)$$

- $\mathbf{X}_{\text{modal}}(\omega_n)$ is the evaluation of the response for ω_n with a modal expansion, which is stated as follows:

$$\mathbf{X}_{\text{modal}}(\omega_n) = \mathbf{X}_N \boldsymbol{\xi}_N(\omega_n)$$

with:

$$\boldsymbol{\xi}_N(\omega_n) = (-\omega_n^2 \mathbf{m}_N + \mathbf{k}_N(\omega_n))^{-1}\boldsymbol{\varphi}_N(\omega_n) \qquad \boldsymbol{\varphi}_N(\omega_n) = \mathbf{X}_N^{\mathsf{T}}\boldsymbol{\Phi}(\omega_n)$$

As shown in Section 6.5, the method proves efficient when the modes are well separated.

As a concluding remark on this topic, Figure 6.12 mentions the evaluation and comparison of various techniques proposed by Rouleau (2013) and Rouleau *et al.* (2014) in the context of damped structural vibrations.

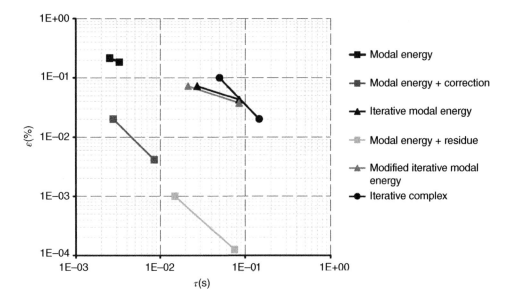

Figure 6.12   Comparison of various modal projection techniques. Rouleau (2013) and Rouleau *et al.* (2014) propose a detailed comparison of various modal projection techniques. The frequency response of a damped cantilever beam when excited by an harmonic load at its free end is calculated. The relative error on the displacement versus the computational time for different truncation criteria is plotted in the graph. The results suggest that the MSE method with a correction offers an optimal compromise between computational cost and accuracy in this case. *Source*: Lucie Rouleau, CNAM, Paris, France, 2013. Reproduced with permission of Lucie Rouleau

Driven by accuracy and optimisation issues in the design of complex mechanical systems, numerical methods based on model reduction techniques lie at the core of many open problems and trigger an intense academic research as well as a growing industrial interest, particularly in the context of vibro-acoustics, as in the example presented in Figure 6.13. Further insights on the topic are proposed, for instance, by Soize and Ohayon (2014).

6.4   Example: Time-Domain Analysis

6.4.1   Accelerated Cantilever Beam with Fluid Coupling

The dynamic response of a simple fluid–structure system subjected to a seismic-type loading is investigated: the geometrical and physical parameters of the problem are defined in Figure 6.14.

The acceleration $\gamma(t)$ imposed on the system is a sine wave defined as follows:

$$\gamma(t) = \gamma_o \sin(2\pi t/\tau_o) \text{ for } t \in [0, \tau_o] \text{ and } \gamma(t) = 0 \text{ for } t \in [\tau_o, \infty[$$

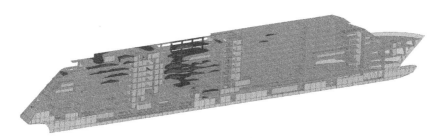

Figure 6.13 Example of industrial computation of vibro-acoustic problems. The acoustic comfort of on-board vessels is a major issue for cruise shipbuilders: numerical simulations with finite element–based models provide the designer with some meaningful data which are useful to understand the behaviour of complex architectures and to assess the vibro-acoustic performance of the ship. The finite element representation of the structure may account for various physics, such as that involved in FSI in the dissipation of energy in various materials (e.g. metallic, composite or visco-elastic), and the size of the numerical model may become large: in such cases, turning to ROMs may prove an interesting option to perform numerous simulations at low computational cost, for instance, when an optimisation strategy is considered. *Source*: Sylvain Branchereau, STX, Saint-Nazaire, France, 2014. Reproduced with permission of STX France

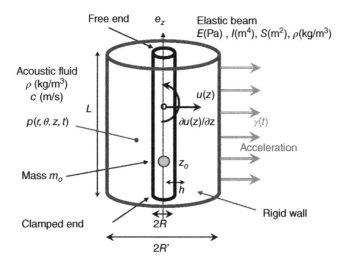

Figure 6.14 Coupled system subjected to seismic-type loading. A straight tube of circular cross section, modelled as an elastic cantilever beam, is coupled to a compressible fluid filling a coaxial cylindrical cavity. The geometrical parameters of the problem are denoted R (inner radius), R' (outer radius), L (length) and h (thickness), while the material properties are denoted ρ (structure or fluid density), E (structure Young's modulus) and c (speed of sound in the fluid). m_o stands for a concentrated mass at position z_o. The acceleration imposed on the system is a sine wave of amplitude γ_o and duration τ_o, as defined in Figure 6.15

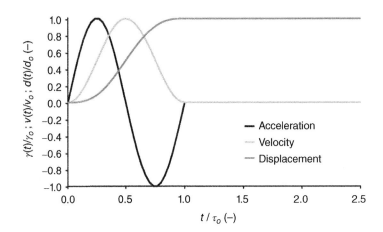

Figure 6.15 Sine wave acceleration. The sine wave acceleration corresponds to a single period of a sine function, with duration τ_o and amplitude γ_o. It is imposed on a system initially at rest. During the first half-period, the system is accelerated: the velocity increases from zero to its maximum value $v_o = \gamma_o \tau_o / \pi$, which is reached at time $\tau_o/2$; during the second half-period, the system is decelerated: the velocity decreases from v_o to zero. The time evolution of the velocity in the interval $[0, \tau_o]$ is $v(t) = v_o(1 - \cos(2\pi t/\tau_o))/2$. Meanwhile, the system moves from its initial position to a new 'at rest' position: it undergoes a total displacement $d_o = \gamma_o \tau_o^2 / \pi$; the time evolution of the displacement in the interval $[0, \tau_o]$ is $d(t) = d_o(t/\tau_o - \sin(2\pi t/\tau_o))$

The equations of motion are formulated in the *reference frame*, using a model which combines the *relative displacement* of the structure and the *absolute pressure* in the fluid; they are stated as follows:

Structure The relative displacement $u(z, t)$ verifies the Euler–Bernoulli bending beam equation in the moving frame, with the imposed acceleration $\gamma(t)$ and the pressure force $\phi(t)$ appearing as source terms in the equation of motion[12]:

$$(\rho S + m_o \delta(z - z_o))\frac{\partial^2 u}{\partial t^2} + EI\frac{\partial^4 u}{\partial z^4} = -(\rho S + m_o \delta(z - z_o))\gamma(t) + \phi(t)$$

The boundary conditions stand for the clamped/free end conditions at $z = 0$ and $z = L$ and are expressed in terms of relative displacement:

$$u|_{z=0} = 0 \qquad \frac{\partial u}{\partial z}\bigg|_{z=0} = 0 \qquad +EI\frac{\partial^2 u}{\partial z^2}\bigg|_{z=L} = 0 \qquad -EI\frac{\partial^3 u}{\partial z^3}\bigg|_{z=L} = 0$$

The fluid force at section z is calculated with the absolute pressure, according to

$$\phi(t) = -\int_0^{2\pi} p(R, \theta, z, t)\cos\theta R d\theta \tag{6.26}$$

---

[12] The passage from the relative to the absolute displacement in this example is similar to that discussed in Remark 2.1.

Fluid The absolute pressure $p(r, \theta, z, t)$ satisfies the Helmholtz equation in the reference frame; with a cylindrical coordinate system, it reads as

$$\frac{1}{c^2}\frac{\partial^2 p}{\partial t^2} - \frac{\partial^2 p}{\partial r^2} - \frac{1}{r}\frac{\partial p}{\partial r} - \frac{1}{r^2}\frac{\partial^2 p}{\partial \theta^2} - \frac{\partial^2 p}{\partial z^2} = 0$$

The acceleration is imposed on the deformable or rigid walls; the corresponding boundary conditions are

$$\frac{\partial p}{\partial z}\bigg|_{z=0} = 0 \qquad \frac{\partial p}{\partial z}\bigg|_{z=L} = 0$$

$$\frac{\partial p}{\partial r}\bigg|_{r=R} = -\rho\left(\frac{\partial^2 u}{\partial t^2} + \gamma(t)\right)\cos\theta \qquad \frac{\partial p}{\partial r}\bigg|_{r=R'} = -\rho\gamma(t)\cos\theta$$

The space discretisation is performed with structural elements and fluid Fourier elements of order $m = 1$, as described in the examples proposed in Chapters 4 and 5, while a $0D$ inertia element accounts for the concentrated mass m_o. The following matrix equations are arrived at:

$$\mathbf{M}_S\ddot{\mathbf{U}}(t) + \mathbf{K}_S\mathbf{U}(t) = -\mathbf{M}_S\mathbf{D}\gamma(t) + \mathbf{R}\mathbf{P}(t)$$

where $\mathbf{D}$ is a unit vector which gives the direction of the prescribed acceleration field:

$$\mathbf{M}_F\ddot{\mathbf{P}}(t) + \mathbf{K}_F\mathbf{P}(t) = -\rho\mathbf{R}^{\mathsf{T}}\mathbf{D}\gamma(t) - \rho\mathbf{R'}^{\mathsf{T}}\mathbf{D'}\gamma(t) - \rho\mathbf{R}^{\mathsf{T}}\ddot{\mathbf{U}}(t)$$

where $\mathbf{D'}$ is a unit vector which specifies the direction of application of the boundary condition on $r = R'$.

The corresponding $(\mathbf{u}, p)$ symmetric condensed formulation reads as follows:

$$\begin{bmatrix} \mathbf{M}_S + \rho\mathbf{R}\mathbf{K}_F^{-1}\mathbf{R}^{\mathsf{T}} & -\mathbf{R}\mathbf{K}_F^{-1}\mathbf{M}_F \\ -\mathbf{M}_F\mathbf{K}_F^{-1}\mathbf{R}^{\mathsf{T}} & 1/\rho\mathbf{M}_F\mathbf{K}_F^{-1}\mathbf{M}_F \end{bmatrix} \begin{Bmatrix} \ddot{\mathbf{U}}(t) \\ \ddot{\mathbf{P}}(t) \end{Bmatrix} + \begin{bmatrix} \mathbf{K}_S & 0 \\ 0 & 1/\rho\mathbf{M}_F \end{bmatrix} \begin{Bmatrix} \mathbf{U}(t) \\ \mathbf{P}(t) \end{Bmatrix}$$

$$= -\begin{bmatrix} \mathbf{M}_S + \rho\mathbf{R}\mathbf{K}_F^{-1}\mathbf{R}^{\mathsf{T}} & -\mathbf{R}\mathbf{K}_F^{-1}\mathbf{M}_F \\ -\mathbf{M}_F\mathbf{K}_F^{-1}\mathbf{R}^{\mathsf{T}} & 1/\rho\mathbf{M}_F\mathbf{K}_F^{-1}\mathbf{M}_F \end{bmatrix} \begin{Bmatrix} \mathbf{D} \\ \rho\mathbf{M}_F^{-1}\mathbf{R'}^{\mathsf{T}}\mathbf{D'} \end{Bmatrix} \gamma(t) \qquad (6.27)$$

Equation (6.27) is of the same form as Equation (6.10) with:

$$\Delta = \begin{Bmatrix} \mathbf{D} \\ \rho\mathbf{M}_F^{-1}\mathbf{R'}^{\mathsf{T}}\mathbf{D'} \end{Bmatrix}$$

$\rho\mathbf{M}_F^{-1}\mathbf{R'}^{\mathsf{T}}\mathbf{D'}$ is the representation of $\mathbf{D}$ in the pressure formulation of the fluid problem. With the displacement/pressure symmetric formulation, FSI is described by the added mass matrix:

$$\mathbf{M}_A = -\begin{bmatrix} \rho\mathbf{R}\mathbf{K}_F^{-1}\mathbf{R}^{\mathsf{T}} & -\mathbf{R}\mathbf{K}_F^{-1}\mathbf{M}_F \\ -\mathbf{M}_F\mathbf{K}_F^{-1}\mathbf{R}^{\mathsf{T}} & 1/\rho\mathbf{M}_F\mathbf{K}_F^{-1}\mathbf{M}_F \end{bmatrix}$$

and by the coupling matrix $-\mathbf{M}_F\mathbf{K}_F^{-1}\mathbf{R}^{\mathsf{T}}$.

6.4.2   System and Excitation Spectra

In the following, the numerical applications are performed with material properties of steel ($\rho = 7,800$ kg/m³, $E = 2.1 \cdot 10^{11}$ Pa) and of water ($\rho = 1,000$ kg/m³, $c = 1,500$ m/s), while the concentrated mass is $m_o = 15$ kg. The geometrical parameters are set as $R = 0.1$ m, $R' = 0.2$ m, $L = 1$ m, $h = 0.01$ m, $z_o = 0.25$ m and the acceleration amplitude and duration are, respectively, $\gamma_o = 1$ m/s², $\tau_o = 1.5$ ms.

The first eigenmode shapes of the system are represented in Figure 6.16, and their scalar properties are reported in Table 6.2.

Fluid–structure coupling is evidenced in a similar manner to that already highlighted in the example studied in Section 5.4. The presence of the concentrated mass slightly modifies the shape of the bending modes, as made conspicuous by comparing Figures 5.11 and 6.16.

The inertial effect is dominant at the first and second modes: the eigenfrequencies are lowered when fluid is present, whereas the effective masses are strongly decreased. Actually, fluid inertia is found to have the paradoxical effect to lighten the structure: this is a consequence of the inertia force which results from the bulk motion of the accelerated fluid. The vibro-acoustic coupling is dominant for the third and fourth modes, which concentrate more than 75% of the system mass. The first four bending modes account for 80% of the structure mass without fluid and for 90% of the structure and fluid mass.

The response spectrum of the excitation is plotted in Figure 6.17, where the square markers serve to identify the eigenfrequencies of the system. A reasonable evaluation of the signal cut-off frequency is $\bar{f} \sim 3,000$ Hz: three modes of the structure have their eigenfrequency below this threshold, while frequencies of the first six modes of the fluid–structure system are within the interval $[0, \bar{f}]$.

From the effective mass criteria and cut-off frequency criteria, it is anticipated that the dynamic response of the system should be described with satisfying accuracy by considering the contribution of the first four to six modes.

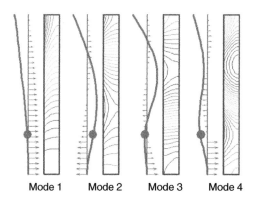

Mode 1   Mode 2   Mode 3   Mode 4

Figure 6.16   Bending modes of the beam with concentrated mass and fluid coupling

Table 6.2 Scalar properties of the bending modes for the beam in the presence of a concentrated mass and fluid coupling (values in brackets indicate the modal characteristics of the structure in vacuo)

n	f_n	μ_n	q_n
1	136 Hz (194 Hz)	1.39 % (52.7 %)	1.47
2	825 Hz (1,115 Hz)	0.88 % (28.3 %)	1.17
3	1,755 Hz (2,730 Hz)	9.95 % (9.63 %)	3.94
4	2,120 Hz (5,794 Hz)	67.7 % (1.75 %)	10.3

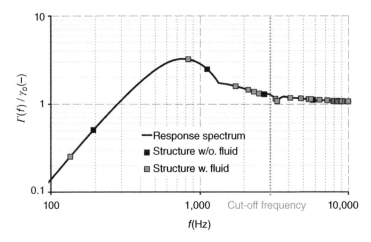

Figure 6.17 **Response spectrum of the sine wave acceleration.** The response spectrum is represented in terms of 'pseudo-acceleration', namely $f \mapsto \Gamma(f)$ with $\omega = 2\pi f$, according to Equation (6.16). The modal analysis of the structure with and without fluid coupling gives the eigenfrequencies $(f_n)_{n\geq1}$ of the system. The corresponding values of pseudo-accelerations $(\Gamma(f_n))_{n\geq1}$ may be identified in the plot as square markers, which allows, in particular, to check the nature of the response with respect to the quasi-inertial, quasi-static or resonant responses to the acceleration γ – as defined in Figure 6.9

6.4.3 Seismic Response: Direct and Modal Methods

The time integration of Equation (6.27) is performed with the Newmark algorithm using the 'standard' parameters $\gamma = 1/2$ and $\beta = 1/4$. Figure 6.18 compares the structural response with and without fluid, in terms of the relative displacement of the beam at the free end $u(z = L, t)$. In the present case, the displacement of the structure with fluid coupling is dramatically reduced on account of inertial coupling: this results from the change in the frequency content and in the participation factors of the eigenmodes.

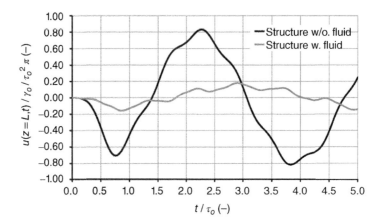

Figure 6.18 Dynamic response of the structure with and without fluid coupling: direct computation (in abscissa, time is normalised by the duration of the sine wave τ_o; in ordinate, displacement is normalised by the maximum displacement triggered in the sine wave $\gamma_o \tau_o^2 / \pi$)

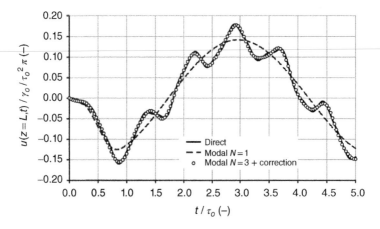

Figure 6.19 Displacement of the beam free end with FSI: comparison of direct and modal computations (time and displacement are normalised here by τ_o and $\gamma_o \tau_o^2 / \pi$)

As far as the contribution of the eigenmodes on the system response is concerned, the number of modes to be retained in the modal method depends on the quantity under concern.

- In most cases, the structure displacement may be computed by retaining a few number of modes solely. For instance, in the present case, three modes are found to produce fairly good results, as may be verified in Figure 6.19. Furthermore, by adding to the model the contribution of the static correction, the accuracy of the computation may be significantly improved.
- The computation of space- or time-derived quantities, such as structure shear force, acceleration or fluid pressure, usually requires more modes. As highlighted by Figure 6.20 in the present case, five modes together with the static correction are necessary to produce accurate results as far as the shear force along the beam axis is concerned.

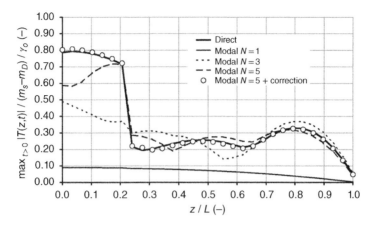

Figure 6.20 Shear force along the beam length: comparison of direct and modal computations (the shear force is normalised here by $(m_S - m_D)\gamma_o$, where $m_S = \rho SL$ is the structure mass and $m_D = \rho \pi R^2 L$ is the fluid-displaced mass)

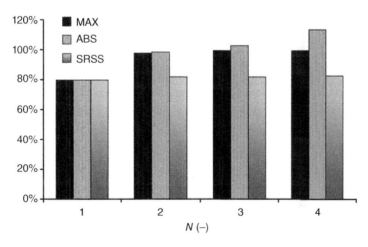

Figure 6.21 Comparison of spectral estimations using the exact and the ABS and SRSS combination rules (MAX refers to the maximum displacement calculated with the modal method in the time domain; N is the order of the modes considered in the analysis)

A short comparison of the modal computation in the time domain and a spectral estimation of the maximum displacement is proposed in Figure 6.21. As evidenced above, a good convergence of the modal computation may be obtained using three modes only.

In the present case, it turns out that the SRSS combination tends to underestimate the system response, while it is slightly overestimated with the ABS combination. Since the modes of the system under concern are sufficiently well separated in frequency, their contribution tend to add-up on the overall response, and, as could be understood, the use of the ABS combination is advisable here, producing a moderate conservatism in the final result.

Remark 6.2 Static calculation – Effective mass *When the frequencies of the system are beyond the cut-off frequency of the excitation, a static calculation allows an accurate approximation of* $\max_{t>0}|\psi(t)|$ *to be performed, where* ψ *may stand for quantity of interest (displacement, strain, stress, etc.). In the present example, a static equivalent formulation of the response is*

$$\begin{bmatrix} \mathbf{K}_S & \mathbf{0} \\ \mathbf{0} & 1/\rho\mathbf{M}_F \end{bmatrix} \begin{Bmatrix} \mathbf{U}(t) \\ \mathbf{P}(t) \end{Bmatrix}$$

$$= -\begin{bmatrix} \mathbf{M}_S + \rho\mathbf{R}\mathbf{K}_F^{-1}\mathbf{R}^T & -\mathbf{R}\mathbf{K}_F^{-1}\mathbf{M}_F \\ -\mathbf{M}_F\mathbf{K}_F^{-1}\mathbf{R}^T & 1/\rho\mathbf{M}_F\mathbf{K}_F^{-1}\mathbf{M}_F \end{bmatrix} \begin{Bmatrix} \mathbf{D} \\ \mathbf{M}_F^{-1}\mathbf{R}'^T\mathbf{D}' \end{Bmatrix} \gamma_o$$

where $\gamma_o = \max_{t>0}|\gamma(t)|$.

A direct method gives a solution to the linear system $\mathbf{KX}_o = -\mathbf{M}\Delta\gamma_o$. *The solution may be approximated using the contribution of N modes, according to*

$$\mathbf{X}_o^N = -\sum_{n=1}^{n=N} \frac{q_n}{\omega_n^2} \mathbf{X}_n \gamma_o$$

In the present case, the modal computation with four modes compares again favourably with the direct computation, as shown in Figure 6.22.

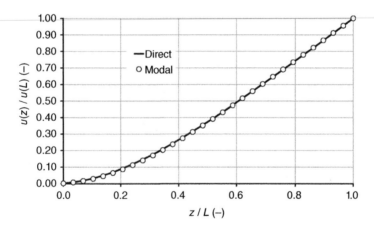

Figure 6.22 *Static response of the beam: modal and direct methods*

As made clear in Figure 6.20, more modes are required to compute the shear force than to compute the displacement field. Indeed, the concentrated mass induces a discontinuity in $T(z)$ *at* $z = z_o$, *and bending of the beam under static acceleration is thus described by*

$$EI\frac{\partial^4 u}{\partial z^4} = -(\rho S + m_o\delta(z_o) + \rho\pi R^2)\gamma_o$$

where $\rho S\gamma_o$ *and* $\rho\pi R^2\gamma_o$ *are the inertial forces associated with the acceleration of the structure and the fluid. Integrating the latter expression between* $z_o - \varepsilon$ *and* $z_o + \varepsilon$ *gives*

$$EI\frac{\partial^3 u}{\partial z^3}\bigg|_{z_o+\varepsilon} - EI\frac{\partial^3 u}{\partial z^3}\bigg|_{z_o-\varepsilon} = -(\rho S\varepsilon + m_o + \rho\pi R^2\varepsilon)\gamma_o$$

The shear force at z is $T(z) = -EI\frac{\partial^3 u}{\partial z^3}$, so that with $\varepsilon \mapsto 0$, the following relation is retrieved:

$$T'(z_o^+) - T(z_o^-) = m_o \gamma_o$$

As may be verified in Table 6.3, the jump on the shear force at $z = z_o$ may be accurately computed using more than five modes.

Table 6.3 Direct and modal computations of the shear force discontinuity

Analytical $T(z_o^+) - T(z_o^-)/\rho SL$ $= m_o \gamma_o / \rho SL$	Direct	Numerical Modal $N = 1$	$N = 5$	$N = 25$	$N = 100$
0.32	0.32	0.01	0.28	0.31	0.32

As for the reaction forces at the system's interfaces, they may be evaluated as follows:

- The reaction force at the beam interface (clamped end on $z = 0$) is calculated as follows:

$$\phi = \langle D, 0 \rangle KX_o = -\langle D, 0 \rangle M \Delta \gamma_o = -(m_S - m_D)\gamma_o$$

$m_S \gamma_o$ is the force proceeding from the structure deformation under acceleration, and $+m_D \gamma_o$ is the pressure force which arises from the accelerated fluid (m_D is the fluid mass displaced by the inner cylinder modelling the beam). Depending on its magnitude, the effect of the inertial force $m_D \gamma_o$ may be to alleviate, annihilate or even enhance the in vacuo value of the reaction force.

- The fluid force on the outer rigid wall ($r = R'$) is

$$\phi' = \langle D', 0 \rangle KX_o = \langle D', 0 \rangle M \Delta \gamma_o = -m_D' \gamma_o$$

with m_D' the fluid mass displaced by the outer cylinder (this inertial force opposes the acceleration γ_o on the outer boundary).

- The total reaction force is

$$\Phi = \phi + \phi' = -\langle D, D' \rangle M_A \Delta \gamma_o = -(m_S + m_F)\gamma_o$$

where $m_F = m_D' - m_D$ is the mass of the fluid filling the cavity.
The modal expansion of Φ reads as follows:

$$\Phi = -\sum_{n \geq 1} \frac{X_n^T M \langle D, D' \rangle^T \, X_n^T M \Delta}{X_n^T M X_n} \gamma_o = -\sum_{n \geq 1} \mu_n \gamma_o$$

Restricting this series to N modes gives an approximation of Φ, denoted Φ_N. It is calculated as $\Phi_N = \sum_{n=1}^{n=N} \mu_n \gamma_o$. Hence,

$$\Phi - \Phi_N = \sum_{n > N} \mu_n \gamma_o$$

which may be understood as another interpretation of Equation (2.27). The effective mass summation is related to the truncation error in the evaluation of the total reaction force, as illustrated in Figure 6.23.

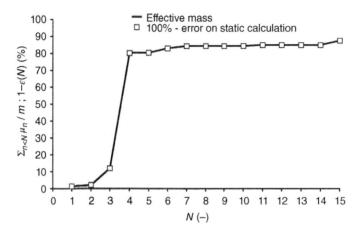

Figure 6.23 *In a static calculation, the summation of the effective masses may be understood as the accuracy of the modal method for evaluating the total reaction force (N denotes the number of modes retained in the expansion of* Φ*;* $\sum_{n<N} \mu_n$ *is the accumulated effective mass up to mode* $\mathbf{X}_N$*;* $\varepsilon(N)$ *stands for the relative error in the computation of the reaction force:* $\varepsilon(N) = |\Phi - \Phi_N|/|\Phi|)$

■

6.5   Example: Frequency-Domain Analysis

6.5.1   *Acoustic Radiation of a Damped Structure Immersed in a Fluid*

The vibrations of a damped structure immersed in a fluid are considered in Figure 6.24 as a simplification of actual configurations, which may be of engineering relevance for instance in naval shipbuilding. The elastic shell is modelled as a two-dimensional ring vibrating in its own plane, according to the simplification introduced in Figure 4.28, and it is coupled to an inner fluid and to an outer unbounded fluid. The vibrations of the structure are supposed to be damped out by using a visco-elastic material: an appropriate modelling of such a system, including the dissipative effect in the visco-elastic layer, is proposed for, instance, by Rouleau *et al.* (2012).

Without loss of generality regarding the method discussed here, a simplified model is adopted here, taking for granted a material with a frequency-dependent elasticity modulus:

$$E^*(\omega) = E(1 + i\eta(\omega))$$

where E stands for Young's modulus of the main structure, which is assumed constant, and $\eta(\omega)$ is the loss factor of the dissipative layer[13], accounting for dissipative effect – it is further supposed that $\eta(\omega)$ is derived from the Zener-type model, as represented in Figure 2.6.

The equations of motion of the structure are detailed in Section 2.4. The wave equation describes the pressure in the sound waves radiated away from the vibrating structure in the

---

[13] This assumption allows for the system's stiffness to be accounted for by E, while its damping is represented by the evolutions of η with ω. It leads to a crude approximation of the actual behaviour of the system but allows a straightforward illustration of some numerical methods used in frequency-domain analyses.

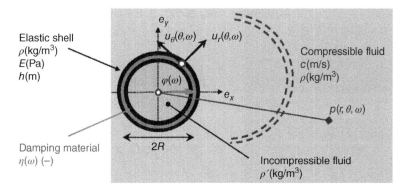

Figure 6.24   Damped elastic ring immersed in a fluid. R and h stand for the radius and thickness of the ring. Material properties of the main structure are ρ (density) and E (Young's modulus), while $\eta(\omega)$ stands for the loss factor of the damping material. The inner and outer fluid density are denoted ρ' and ρ; the inner fluid is supposed incompressible, while the outer fluid is compressible (speed of sound in the fluid is c). $u_r(\theta, \omega)$ and $u_\theta(\theta, \omega)$ stand for the radial and ortho-radial displacement of the shell, while $p(r, \theta, \omega)$ is the fluid pressure. A unit pointwise force $\varphi(\omega)$ is applied on the structure at $\theta = 0$. Frequency response of the system is calculated over the range $0 - 1,500$ Hz

outer fluid, as in Section 3.8, while the inner fluid pressure verifies $\Delta p = 0$ with the coupling condition $\dfrac{\partial p}{\partial r} = \rho'\omega^2 u_r$ at $r = R$.

An analytical model is used for the inner fluid, while a finite element model yields the fluid pressure for the outer fluid. Using a Fourier series expansion of the displacement and the pressure, the frequency response of the system is obtained by a linear superposition of each component response. The latter is obtained by solving the following equation, duplicated here from Section 5.5 for convenience:

$$\left(-\omega^2 \begin{bmatrix} \mathbf{M}_S^m + \mathbf{M}_A^m & \mathbf{0} \\ \rho \mathbf{R}^{\mathsf{T}} & \mathbf{M}_F^m \end{bmatrix} + i\omega \begin{bmatrix} \mathbf{0} & \mathbf{0} \\ \mathbf{0} & \mathbf{C}_F^m \end{bmatrix}\right.$$

$$\left. + \begin{bmatrix} \mathbf{K}_S^m(\omega) & -\mathbf{R} \\ \mathbf{0} & \mathbf{K}_F^m \end{bmatrix}\right) \left\{ \begin{matrix} \mathbf{U}(\omega) \\ \mathbf{P}(\omega) \end{matrix} \right\} = \left\{ \begin{matrix} \boldsymbol{\Phi}^m(\omega) \\ \mathbf{0} \end{matrix} \right\} \qquad (6.28)$$

The contribution of the inner fluid to FSI is represented with the added mass matrix:

$$\mathbf{M}_A^m = \begin{bmatrix} \rho' R \mu_m & 0 \\ 0 & 0 \end{bmatrix}$$

where the added mass coefficient is $\mu_m = \dfrac{1}{m}$, for all $m \neq 0$. The contribution of the Fourier component $m = 0$ is therefore discarded since the deformation of this type does not comply with the incompressibility constraint $\displaystyle\int_0^{2\pi} u_r(\theta)R d\theta = 0$. The contribution of the outer fluid on FSI is represented with the coupling matrix $\mathbf{R}$ and the damping matrix $\mathbf{C}_m$.

The geometrical and material properties of the problem are the same as that in the second example of the preceding chapter, Section 5.5; the inner and outer fluids are identical. The parameters of the Zener model are supposed to be $E_o = 7.5$ MPa, $E_\infty = 32.5$ MPa, $\tau = 5.0 \times 10^{-5}$ s and $\alpha = 0.65$.

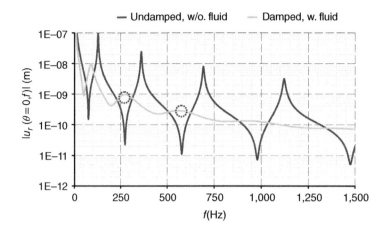

Figure 6.25 Frequency response: undamped structure without FSI and damped structure with FSI

Figure 6.25 gives the frequency response of the system to the external load in terms of the radial displacement at $\theta = 0$ for the undamped structure without fluid on the one hand and for the damped structure immersed in the fluid on the other hand. The system's response is calculated for frequency ranging from 0 to 1, 500 Hz with a 1-Hz frequency step, solving Equation (6.28) with a direct method. The time elapsed to calculate the response with damping and FSI serves as reference to estimate the computational cost.

The combined effect of dissipative damping and FSI is apparent on the system's response. In the present example, both structural and radiative damping are accounted for; however, the former tends to dominate the latter.

Figure 6.26 gives the displacement of the structure and the corresponding pressure angular pattern in the fluid at the second and third resonances, namely, for $f = 275$ Hz and $f = 575$ Hz. The attenuation of the vibro-acoustic response is revealed, for instance, by comparing the pressure level in the damped and undamped cases, in conjunction with a reduction in the structural peak response at resonances, as evidenced in Figure 6.25.

6.5.2 Frequency Response: Direct and Modal Methods

The computation of the frequency response with a ROM is carried out by making use of a symmetric formulation of the coupled problem[14], a symmetric formulation of the coupled problem is used.

With the $(\mathbf{u}, \chi)$ formulation, the equivalent of Equation (6.28) for each Fourier component m reads as follows:

$$(-\omega^2 \mathbf{M}_m + i\omega \mathbf{C}_m + \mathbf{K}_m(\omega))\mathbf{X}_m(\omega) = \mathbf{F}_m(\omega) \qquad (6.29)$$

where $\mathbf{X}_m = \begin{Bmatrix} \mathbf{U}_m(\omega) \\ \chi_m(\omega) \end{Bmatrix}$ and $\mathbf{F}_m = \begin{Bmatrix} \mathbf{\Phi}_m(\omega) \\ \mathbf{0} \end{Bmatrix}$.

---

[14] A Fourier representation of the system is a first step towards model reduction. It offers potentialities, such as parallel computing, which are not exploited here; additional reduction in the computational cost is achieved with modal projection.

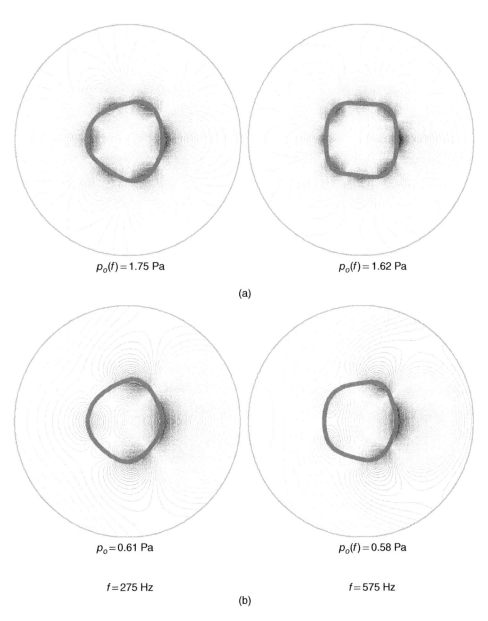

$p_o(f) = 1.75$ Pa $\qquad\qquad\qquad$ $p_o(f) = 1.62$ Pa

(a)

$p_o = 0.61$ Pa $\qquad\qquad\qquad$ $p_o(f) = 0.58$ Pa

$f = 275$ Hz $\qquad\qquad\qquad$ $f = 575$ Hz

(b)

Figure 6.26 Acoustic radiation of an immersed structure: structure deformation and fluid pressure ($p_o(f) = \max_{r\in[R,3R],\theta\in[0,2\pi]} |p(r,\theta,f)|$ refers to the maximum pressure in the fluid at frequency f). (a) Undamped structure. (b) Damped structure

Matrices $\mathbf{M}_m$, $\mathbf{C}_m$ and $\mathbf{K}_m$ appearing in Equation (6.29) are symmetric; they are expressed as follows:

$$\mathbf{M} = \begin{bmatrix} \mathbf{M}_S^m + \mathbf{M}_A^m & \mathbf{0} \\ \mathbf{0} & -\rho\mathbf{M}_F^m \end{bmatrix} \quad \mathbf{C} = \begin{bmatrix} \mathbf{0} & \rho\mathbf{R} \\ \rho\mathbf{R}^\mathsf{T} & -\rho\mathbf{C}_F^m \end{bmatrix}$$

and:

$$\mathbf{K} = \begin{bmatrix} \mathbf{K}_S^m(\omega) & \mathbf{0} \\ \mathbf{0} & -\rho\mathbf{K}_F^m \end{bmatrix}$$

For each Fourier component, a projection basis is obtained here in a two-step method, as detailed by Algorithm 6[15].

- First, a projection basis $\mathbf{U}_N$ for the displacement of the damped structure with the inertial contribution of FSI is calculated, using one of the methods discussed in Section 6.3.2[16]. $(\omega_n)_{n\in 1,N}$ are the corresponding pulsations and χ_N is the analogous basis for the fluid velocity potential, derived from the displacement:

$$\chi_n = i\omega_n \mathbf{K}_F^{-1}\mathbf{R}^\mathsf{T}\mathbf{U}_n \qquad \forall n \in [1, N]$$

As may be emphasised by looking at Figure 6.27, the computational cost for the extraction of the structure projection basis does not exceed a few percent of the reference computation.
- Second, the projection basis $\mathbf{X}_N = \langle \mathbf{U}_N, \chi_N \rangle$ is enriched with the error of the response at pulsation ω_n:

$$\mathbf{X}_n^* = [(-\omega_n^2\mathbf{M} + i\omega_n\mathbf{C} + \mathbf{K}(\omega_n))^{-1}$$
$$-\mathbf{X}_N(-\omega_n^2\mathbf{m}_N + i\omega_n\mathbf{c}_N + \mathbf{k}_N(\omega_n))^{-1}\mathbf{X}_N^\mathsf{T})]\mathbf{\Phi}(\omega_n)$$

For each Fourier component, the reduced-order problem is

$$(-\omega^2\mathbf{m}_N^* + i\omega\mathbf{c}_N^* + \mathbf{k}_N^*(\omega))\boldsymbol{\xi}_N^*(\omega) = \boldsymbol{\varphi}_N^*(\omega)$$

The numerical efficiency and the accuracy of the reduced-order modelling for large-scaled systems is evidenced with the present example. The frequency response is calculated with $M = 15$ Fourier components: as may be verified by looking at Figure 6.28, the frequency response is rather well approximated by the modal computation.

For each Fourier component m, the size of the ROM is 2, while that of the complete FEM model is $\mathcal{O}(I)$, with I the number of fluid finite elements. As depicted in Figure 6.29, the computational cost for the modal method remains rather steady when I increases, while it tends to increase steeply with the size of the problem when making use of the direct method.

---

[15] The algorithm proposed here is adapted to the case under concern and therefore has an illustrative purpose and may not prove suited for more general applications.
[16] The displacement basis $\mathbf{U}_N$ is calculated here with the MSE method.

Algorithm 6 Computation of the frequency response of a damped structure immersed in a fluid and subjected to an external load – dissipative damping and radiative damping are accounted for, respectively, in the structure and in the fluid models

---

Assemble structure and fluid matrices $\mathbf{M}_S$, $\mathbf{K}_S(\omega)$, $\mathbf{M}_F$, $\mathbf{C}_F$, $\mathbf{K}_F$ and $\mathbf{R}$.

Calculate the added mass matrix $\rho \mathbf{R} \mathbf{K}_F^{-1} \mathbf{R}^\mathsf{T}$.

Solve the non-linear eigenvalue problem:

$$(-\omega^2 (\mathbf{M}_S + \rho \mathbf{R} \mathbf{K}_F^{-1} \mathbf{R}^\mathsf{T} + \mathbf{K}_S(\omega))\mathbf{U} = \mathbf{0}$$

Store the N structure pseudo-modes in terms of displacement $(\mathbf{U}_n)_{n \in [1,N]}$.

Store the corresponding eigenpulsations $(\omega_n)_{n \in [1,N]}$.

Calculate the N corresponding fluid pseudo-modes in terms of velocity potential:

$$\chi_n = i\omega_n \mathbf{K}_F^{-1} \mathbf{R}^\mathsf{T} \mathbf{U}_n$$

Assemble the projection basis $\mathbf{X}_N = \langle \mathbf{X}_n \rangle_{n \in [1,N]}$ with $\mathbf{X}_n = \left\{ \begin{matrix} \mathbf{U}_n \\ \chi_n \end{matrix} \right\}$ for all n.

Assemble fluid–structure matrices in $(\mathbf{u}, \chi)$ symmetric formulation:

$$\mathbf{M} = \begin{bmatrix} \mathbf{M}_S & \mathbf{0} \\ \mathbf{0} & -\rho \mathbf{M}_F \end{bmatrix} \qquad \mathbf{C} = \begin{bmatrix} \mathbf{0} & \rho \mathbf{R} \\ \rho \mathbf{R}^\mathsf{T} & -\rho \mathbf{C}_F \end{bmatrix} \qquad \mathbf{K} = \begin{bmatrix} \mathbf{K}_S(\omega) & \mathbf{0} \\ \mathbf{0} & -\rho \mathbf{K}_F \end{bmatrix}$$

Calculate the reduced-order mass and damping matrices $\mathbf{m}_N$ and $\mathbf{c}_N$ with projection basis $\mathbf{X}_N$.

for $n \in [1, N]$ **do**

  Evaluate $\mathbf{k}_N(\omega_n)$ and $\mathbf{K}(\omega_n)$.

  Compute the error on the response at pulsation ω_n:

$$\mathbf{X}_n^* = [(-\omega_n^2 \mathbf{M} + i\omega_n \mathbf{C} + \mathbf{K}(\omega_n))^{-1}$$

$$-\mathbf{X}_N (-\omega_n^2 \mathbf{m}_N + i\omega_n \mathbf{c}_N + \mathbf{k}_N(\omega_n))^{-1} \mathbf{X}_N^\mathsf{T})] \Phi(\omega_n)$$

  Update the projection basis $\mathbf{X}_N^* = \langle \mathbf{X}_N, \mathbf{X}_n^* \rangle$ – if needed with an ortho-normalisation procedure.

end for

Calculate the reduced-order mass and damping matrices $\mathbf{m}_N^*$ and $\mathbf{c}_N^*$ with the enriched projection basis $\mathbf{X}_N^*$.

Calculate the frequency response with the ROM:

$$(-\omega^2 \mathbf{m}_N^* + i\omega \mathbf{c}_N^* + \mathbf{k}_N^*(\omega)) \xi_N^*(\omega) = \varphi_N^*(\omega)$$

Calculate the frequency response at location $\mathbf{x}$ for pulsation ω:

$$\mathbf{X}(\omega, \mathbf{x}) = \mathbf{X}_N^*(\mathbf{x}) \xi_N^*(\omega)$$

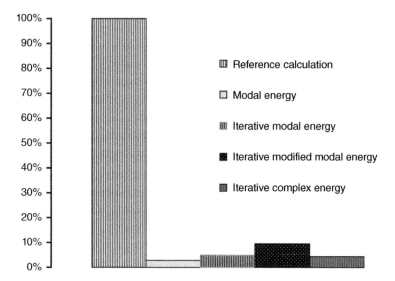

Figure 6.27   Computation of the projection basis. The computational cost for the computation of the projection basis using the MSE, IMSE, IMMSE or ICE methods are compared. The reference corresponds to the direct computation of the response for the damped structure, over the frequency range of interest (between 0 and 1, 500 Hz) and with a narrow-band analysis (with a frequency step of 1 Hz). In the present example, calculating the projection basis represents only a few percent of the computational effort required in the case the direct approach is used

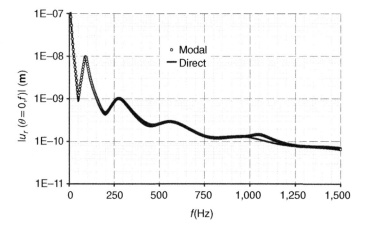

Figure 6.28   Direct and modal computations of the frequency response: the modal approach produces results which are in good agreement with the direct computation

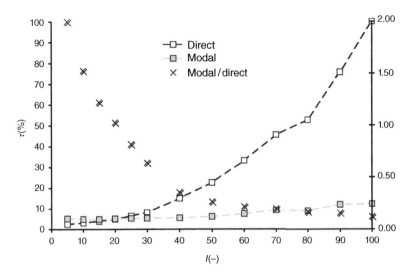

Figure 6.29   Computational cost for the frequency response calculation with direct/modal methods

Remark 6.3 A comparison of two ROM strategies for vibro-acoustic coupling *The numerical method presented in Algorithm 6 is simple enough to be adapted to academic cases such as depicted in Figure 6.24, but it may however not be suited to more complex problems. A detailed overview of algorithms which yield a projection basis for Equation (6.20), as well as the techniques to control the accuracy and efficiency of the ROM derived from the projection, would occupy an entire volume in itself – if not many! – and is beyond the scope of the introduction provided in this book. As a concluding remark on the topic however, it is found relevant to present an application of reduced-order modelling for a real damped structure immersed in a fluid by making use of two different strategies.*

The example is concerned with a four-blade propeller made of steel, which is immersed in water and whose vibrations are damped out by visco-elastic patches embedded within the blades. A pointwise force is applied at the tip of the upper blade as a dynamic excitation with unit amplitude throughout the frequency range of interest.

The dynamic response of the propeller is computed by coupling fluid and structure elements: the corresponding finite element model is produced with the FEM code Code-Aster[17] *and it is represented in Figure 6.30[18]. The finite element mesh is composed of some* 265, 000 *tetrahedral elements with quadratic approximation of both the structure and fluid fields, resulting in a matrix system of about* 850, 000 *equations, this example is thereby holding the complexity of an industrial problem.*

The fluid volume is bounded with a sphere of radius $R_\infty = 2.0$ m *from the origin of the spherical coordinate system, which is placed in the vicinity of the centre of gravity of the propeller. Radiative damping is accounted for with an acoustic impedance* $Z(\omega)$ *expressed as*

---

[17] See for instance www.code-aster.org.
[18] A detailed model of this may be found in the study by Leblond and Sigrist (2015). The diameter of the propeller is 1.8 m, the lengths of the axis and of the blades are, respectively, 0.33 and 1.0 m, while the thickness of the patches is about 0.003 m. The dissipative properties of the visco-elastic material are accounted for again here using the Zener model.

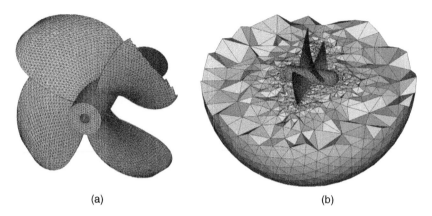

(a)   (b)

Figure 6.30 *Finite element model of the immersed propeller with blades damped by visco-elastic patches. (a) Blades with visco-elastic patches. (b) Propeller immersed in a fluid domain*

a BGT condition of the first order, as discussed in Section 3.2.2:

$$Z(\omega) = \frac{\rho c}{1 + c/(i\omega R)}$$

A symmetric $(\mathbf{u}, p, \varphi)$ *coupled formulation is proposed here to model vibro-acoustic coupling, so that the corresponding finite element matrix system reads as follows:*

$$(-i\omega^3 \mathbf{C} - \omega^2 \mathbf{M} + \mathbf{K}(\omega))\mathbf{X}(\omega) = \mathbf{F}(\omega) \tag{6.30}$$

where the mass, damping and stiffness matrices are, respectively, defined as follows:

$$\mathbf{M} = \begin{bmatrix} \mathbf{M}_S & \mathbf{0} & \rho\mathbf{R} \\ \mathbf{0} & \mathbf{0} & \mathbf{M}_F \\ \rho\mathbf{R}^T & \mathbf{M}_F^T & -\rho\mathbf{K}_F \end{bmatrix} \qquad \mathbf{C} = \begin{bmatrix} \mathbf{0} & \mathbf{0} & \mathbf{0} \\ \mathbf{0} & \mathbf{0} & \mathbf{0} \\ \mathbf{0} & \mathbf{0} & -\rho\mathbf{C}_F \end{bmatrix}$$

and:

$$\mathbf{K}(\omega) = \begin{bmatrix} \mathbf{K}_S(\omega) & \mathbf{0} & \mathbf{0} \\ \mathbf{0} & 1/\rho\mathbf{M}_F & \mathbf{0} \\ \mathbf{0} & \mathbf{0} & \mathbf{0} \end{bmatrix}$$

$\mathbf{X}^T = \langle \mathbf{U}^T, \mathbf{P}^T, \mathbf{\Phi}^T \rangle$ *gathers the structure and fluid degrees-of-freedom, respectively, stated in terms of displacement, pressure and displacement potential.*

In the above expression, matrices $\mathbf{K}_F$ *and* $\mathbf{C}_F$ *are standing for the finite element discretisation of the following integral terms, respectively:*

$$\int_\Omega \frac{\partial\varphi}{\partial x_i}\frac{\partial\delta\varphi}{\partial x_i}\,d\Omega + \int_{\Gamma_\infty} \frac{\varphi\delta\varphi}{R}\,d\,\Gamma_\infty \rightarrow \delta\mathbf{\Phi}^T\mathbf{K}_F\mathbf{\Phi} \qquad \int_{\Gamma_\infty} \frac{\varphi\delta\varphi}{c}\,d\,\Gamma_\infty \rightarrow \delta\mathbf{\Phi}^T\mathbf{C}_F\mathbf{\Phi}$$

The above expressions indicate that radiative damping is accounted for by matrix $\mathbf{C}_F$ *and by the second term appearing in the definition of matrix* $\mathbf{K}_F$. *On the other hand, the dissipative properties of the visco-elastic patches are modelled with the frequency-dependent stiffness matrix* $\mathbf{K}_S(\omega)$.

An ROM is obtained here using two different numerical strategies.

- *In the first one, the ROM is derived from the projection of Equation (6.30) onto a set of vectors obtained with the MSE method: the projection basis is composed of the natural modes of vibration of the undamped structure, namely, the first N eigenvectors of the following eigenvalue problem:*

$$(-\omega^2 \mathbf{M} + \mathbf{K}(0))\mathbf{X} = \mathbf{0}$$

  This projection basis does not take into account any dissipation within the structure and also partly *accounts for FSI (in particular radiative damping is not fully represented in the above eigenvalue problem, as the matrix* **C** *is not present in the formulation). The dissipative properties of the structure and the coupling with the fluid are expected to be modelled with the ROM computation. Some of these natural modes of vibration are represented in Figure 6.31* (a) *in terms of the pressure field projected on the deformed wetted surface of the propeller.*
- *In the second one, referred to in the following using the generic term* Reduced Basis *(RB) without stepping into a detailed discussion on this denomination, the solution* $\mathbf{X}(\omega)$ *of the vibro-acoustic equation is expanded as a separated space-frequency series of order N, which reads as follows:*

$$\mathbf{X}_N(\omega) = \sum_{n=1}^{n=N} \xi_n(\omega)\mathbf{X}_n$$

$(\mathbf{X}_n)_{n\in[1,N]}$ *are vectorial functions depending on the space variable and* $(\xi_n)_{n\in[1,N]}$ *are scalar functions of the frequency* ω; *N is supposed to be small compared to the size of the matrix system in Equation (6.30). The reduced basis* $\mathbf{X}^N = (\mathbf{X}_n)_{n\in[1,N]}$ *is obtained according to Algorithm 7, as detailed in Leblond and Sigrist (2015).*

In contrast with the MSE method that produces a set of vectors with a physical significance (in the present case, they are the natural modes of vibration of the undamped propeller with fluid inertial effect), which may however not form a complete *projection basis for the problem under concern (meaning that the convergence of the ROM computation towards the actual solution of the vibro-acoustic equation as the number of modes is increased may not be observed), the projection basis derived from the RB method is shown to be 'more complete', resulting in a better and more efficient convergence towards the solution.[19]*

Figure 6.31 (b) *gives a representation of the first vectorial functions of the projection basis, again in terms of the pressure field projected on the deformed wetted surface of the propeller: an observation of the vector shape reveals the influence of the external excitation visible at the top of the upper blade and thereby highlights the difference in nature between the projection basis obtained with the MSE and RB methods.*

The frequency response of the vibro-acoustic system may be computed over an extended frequency range (in the present case $0 - 600$ *Hz) with a fine resolution (typically with a frequency step of 1 Hz) at low computational cost and with satisfying accuracy, in first approximation at least, using the ROM obtained with the MSE method, while a direct resolution of Equation (6.30) in the same conditions would not even be tractable with the present finite element model.*

---

[19] The projection basis bears also a physical significance, because it is built from the solution of the 'complete' problem, as stated by Equation (6.30), for certain values of the pulsations $\omega \in [\omega^-, \omega^+]$.

Algorithm 7 Computation of the Reduced Basis model with the so-called 'Greedy Algorithm' to solve the vibro-acoustic problem modelled by Equation (6.30)

Require: Finite element matrices $\mathbf{M}$, $\mathbf{C}$ and $\mathbf{K}(\omega)$ and vector $\mathbf{F}(\omega)$ of the complete vibro-acoustic problem.

Ensure: Reduced basis $\mathbf{X}^N = (\mathbf{X}_n)_{n \in [1,N]}$.

Set $\mathbf{X}^0 = \emptyset$ *and* $\xi_0(\omega) = \emptyset$

for $n \in [1, N]$ **do**

Generate a random sampling of L values of the pulsations ω in the range of interest:

$$\mathcal{W}_L = \{\omega_l\}_{l \in [1,L]} \qquad \omega_l \in [\omega^-, \omega^+] \quad \forall l \in [1, L]$$

If $\mathbf{X}^{n-1} \neq \emptyset$, solve the reduced-order model obtained from the projection of Equation (6.30) onto $\mathbf{X}^{n-1}$, for all pulsations in the random sampling $\mathcal{W}_L$:

$$(-i\omega_l^3 \mathbf{c}_{n-1} - \omega_l^2 \mathbf{m}_{n-1} + \mathbf{k}_{n-1}(\omega_l))\xi_{n-1}(\omega_l) = \varphi_{n-1}(\omega_l) \qquad \forall l \in [1, L]$$

Identify the pulsation value in $\mathcal{W}_L$ associated with the largest norm of the residual:

$$\omega^* = \arg \sup_{\omega \in \mathcal{W}_L} \|\mathbf{R}_{n-1}(\omega)\|$$

where $\mathbf{R}_{n-1}(\omega)$ is:

$$\mathbf{R}_{n-1}(\omega) = (-i\omega^3 \mathbf{C} - \omega^2 \mathbf{M} + \mathbf{K}(\omega))\mathbf{X}_{n-1}(\omega) - \mathbf{F}(\omega)$$

with $\mathbf{X}_{n-1}(\omega) = \sum_{\mu=1}^{\mu=n-1} \xi_\mu(\omega)\mathbf{X}_\mu$.

Evaluate the exact error at this parameter value ω^*:

$$\mathbf{e}_{n-1}(\omega^*) = \mathbf{X}(\omega^*) - \mathbf{X}_{n-1}(\omega^*)$$

Enrich the projection basis $\mathbf{X}^{n-1}$ with the additional vector $\mathbf{X}_n$:

$$\mathbf{X}_n = \frac{\mathbf{e}_{n-1}(\omega^*)}{\|\mathbf{e}_{n-1}(\omega^*)\|} \qquad \mathbf{X}^n = \langle \mathbf{X}^{n-1}, \mathbf{X}_n \rangle$$

end for

Perform an accuracy check by computing the maximum error ε with given a new random sampling $\mathcal{W}'_L = \{\omega'_l\}_{l \in [1,L]}$:

$$\varepsilon = \sup_{\omega \in \mathcal{W}'_L} \left\| \mathbf{X}(\omega) - \sum_{n=1}^{n=N} \xi_n(\omega)\mathbf{X}_n \right\|$$

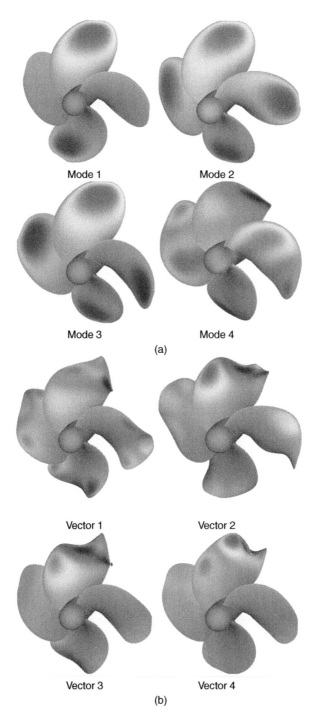

Mode 1   Mode 2

Mode 3   Mode 4

(a)

Vector 1   Vector 2

Vector 3   Vector 4

(b)

Figure 6.31 *Modal shape in terms of displacement for the first modes of the projection basis obtained with the (a) modal strain energy method and the (b) reduced basis method proposed by Leblond and Sigrist (2015)*

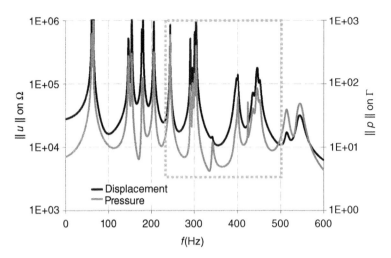

Figure 6.32 Frequency response. *The frequency response of the immersed propeller with damped blades is computed with a reduced-order model obtained with the MSE method. It is plotted in terms of displacement in the structure volume Ω and of pressure on the fluid boundary Γ versus the frequency f. ‖u‖ and ‖p‖ denote the norms on $L^2(\Omega)$ and $L^2(\Gamma)$: they are calculated here according to $\|u\|^2 = \int_\Omega |u|^2\, d\Omega$ and $\|p\|^2 = \int_\Gamma p^2\, d\Gamma$ using the finite element approximations of u and p*

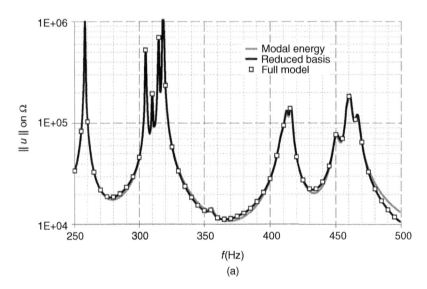

Figure 6.33 Full and reduced-order computations of the frequency response. *The frequency response of the vibro-acoustic system (damped propeller immersed in water) is plotted in the frequency range of 250 − 500 Hz with a frequency step of 1 Hz for the reduced model and 5 Hz for the full model, in terms of (a) displacement in the structure volume Ω and in terms of (b) pressure at the fluid boundary Γ. The ROM computation obtained with the MSE and RB methods involve 35 modes; they are compared with the direct resolution of Equation (6.30) performed with the full finite element model*

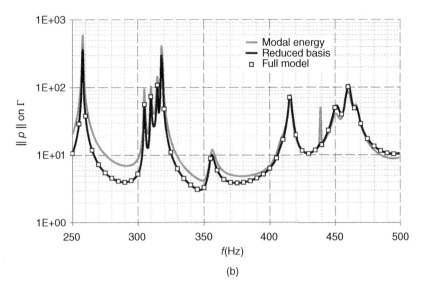

(b)

Figure 6.33   (continued)

The ROM involved in the present computation some N = 35 modes or vectors, which corresponds to a spectacular reduction of the size of the problem.

Figure 6.32 plots the frequency response to the pointwise excitation, both in terms of displacement in the structure volume Ω and pressure at the fluid boundary Γ, and reveals that the dynamic behaviour of the system is dominated by structural elasticity with fluid inertia at frequencies below 250 Hz, while dissipative damping and radiative damping have a marked influence above 325 Hz. It should be noted in addition that the frequency content of the system is rather dense in the frequency range of interest, the resonances being well identified individually but involving however close peaks of response.

A detailed comparison of the ROM and full finite element models, the latter involving a direct resolution of the vibro-acoustic equation and serving as reference for the computation of the frequency response, is proposed in Figure 6.33 for the frequency range of 250 − 500 Hz for which the dissipative properties of the visco-elastic patches come into play in the dynamic behaviour of the damped propeller.

It is observed that the numerical results obtained with the RB method are in full agreement with the reference computation, whereas the accuracy of the MSE is rather poor in some cases, as evidenced in particular when it comes to the computation of the pressure in the fluid.[20]

The accuracy and efficiency of the proposed RB method make it particularly interesting for engineering applications: involving a low computational cost, the method is able to perform the vibro-acoustic response of a mechanical system of industrial complexity, that is, in

---

[20] A detailed analysis of the convergence properties of the RB and MSE methods is proposed by Leblond and Sigrist (2015) and shows in addition that the convergence of the MSE method is much slower than the convergence observed with the RB method: with 35 modes indeed, mean relative errors of 50% and 150% are obtained, respectively, for the displacement and pressure norms with the MSE method, while the corresponding relative errors are insignificant when using the RB method.

the present case, a system with relatively high modal density, fluid–structure coupling and dissipative properties.   ∎

References

Balmès E 1996 Parametric families of reduced-order finite element models: theory and applications. *Mechanical Systems and Signal Processing*, **10**, 381–394.

Bathe KJ 1982 *Finite Element Procedures in Engineering Analysis*. Prentice-Hall.

Bathe KJ and Wilson EL 1973 Stability and accuracy of direct integration methods. *Earthquake Engineering and Structural Dynamics*, **1**, 283–291.

Belytschko T, Liu WK, and Morand B 2000 *Non-Linear Finite Element for Continua and Structure*. John Wiley & Sons, Ltd.

Biot MA 1941 A mechanical analyzer for the prediction of earthquake stresses. *Bulletin of the Seismological Society of America*, **31**, 151–171.

Dokainish MA and Subbaraj K 1989 A survey of direct time-integration methods in computational structural dynamics-I. Explicit methods. *Computers and Structures*, **32**, 1371–1386.

Duigou L, Daya M, and Potier-Ferry M 2003 Iterative algorithms for non-linear eigenvalue problems. *Computational Methods in Applied Mechanics and Engineering*, **192**, 1323–1335.

Géradin M and Rixen D 1994 *Mechanical Vibrations: Theory and Application to Structural Dynamics*. John Wiley & Sons, Inc.

Gupta AK 1990 *Response Spectrum Method in Seismic Analysis and Design of Structures*. CRC Press.

Harris CM (ed.) 1998 *Shock and Vibration Handbook*. McGraw-Hill.

Hibbit HD 1979 Some follower forces and load stiffness. *International Journal for Numerical Methods in Engineering*, **14**, 937–941.

Hilber HM, Hughes TRJ, and Taylor RL 1977 Improved numerical dissipation for time integration algorithms in structural dynamics. *Earthquake Engineering and Structural Dynamics*, **5**, 283–292.

Hughes TRJ and Belytschko T 1983 A précis of developments in computational methods for transient analysis. *Journal of Applied Mechanics*, **50**, 1033–1041.

Leblond C and Sigrist JF 2015 Parametric reduced order modeling for the low frequency response of submerged visco-elastic structures. Submitted for publication.

Lion RH and Dejong RG 1995 *Theory and Application of Statistical Energy Analysis*. Butterworth-Heinemann.

Merlette N, Pagnacco E, and Ladier A 2012 Recent developments in Code Aster to Compute FRF and modes of VEM with frequency dependent properties. In *Proceedings Acoustics 2012*.

Moreira R and Rodrigues J 2006 Partial constrained viscoelastic damping treatments of structures: a modal strain energy approach. *International Journal of Structural Stability and Dynamics*, **6**, 397–411.

Moumni Z and Axisa F 2004 Simplified modelling of vehicle frontal crashworthiness using modal approach. *International Journal of Crashworthiness*, **9**, 285–297.

Newmark MN 1959 A method of computation for structural dynamics. *Journal of Engineering Mechanics*, **85**, 67–94.

Ohayon R and Soize C 2012 Advanced computational dissipative structural acoustics and fluid-structure interaction in low- and medium-frequency domains. Reduced-order models and uncertainty quantification. *International Journal of Aeronautical and Space Sciences*, **13**, 14–40.

Pettigrew MJ 2003a Vibration analysis of shell and tube heat exchangers – Part 1: flow, damping, fluidelastic instability. *Journal of Fluids and Structures*, **18**, 469–483.

Pettigrew MJ 2003b Vibration analysis of shell and tube heat exchangers – Part 2: vibration response, fretting-wear, guidelines. *Journal of Fluids and Structures*, **18**, 485–500.

Pettigrew MJ, Rogers RJ, and Axisa F 2011 Damping of heat exchanger tubes in liquids: review and design guidelines. *Journal of Pressure Vessel Technology*, **133**, Paper #014002.

Plouin AS and Balmès E 1998 Pseudo-modal representations of large models with viscoelastic behavior. In *Proceedings of the 16th International Modal Analysis Conference (IMAC XVI)*.

Rebouillat S and Liksonov D 2010 Fluid-structure interaction in partially fluid-filled liquid containers: a comparative review of numerical approaches. *Computers & Fluids*, **39**, 739–746.

Rouleau L 2013 Modélisation vibro-acoustique de structures sandwich munies de matériaux visco-élastiques. PhD thesis, Conservatoire National des Arts & Métiers, Paris.

Rouleau L, Deü JF, and Legay A 2014 Review of reduction methods based on modal projection for highly damped structures. In *Proceedings of the* 11th *International Conference on World Congress on Computational Mechanics (WCCM 2014)*.

Rouleau L, Deü JF, Legay A, and Sigrist JF 2012 Vibro-acoustic study of a viscoelastic sandwich ring immersed in water. *Journal of Sound and Vibration*, **331**, 522–539.

Soize C and Ohayon R 1998 *Structural Acoustics and Vibrations: Mechanical Models, Variational Formulations and Discretisation*. Academic Press.

Soize C and Ohayon R 2014 *Advanced Computational Vibroacoustics: Reduced-Order Models and Uncertainty Quantification*. Cambridge University Press.

Subbaraj K and Dokainish MA 1989 A survey of direct time-integration methods in computational structural dynamics-II. Implicit methods. *Computers and Structures*, **32**, 1387–1401.

Trindade M, Benjeddou A, and Ohayon R 2000 Modeling of frequency-dependent viscoelastic materials for active-passive vibration damping. *Journal of Vibration and Acoustics*, **122**, 169–174.

Vasques C, Moreira R, and Rodrigues J 2010 Viscoelastic damping technologies. Part I: modeling and finite element implementation. *Journal of Advanced Research in Mechanical Engineering*, **1**, 76–95.

Wilson EL, Kiareghian AD, and Bayo ER 1981 A replacement for the SRSS method in seismic engineering. *Earthquake Engineering and Structural Dynamics*, **9**, 187–192.

Wriggers P 2008 *Nonlinear Finite Element Methods*. Springer-Verlag.

Index

Fluid–Structure Interaction: An Introduction to Finite Element Coupling, First Edition. Jean-François Sigrist.
© 2015 John Wiley & Sons, Ltd. Published 2015 by John Wiley & Sons, Ltd.
Companion Website: www.wiley.com/go/sigrist

Printed and bound by CPI Group (UK) Ltd, Croydon, CR0 4YY

16/04/2025

14658560-0001